THE CITY

MOVEMENT IN CITIES

URBAN GEOGRAPHY

MOVEMENT IN CITIES

Spatial Perspectives on Urban Transport and Travel

P.W. DANIELS & A.M. WARNES

LONDON AND NEW YORK

First published in 1980

This edition first published in 2007 by
Routledge
2 Park Square, Milton Park, Abingdon, Oxon OX14 4RN

Simultaneously published in the USA and Canada
by Routledge
711 Third Avenue, New York, NY 10017

Transferred to Digital Printing 2007

Routledge is an imprint of the Taylor & Francis Group, an informa business

First issued in paperback 2013

© 1980 P.W. Daniels and A.M. Warnes

All rights reserved. No part of this book may be reprinted or reproduced or utilized in any form or by any electronic, mechanical, or other means, now known or hereafter invented, including photocopying and recording, or in any information storage or retrieval system, without permission in writing from the publishers.

The publishers have made every effort to contact authors and copyright holders of the works reprinted in the *The City* series. This has not been possible in every case, however, and we would welcome correspondence from those individuals or organisations we have been unable to trace.

These reprints are taken from original copies of each book. In many cases the condition of these originals is not perfect. The publisher has gone to great lengths to ensure the quality of these reprints, but wishes to point out that certain characteristics of the original copies will, of necessity, be apparent in reprints thereof.

British Library Cataloguing in Publication Data
A CIP catalogue record for this book
is available from the British Library

Movement in Cities

ISBN13: 978-0-415-41759-4 (hardback)
ISBN13: 978-0-415-86039-0 (paperback)
ISBN13: 978-0-415-41318-3 (set)

Routledge Library Editions: The City

MOVEMENT IN CITIES

Spatial perspectives on
urban transport and travel

P. W. Daniels
and
A. M. Warnes

METHUEN
LONDON AND NEW YORK

First published in 1980 by
Methuen & Co. Ltd
11 New Fetter Lane, London EC4P 4EE
University Paperback edition published in 1983

Published in the USA by
Methuen & Co.
in association with Methuen, Inc.
733 Third Avenue, New York, NY 10017

© 1980 P. W. Daniels and A. M. Warnes

Printed in Great Britain at
the University Press, Cambridge

All rights reserved. No part of this book may be reprinted or reproduced or utilized in any form or by any electronic, mechanical or other means, now known or hereafter invented, including photocopying and recording, or in any information storage or retrieval system, without permission in writing from the publishers.

British Library Cataloguing in Publication Data
Daniels, P. W.
Movement in cities. – (University paperbacks)
1. Urban transportation – Great Britain
I. Title   II. Warnes, A. M.
338.4′0941   HE311.G7

ISBN 0-416-35620-6

FOR
OUR CHILDREN
Andrew, Charlotte,
Paul and Sarah

Contents

List of Figures   *ix*
Acknowledgements   *xiii*
Preface   *xv*
1   The transport revolution and urban growth   *1*
2   The activities of urban populations and their relationship to urban movement   *17*
3   The aspatial characteristics of movement   *54*
4   Goods movements within towns   *77*
5   Spatial patterns of urban movement: the organizing principles   *114*
6   Spatial patterns of urban movement: empirical evidence   *136*
7   Analysis and prediction of travel patterns   *175*
8   The development of legislation concerning urban transport policies and planning   *229*
9   Management of urban travel demands   *263*
10 Some topical issues in urban transport decision making   *306*
11 The geographical perspective on urban movement   *349*
    Bibliography   *356*
    Indexes   *387*

List of Figures

Figure 1.1 Schematic relationship between urban form and transport
Figure 2.1 Activity profiles of six population groups in three West German towns
Figure 3.1 The half-hourly distribution of internal person trips in Northampton, 1964
Figure 3.2 The hourly distribution of trips in Norwich, 1967
Figure 3.3 The variation from day to day of personal trip making for different purposes: Great Britain, 1964
Figure 3.4 The variation between residential densities and trip rates and distances: Great Britain, 1972/3
Figure 3.5 Mode of travel by area: Great Britain, 1975/6
Figure 3.6 The mode of travel to work in the West Midlands, 1964
Figure 3.7 The mode of personal travel according to trip purpose: Great Britain, 1972/3
Figure 4.1 Screenlines and cordons used in commercial traffic surveys of London and Liverpool
Figure 4.2 Monthly traffic flows on urban roads in Great Britain
Figure 4.3 The hourly distribution of commercial vehicle trips in London, 1960
Figure 4.4 Flows of over 500 trips between transportation area study zones in the West Midlands, 1964
Figure 4.5 Number of internal goods vehicle trips within and between study zones in Liverpool, 1966
Figure 5.1 The distance decay of urban personal journeys
Figure 5.2 Distance and mode of travel of urban personal journeys, 1975/6
Figure 5.3 The exponential density gradient of urban population densities
Figure 5.4 Schematic representation of the distribution of short-distance personal journeys within a city
Figure 5.5 Principal features of intra-suburban personal movement patterns
Figure 6.1 Population and employment density gradients, and their commuting implications

Figure 6.2 The employment distribution of Greater London, 1962
Figure 6.3 The employment distribution of the West Midlands, 1964
Figure 6.4 Population and employment densities by 2km zones in Greater London and Greater Manchester, 1966
Figure 6.5 The relationship between distance from the centre of Manchester and the strength (centripetal vector) of commuting movements towards the centre
Figure 6.6 Commuting movements from selected local authorities of the Greater Manchester Region, 1966
Figure 6.7 Factors influencing the separation of homes from workplaces
Figure 6.8 Shopping centres and their use: Watford, 1969
Figure 6.9 Shopping movements in Coventry, 1972
Figure 6.10 Shopping centres and household survey districts in Swansea
Figure 7.1 Growth in the number of road vehicles: Great Britain 1904–75
Figure 7.2 Main stages in the transportation study process
Figure 7.3 Variations in commercial-vehicle trip generation according to type of industry, Medway Towns
Figure 7.4 Some factors affecting choice of mode for the journey to work
Figure 7.5 Schematic model of binary choice of travel mode for urban travel
Figure 7.6 Trip-time ratios and their effect on travel-mode choice, Coventry
Figure 7.7 Relationship between travel-mode choice and cost ratio between bus and car, Coventry
Figure 7.8 Basic data for distributing trips between zones using Fratar method
Figure 7.9 Basic data for trip distribution using a gravity model
Figure 7.10 Observed and predicted patterns of trip distribution for work and shop purposes using a gravity model, Dublin, 1970
Figure 7.11 Observed and predicted distribution of trips using intervening opportunities model, Harlow, 1966
Figure 7.12 Growth rates of cars per head, Great Britain, 1965–70
Figure 7.13 Cars per head, Great Britain, 1950–71 and forecasts for 1972–2010
Figure 8.1 New arterial road construction around London, 1900–39
Figure 9.1 Effect of congestion on vehicle operating costs
Figure 9.2 Average journey speeds at peak and off-peak periods in relation to distance from the city centre
Figure 9.3 Average trip times for car and bus travel within central areas of cities according to number of commuters and the proportion travelling by car
Figure 9.4 Patronage of Chells and St Nicholas 'Superbus', Stevenage, 1971–4

LIST OF FIGURES

Figure 9.5 Busway network in Runcorn New Town
Figure 9.6 Idealized bus system for a medium-to-large British city
Figure 9.7 Nottingham's traffic 'collar'
Figure 9.8 Alternative methods of charging for the use of urban roads
Figure 9.9 Parking demand by trip purpose, Croydon
Figure 9.10 Central area traffic control: Leeds, 1981
Figure 9.11 Parking in central Coventry, 1972–6
Figure 10.1 Some relationships between the car, the origins of vehicle noise and its control
Figure 10.2 Distance from an elevated urban motorway and the impact of noise
Figure 10.3 Areas of pedestrian vulnerability, Coventry
Figure 10.4 Pedestrian and traffic segregation in the centre of Liverpool
Figure 10.5 Environmental areas and their relationship to primary distributor roads
Figure 10.6 The loop-and-link underground system on Merseyside.

Acknowledgements

We are indebted to Professor R. Lawton of the Department of Geography, University of Liverpool, for making the original suggestion that we tackle a book of this kind. The subsequent assistance with travelling expenses, cartographic and photographic work provided by the Departments of Geography at the Universities of Liverpool and Salford and at King's College London has been invaluable. For preparing the maps we would particularly like to thank Miss Joan Treasure of the Cartography Unit, Department of Geography, University of Liverpool, and Miss Roma Beaumont, Miss Alison Hine and Mr Gordon Reynell of the Department of Geography, King's College London. A number of unsuspecting students sampled parts of the manuscript and their reactions proved instructive, while particular thanks for reading and making useful comments on individual chapters are due to Peter Brown, Department of Civic Design and Transport Studies, University of Liverpool.

We are most grateful for all the constructive guidance we have received at various times during the gestation of the manuscript but we accept full responsibility for the final form of the views expressed therein. Mrs Patricia Hobson, Mrs Joan Stevenson Mrs Anne Rodgers, Miss Anne Richardson, Miss Bridget O'Donnell, Mrs Joy Barnett and Miss Hilary Salter all dealt expeditiously and in a good-humoured way with the interpretation and typing of our sometimes illegible draft manuscripts. Miss Pat Aylott assisted with the reproduction of diagrams at King's College. We also thank all those who have given us encouragement, help and understanding.

We should like to thank the following copyright holders for permission to reproduce material: HMSO (tables 2.2, 2.6, 2.7, 3.3, 3.4, 3.5, 3.6, 3.8, 4.5, figures 3.1, 3.3, 3.4, 4.2, 10.5), Newcastle City Council (table 2.5), Norfolk County Council (tables 2.6, 2.7, figure 3.2), the Editor, *Urban Studies* (figure 2.1), Policy Studies Institute (tables 2.12, 2.13), West Midlands County Council (figures 3.6, 4.4, table 4.15), The American Society of Civil Engineers (tables 4.1, 4.2), Greater London Council (table 4.4, figures 4.1, 6.2, 6.3), Merseyside County Council (tables 4.4, 4.16, figure 4.1), the Editor, *Traffic Engineering and Control* (tables 4.4, 4.13, figures 4.3, 9.2, 9.7,

9.9), Liverpool City Council (tables 4.4, 4.16, figure 4.1), The Institute of British Geographers (table 6.4, figure 6.5), Dr R. L. Davies (figure 6.9), The British Road Federation (table 7.1), Mclaren Publishing (figures 7.6, 7.7), the Editors, *Progress in Planning* (table 8.1), Penguin Books (table 9.5, figure 9.1), the Editors, *Journal of Transport Economics and Policy* (figure 9.3), *New Society* (figure 9.6), Butterworth (figure 9.8), The Royal Statistical Society (table 9.3), Gordon and Breach (figure 9.10), Coventry City Council (figures 9.11, 10.3), the Editors, *Town Planning Review* (figure 10.2).

Preface

This book has been written principally for undergraduates who are taking courses in urban or transport geography or who otherwise have an interest in the spatial aspects of transport and travel in cities. Because there are so many books on transport, it might be argued that there is little reason for writing another. As well as being numerous, British transport books feature a great diversity of topics, from the briskly-selling illustrated reminiscences for railway, tram, bus, canal and other enthusiasts, to the technical manuals written by and for the exclusive professional use of traffic scientists or public service vehicle managers. Existing geographical transport texts have many strengths, but we believe that absent from these is a satisfactory account of movement in urban areas.

Academic transport study has itself stimulated a great variety of texts, with strong contributions from engineers, historians, economists and transport planners. Indeed, it has commonly been the case that geography students embarking upon transport courses have been directed in the first instance to a number of excellently written economic histories, such as Savage (1974), Dyos and Aldcroft (1971) and Bagwell (1974). This seems to have been partly because most geographical texts have dealt only with specialized transport features, although two highly contrasted exceptions are Appleton (1962) and Haggett and Chorley (1969). Both of these introduced and tenaciously applied a distinctive spatial approach to transport and movement study, and arguably should have led to rather more derivative case studies and re-evaluations than have actually appeared. (Among the few works to follow Appleton's ideas is Farrington, 1973. Two recent texts which at least partly follow network approaches are Hay, 1973 and Fullerton, 1975. However, the continued vigour of network approaches was recently demonstrated by Leinbach, 1976.) Appleton elucidates a number of relationships between the morphology of routes, the physiography of the territory through which they pass, and the technology and operating characteristics of each transport type. Haggett and Chorley, on the other hand, explore the potential for analysing route and flow networks in terms of their simplified structure or topology. These attempts to apply a consistent approach to the geography of movement are still far outnumbered by books with more narrowly defined subjects. For example, from among British geographers, Bird (1963, 1968, 1971) has written several original accounts of the evolution of ports, Sealy (1966, 1976) has provided the geography student with texts on airports and air traffic, and O'Dell and Richards (1971) and H. P. White (1973) among others, have brought a geographical interpretation to bear on railway history. Most recent in this genre has been Townroe's work (1974) on the social and political consequences of the motor car. A usual characteristic of these volumes, and also of many of the more recent monographs with applied aims and topics, has been their focus on international, national, regional and inter-urban scales of movement analysis, and their relative neglect of shorter movements, particularly those taking place within towns (see, for example, Chisholm and O'Sullivan, 1973; Hay and Smith, 1970). The same limitation applies to the few other broad-based geographical accounts of transport that have appeared, none of which, incidentally, deal with British conditions (Taaffe and Gauthier, 1973; Wolkowitsch, 1973; Becht, 1970; Lowe and Maryadas, 1975). As the majority of personal journeys, and probably also of goods movements, occur at this scale, there is a good case for attempting to redress the balance.

Short distance or intra-urban journeys are individually mundane but in total they are of great practical importance to the structure and development of urban areas and in terms of the resources that they consume. They should also be a sure source of stimulation to the geographer's spatial interests. If, in the past, it has been the static and moving landscape of transport routes and vehicles which has most frequently excited the geographer-enthusiast, changing times and concerns should now be shifting the focus of most students' inquistiveness. We hope that by attempting to synthesize in this book a small part of the available information on the social and economic implications of urban travel and transport, on the formulation and execution of government and local planning policies, and on the individual's response to the available transport facilities and functions, we can provide a helpful introduction to an enormously wide and interesting field of study.

It can also be argued that those studying urban geography deserve rather different and fuller accounts of the role of transport in serving and shaping towns. At one time or other all urban residents and economic activities must resort to travel between two points within the same city in order to help create or to consume a good or service. Social contacts, many recreational pursuits, organized worship and education, to name a few activities, would also be impossible without movement through space. Movement is therefore fundamental to the creation and maintenance of the spatial structure of cities, and it is important also in promoting urban change through time. Yet many accounts recommended to the student oversimplify the relationship between urban growth and transport development. It is our view that many currently favoured topics in urban geography also demand some reexamination in the context of movement in cities. As one example, studies of travel for shopping purposes have not only tended to ignore the mass of short-distance walking journeys, but also appear to have been more concerned to prove or refute the existence of hierarchies of shopping centres or other hypotheses than to establish fully the characteristics of consumer travel and the problems associated with them. While it must be accepted that standard urban geographies cannot comprehensively cover their subject because only selected features have been researched, such as the distribution of land use, economic activity, and certain social patterns, it is also true that the selection of topics is not only influenced by the constraints of data and information availability, but also by the preferences, bias and tradition of urban geography. Hence, for example, the location of industry in urban areas has received much less attention than residential patterns or retail structure. Urban transport seems to suffer the same fate; while its influence on the location of land uses in urban areas is frequently acknowledged, this is usually in passing rather than in any depth.

Recently it has been acknowledged that urban geographers have been too exclusively interested in spatial patterns or outcomes and have not paid sufficient attention to the processes which create these distributions. It is these processes which structure a town and its inhabitants' lives and among them the available kinds of movement facilities – and their cost, speed, convenience, and network characteristics – are of great importance. We have attempted, therefore, not only to describe movement patterns but also to elucidate the important influence of transport and travel upon many aspects of urban geography. The book also concerns itself at times with existing urban transport policies and problems, not only because an awareness of politics, policies and planning is essential to understand urban movement, but also because we argue that the geographer can contribute much more substantially to the improvement of urban areas by studying movements of different kinds and by contributing to transport policy formulation.

The structure of the book can be broadly divided into three parts which represent a continuum rather than discrete units. The first section comprises five chapters which describe and explain the causes of urban travel: why it is necessary, what determines the frequency of travel, what distances are travelled for different purposes and which modes and routes are used. The objective here is to identify and explain patterns of movement both within cities and between them. The section opens with a discussion of the impact of changes in nineteenth- and early twentieth-century transport technology on urban growth and travel patterns. An attempt is made in Chapter 5 to develop this essentially static and disaggregative descriptive approach into an aggregated spatial account of movement in cities.

Static, cross-sectional approaches to the analysis of urban travel are useful, but tend to miss the fundamental dynamic influence of economic and social change on patterns of movement. For example, urban geographers can acquire only a limited understanding of city retail hierarchies or of residential location behaviour if they are not aware of the trends in travel behaviour which sustain these patterns. Hence, the second part of the book, consisting of two chapters, discusses the reasons for and directions of change, and examines the techniques which have been widely used to monitor and predict urban travel.

Both the patterns of movement and the ever-present changes give rise to the many urban transport problems which exercise national and local government. Geographers have amassed an expanding range of research skills and capacity which equip them to offer a useful evaluation of policies and of transport management practices. These applied aspects are discussed in Chapters 8 to 10, where an effort is also made to cast an eye towards the future.

Most of the material that has been used has been derived from studies of transport in British cities. Forays into the very extensive American literature on this subject have been restricted, in the belief that we have been too ready to rely on this source in the past. This may indeed have resulted in an inappropriate application of American generalizations and solutions to transport and its problems in British cities, which, it is sometimes forgotten, have higher densities of development, less extensive areas, lower standards of living and levels of car ownership, and a different political, administrative and planning environment. Quite deliberately, therefore, we have often emphasized the distinctive features of British conditions but nevertheless our dominant approach has been nomothetic rather than idiographic. A large proportion of our comments will appertain in whole or part to transport and travel in other highly industrialized nations. It is indeed possible that the characteristics and conditions of travel in Cambridge, Carlisle or Coventry are more similar to those in a large number of modern cities than the advocates of a diffusion of Chicago or Californian approaches would have us believe. To this end, we hope that this book serves as a useful starting point for teaching and research outside Britain.

1
The transport revolution and urban growth

Person and vehicle movements in today's cities are as much guided by the events of the past as they are shaped by the requirements of the present. The process of evolution and its effects on the growth of cities was slow, but the momentum increased quickly during the eighteenth century as a result of the stimulation provided by the industrial revolution. More efficient methods of moving larger quantities of raw materials and goods were needed and increased industrial and commercial activity generated growing demands for people to travel within and between cities. Migration from rural areas to the cities also encouraged more advanced transport facilities and services beginning with the stage coaches of the late eighteenth and early nineteenth centuries. Probably the most significant feature in the evolution of transport during this period was that it became possible to travel increased distances without consuming more time. This was crucial in the context of urban development because it allowed the residences and workplaces of urban dwellers to be separated to an increasingly significant degree and changed the traditional 'compact' city to a more dispersed form. Cities which were proving particularly attractive to industrial entrepreneurs and to rural migrants seeking work could begin to handle these demands more effectively as well as to sustain the effectiveness of their urban functions. In addition, new methods of movement conferred different accessibility advantages on intra-urban locations and encouraged functional segregation of land uses and the appearance of some new uses, such as higher-order retail functions. This would not have been possible without the liberalizing influence of urban transport technology in the sense that it promoted outward expansion of cities, with residential development leading the way, while at the same time sustaining the accessibility advantages of the centre.

The golden era of urban transport evolution was undoubtedly the nineteenth century. The pace of transport development was more rapid than at any other time and its consequences were not confined to changes in city form and growth but extended to increasing the volume of movement, its composition, and the purposes of urban trip-making. These structural changes in urban travel have continued to take place and have presented cities with a wide range of transport-related problems with origins in the events of a century and more ago. Some of the more important nineteenth-century developments are therefore worthy of closer examination as part of the explanation for the existing characteristics of movement in cities which occupy such a prominent place in this book.

Urban transport and travel before 1830

Up to the beginning of the nineteenth century the structure of cities was in part dependent upon the distances which residents were prepared to cover on foot in order to conduct their work, recreation, social activities or business transactions. The planning and arrangement of cities was undertaken in a way which would minimize the need for internal transport, both in terms of the number of trips necessary and the length of these trips (Schaeffer and Sclar, 1975, pp. 8–17). City space was therefore characterized by extremely high densities with narrow streets, buildings built as high as contemporary construction techniques would allow and little or no wasted land except where ground conditions made building impossible – housing even spread across the bridges over the Thames in London. Such 'foot cities' lacked the distinctive social and functional segregation of land uses which are well-developed in modern cities (see also Boal, 1970, pp. 73–80). Buildings served not only as family homes but as workshops for the family so that the tanners, bakers, candlestick-makers or moneylenders operated from the same premises as their family residence. In addition, they often provided accommodation on the same premises for apprentices, housekeepers, relatives and others needed to keep the business or the household in successful operation. With employers and employees resident on the same site the journey to work requirement was much more limited than today and, depending on the self-sufficiency and the size of individual households, most of the trips generated were in connection with businesses or shopping. While congestion was not unknown in such towns and cities, the street system was perfectly adequate until technological changes led to an erosion of the household organization as economies of scale encouraged larger production and business units and, eventually, the separation of homes from workplaces. This could not have taken place without the aid of developments in transport technology which took place during the nineteenth century. Although it is difficult to prove whether increasing separation of homes from workplaces was the main stimulus for transport development or vice versa; given that most entrepreneurs could not provide a transport service ahead of demand it seems reasonable to suggest that changes in urban structure preceded transport improvements. These substituted the dependence on foot transport with vehicles which permitted, in relative terms, rapid exchange of goods and services as well as larger quantities than could

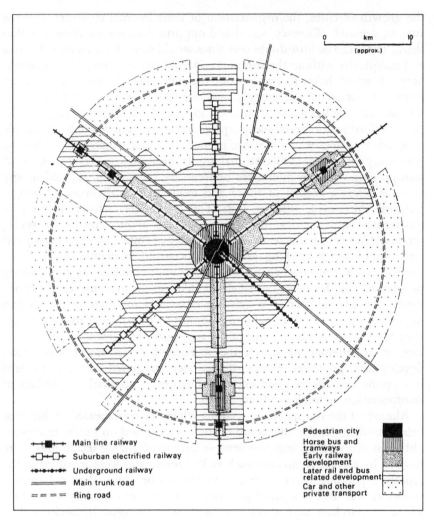

Figure 1.1 *Schematic relationship between urban form and transport*

be moved by one man or by pack animals. 'Foot cities' could rarely expand beyond a population of 50,000. The nineteenth century transport revolution allowed this to be increased many times over without apparently impairing the efficiency of the city. The link between transport and urban development was firmly forged during the century and the 'foot city' was soon replaced by the 'tracked city' and, during the twentieth century, the 'rubber' or 'highway' city (Schaeffer and Sclar, 1975, pp. 18–60).

Therefore, major changes in transport technology which began to emerge during the first quarter of the nineteenth century had a major influence on the growth of cities, the organization of their internal structure, and the supply, demand, efficiency, speed and opportunities for movement within them. There can be little doubt that cities would never have increased in size and complexity without the stimulus and guidance provided by a transport network which helps to smooth out the differences between locations in terms of locational advantage/disadvantage as well as reducing the time/distance between sets of points in an area of urban growth (figure 1.1). As Mumford has observed the 'purpose of transportation is to bring people or goods to places where they are needed' in a way which will 'widen the possibility of choice without making it necessary to travel. A good transportation system minimizes unnecessary transportation; and in any event it offers a change of speed and mode to fit a diversity of human purpose' (Mumford, 1964, p. 178). This is precisely what took place in nineteenth-century cities in Britain. Marked improvements in socio-economic indices such as real income, security of employment, occupational status and shorter working hours also occurred in parallel with, if not partially as a consequence of, transport improvements as the century progressed and this helped to swell the demand for movements of various kinds. Indeed, the way in which our cities and towns have grown and organized themselves internally in response to transport changes through time has helped to mould some of the transport problems which they now face in an attempt to match the demand for movement and urban development processes. Hence, it has been suggested that the 'modern metropolis in both its good and bad features is peculiarly a product of transportation technology' (Rae, 1968).

Almost all the transport innovations which led to increased mobility of urban residents in the nineteenth century were based on public transport, although many of the improvements in carriage-building technology were devised by private companies such as Pickfords who were involved in the rapidly expanding carrying trades in the major cities. Personal private transport was mainly a privilege enjoyed by the very wealthy who could afford to purchase and operate horse-drawn carriages. Intra-city public transport was mainly a product of the period after 1830 – prior to this time public transport had been almost exclusively inter-city. The long-distance stagecoach dominated these long-distance routes, to be followed and eventually replaced by the railways during the first half of the century (Dickinson, 1959). The well-developed canal network between settlements was the main method for moving goods and raw materials, while within cities movement of goods was highly inefficient, depending mainly on horses and some horsedrawn carts. Some cities, such as Birmingham, did have quite well-developed urban canal systems, but this was exceptional and, in general, there was little scope for enhanced industrial and commercial location in towns and cities until superior transport networks were introduced. The canals were insignificant for person-movements but cities with navigable rivers, such as London, Liverpool and Glasgow, did have ferry services in operation at the end of the eighteenth century. Apart from their contribution to movement into and within these cities, the ferries helped to open up new areas for urban development, such as Birkenhead across the Mersey from Liverpool, which had previously been inaccessible.

Before the development of public transport most movements in British cities were made on foot, whether for the journey to work, day-to-day business affairs or social activities. This imposed sharp limitations on the outward spread of most urban areas to a maximum of three miles, or approximately one hour's walking distance, and in most cases much less. For the vast majority of the population workplaces and residences were in close proximity with most people working in their immediate locality (see, for example, Warnes, 1970). Functional separation of urban activities was far less obvious than today because movement between activities was time-consuming, and only really essential trips, such as the journey to work, were made. Opportunities for making other trips were restricted by the length of the working day – ten hours and more. At the beginning of the nineteenth century, towns and cities were, therefore, very compact and while the patterns of movement within them may well have been complex, trip purposes were restricted and inflexible.

Cities remained compact until the railways became more interested in providing short-distance services. But during the interim there were a number of other important developments in urban public transport. Short-stage coaches were the first to attempt to operate within urban areas, particularly the larger centres such as London and Birmingham, either providing totally independent services or as feeders to the railways. This traffic had reached considerable proportions in the London area by the early 1820s. Some 600 short-stages a day, making 1,800 journeys, left the City and West End to destinations in the suburbs and outlying villages to the south and north-east of the Thames (Barker and Robbins, 1963, pp. 4–5). Each coach carried four to six persons inside as well as some outside passengers, but journeys were costly; 1s. 6d.–2s. single to the City or West End from Paddington. Only the wealthier classes could afford to use them.

The rapid evolution of road transport techniques

But the short-stage coaches were comparatively short-lived and quickly usurped by the larger, more comfortable horse omnibus which came into popular use in London in 1829 on a service from Paddington to the Bank (Dyos and Aldcroft, 1971, pp. 219–20). The journey time was scheduled as one hour but the five-mile journey sometimes only took forty minutes (Barker and Robbins, 1963, p. 22). The capacity of the early omnibuses was twelve persons but in later versions some were able to carry twice this number. The original fare was 6d. per trip but later, when competition became more severe, charges fell to a standard fare of 1d. for each journey. This made the omnibuses more accessible to a wider cross-section of the population. Although the early services were along set routes, there were no official picking-up and dropping-off points – these did not become legal in London until 1832. But by the mid-nineteenth century the horse omnibus was a more important urban carrier than the railways and by 1860 the London General Omnibus Company, which alone owned some 600 vehicles of the 850 on the streets at the time, was carrying 40 million passengers a year on sixty-three distinct services in and around London (Dyos and Aldcroft, 1971, p. 220; Savage, 1959, pp. 87–9). The diffusion of the horse omnibus to provincial cities occurred some years later – to Leeds in 1839, Birmingham in 1860, and Bradford in 1870 – but the evolution of the service network, fare structures, and the contribution to increased person movement and, to a more limited degree, goods movement, was much the same. Horse omnibuses were expensive to run, however, and therefore steam carriages were tried in London between 1821 and 1840 but mechanical problems rendered a promising future uncertain and they only offered short-lived competition to the omnibuses.

By the end of the 1860s the demand for horse omnibus journeys was growing steadily each year and signs of peaking in person and vehicle movements were already beginning to appear as the clerks and artisans able to afford the fares for suburban journeys lived in increasing numbers outside the overcrowded city centres. Much of the congestion was caused by the limited capacity of the horse omnibuses, which meant that large numbers were required to satisfy demand. This congestion was exaggerated by the limited range of the omnibus and, in cities which were still essentially compact, journey times were particularly slow in the central areas. A more rapid method of transport with a larger carrying capacity would soon begin to test the dominance of the horse bus. In their heyday, fares on the horse omnibuses had come down to less than 1d. per mile in the larger cities, such as Sheffield, Leeds and London, and the average speed, including stops, of town services was five miles per hour (Savage, 1959, p. 89).

It has already been suggested that the railways were slow to emerge as serious competitiors to the horse omnibuses. This allowed horse trams, introduced commercially in London and Birmingham in the late 1860s, to share fully in the ever-growing demand for public transport during the second half of the century. The main advantages of the horse trams were their greater carrying capacity (almost double that of the horse bus), greater efficiency because there was less friction encountered by the horses pulling the trams, the ease with which passengers could climb on and off, and the ride which was smoother and safer than the horse omnibuses. The first urban tramway in Britain was constructed in Birkenhead in 1860 and ran between Woodside and Birkenhead Park, but it was not a great success because of a number of accidents which led to adverse publicity and great public uncertainty about the safety of the trams (Savage, 1959, pp. 89–90). The same entrepreneur, G. F. Train, opened two short experimental lines in central and inner London in 1861, but again with comparatively little success, partly because he insisted on using elevated tracks which naturally interfered with movement of other traffic on the roads. It was another ten years before tramways became an accepted form of transport to and between the suburban areas of our major cities such as Liverpool, Hull, Glasgow and London.

There was much initial hostility from residents in affected streets, from some local authorities and government departments, but a tramway finally became fully operational in Liverpool in 1868, and in London in 1870. Expansion of networks was slow, however, mainly because of restrictive legislation and the tight control on the operation of tramway undertakings which were exclusively private companies. Initially, therefore, the horse tramways had a very limited route mileage so that they in no way totally replaced the horse bus which frequently used similar routes up to five miles from the centre of towns. In 1878 there were only 237 miles of tramway in the whole country but by the turn of the century this had increased to 1,000 miles and by 1911 to 2,500 miles (Savage, 1959, p. 90). Prior to 1904 most of the trams were horse-drawn and they suffered the same basic disadvantages as the horse omnibuses and did not greatly extend the length of urban trips. There were some steam trams operating in the Black Country, the North of England and Scotland, but substantial improvements in journey times and the length of networks only became possible following the introduction of electric traction. Leeds began to install electric trams in 1891 but they did not enter general service throughout the country until the early years of this century. By 1914 there were some 12,000 electric trams in operation, mainly in urban areas, and they carried some 3,300 million passengers. Municipalization of tramway systems also began to take place at the end of the nineteenth century and this permitted the introduction of cheaper fares which then allowed the working classes to live further from their places of work than before (for a full discussion, see Dickinson and Langley, 1973).

Both the horse omnibus and the tram benefited from the general increase in the propensity of urban residents to travel. This was encouraged by the lower fares which resulted from the intense competition between the two modes along the major arterials. The tramways could achieve less comprehensive coverage of outer areas than the omnibuses because they were tied to tracks which clearly did not allow easy introduction of new routes or the operation of experimental services in untapped areas of the suburbs. They were also almost totally excluded from the central area of London, which remained the domain of cabs and horse omnibuses almost until the end of the century. Elsewhere, such as Manchester, this was not the case and virtually all the city's tramway routes converged on Market Street and made the area an ideal location for the growth of commerce and entertainment which was dependent on substantial patronage.

The urban transport role of railways

Throughout this period the railways, as a mode of travel within urban areas, were becoming better-established but only provided very limited coverage of the existing areas of urban development or potential areas for urban expansion. The railways certainly handled the longer-distance suburban journeys of six miles or more, leaving the omnibuses and trams to handle the shorter-distance traffic, but the opportunities for short-distance rail passenger work in London, for example, were not demonstrated until 1836 and it was another forty years before railways became extensively used for urban trips. It should be remembered, of course, that the scale and extent of British cities was very restricted until late in the nineteenth century – the continuous built-up area of Manchester for example, extended no more than two miles from the Market Place at mid-century. Most of the railway construction activity of the last quarter of the nineteenth century was in fact confined to extensions into the suburbs and the immediate hinterland of the larger cities. The road-based modes of travel often provided feeder services to those stations which did exist within ten to fifteen miles of the cities, but, in general, intra-urban movement remained confined to a zone within four to six miles' radius of the city centre. The rail services which were used, for the journey to work in particular, depended upon patronage by the wealthy classes – return fares of one shilling per day were an impossibility for the majority of urban residents who earned £1 per week or less.

But there can be no doubt that the railways were the key to much more rapid outward expansion of Victorian cities (see, for example, Kellett, 1969; Perkin, 1970). Clark (1957–8, p. 248) has suggested that they were the 'implement which really chiselled apart the compact Victorian city'. In many cities, however, the railway companies did not possess termini near enough to the centre to provide the kind of access to centres of employment which the trams and omnibuses provided. Neither could they utilize existing facilities for their routes and services – they had to crave routes through existing built-up areas and they eventually came to own eight to ten per cent of central land in cities as well as indirectly influencing the functions of twenty per cent of all urban land (Kellett, 1969, p. 2). Because of their preoccupation with inter-city services the railways had been largely concerned with locating city termini where property values were lowest and this was usually at the edge of the built-up area. Quite commonly these early terminals were subsequently converted to goods depots as the railway companies undertook extremely expensive extensions into and through the central area to serve larger-capacity termini. The goods depots at Oldham Street and Liverpool Street in Manchester, at Edge Hill in Liverpool and at Bricklayers Arms in south-east London are all examples. The London Bridge terminus of London's first suburban railway, the London and Greenwich, was fortunately within easy walking distance of the City and the railway carried more than 650,000 passengers on its frequent trains during the first fifteen months of its existence. The trains travelled at fifteen-minute intervals and when the line was fully extended to Greenwich in 1838 horse omnibuses from Blackheath, Lewisham and Woolwich were scheduled so that they met every train. This had clear implications for urban residential expansion and in 1844 the line carried more than two million passengers (Barker and Robbins, 1963, pp. 44–6). The lines which followed the London and Greenwich were remarkably slow to develop their suburban traffic and the London and Birmingham's first station after its terminus was at Harrow, 11.5 miles; the Great Western's was at Ealing, 5.7 miles; and the Great Northern's at Hornsey, 4.0 miles (Hall, 1964, p. 63).

The railways brought with them a distinctive form of urban expansion which had not been characteristic of the trams and omnibuses (Buchanan, 1958, pp. 59–60). Because steam trains were slow to accelerate it was necessary to have a considerable distance between stations (between one and two miles) in order that trains could gain sufficient headway to utilize their speed potential as efficiently as possible. Hence, beyond the high-density urban areas served by the competing public transport systems, growth took place at individual nuclei surrounding each of the stations along the radial routes out of the major cities (figure 1.1). As the railways attracted more goods and passengers these nuclei spread outward and eventually, with the aid of later improvements in private transport, they coalesced to form more cohesive areas of urban development typified by outer London. Urban railways did not, therefore, cause indiscriminate scatter of urban growth and travel; rather they introduced considerable rigour and coherence into expansion, as well as regularity into movement, which had been less obvious during earlier periods.

Congestion of movement within central areas was also eased by the railways, especially in central London where it was necessary to transfer to road transport to complete journeys to destinations in the West End and most parts of the City. The situation was eased by the construction of the Metropolitan Railway during the 1870s and the Inner Circle Line which linked the major termini around central London. These were cut-and-cover lines, however, which were unable to penetrate the heart of the central area because of the exorbitant cost of compensation for buildings displaced and construction costs – some of the later extensions of the Metropolitan had cost more that £1 million a mile (Hall, 1964, p. 70; Lee, 1972). Improved tunnelling methods eventually allowed continuous tunnels at deep levels and at lower costs than cut-and-cover and this, combined with electrification, provided the ideal conditions for providing rail services which crossed the central area and which filled an important gap in the transport services of the area. Electrification of the underground was essential if it was to compete successfully with trains for inner urban traffic; the latter would be preferred for their cleanliness alone as long as the steam trains on the underground continued to create dirt and smell of the worst order. The first steam underground railway to be electrified (1903) was on Merseyside on the route linking Liverpool and Birkenhead under the Mersey (Ellis, 1959 pp. 110–16). This was followed by similar conversions in Glasgow and Newcastle. These lines provided the first real opportunity to contain the serious vehicular congestion in central London, especially during peak hours, which was an inevitable consequence of a transport system which did not allow 'through' trips to be made from, say, the South Coast to the Midlands, without the need to transfer between London termini. Similarly, there were serious commuter distribution problems once they reached the central London rail termini, particularly if they worked in the West End. The difficulties were exacerbated by the complete absence of direct rail links between the City and West End, a role only very partially served by the Circle Line which linked the main-line termini, even though the two areas were far from being independent economic systems. The City and South London, opened in 1890, was the first tube railway in London and, in common with the Waterloo and City Line opened in 1898, it connected the City with the south bank of the Thames. This reflected much of the demand from commuters but did nothing to ease the central area's basic congestion and communication problems. This opportunity did not occur until the Central London line was opened between Shepherds Bush and Bank in 1900 and this was the first line to cut across the West End from west to east and to link it to the City. Three other lines followed in quick succession in 1906–7 and they all crossed the West End. The first two decades of tube railway development were largely a response to central London's congestion problems which the motor buses, just beginning to appear at the end of this period, could only hope to resolve marginally. Underground lines across the central area were profitable for obvious reasons as the railway entrepreneurs had shown little interest in servicing the needs of the suburban travellers who were able to rely on the expanding surface network of the underground, the electric tramways and the omnibus services.

As a consequence the model of the relationship between transport and urban growth (figure 1.1) was modified in the London case because the underground trains offered superior headways over the conventional railways and therefore allowed more frequent stations to be provided. There was an obvious potential for opening up areas for development still relatively near the centre but inaccessible from the more widely spaced railway stations (Howson, 1967). The pace of urban growth was accelerated as a result, with the expansion of nuclei and underground stations following a similar cycle to that found along the non-electrified railways.

The technological developments in railways and other forms of urban transport during the nineteenth century were a necessary but an insufficient condition for the increases in personal mobility and the acceleration of urban growth which began during this period. Often the relationship between transport and urban development has been interpreted too simply and directly. As important as the developing technological possibilities was the declining real cost of transport, and indeed of housing, both of which were associated with the increasing standard of living of the population. It is true that the declining real cost of public transport fares can be partly attributed to the increasing efficiency and higher carrying capacities of successive technologies, notably with the much greater attractiveness of electric trams over horse-drawn road transport. But if electric trams became in the words of Richard Hoggart 'gondolas of the people', and the first technological form to be widely used by most income groups, it was probably the increased ability of urban populations to pay for personal transport in the last decades of the century which accounts for this development. Likewise, on the railways, it would be simplistic to see the rapid growth of commuter traffic towards the end of the century solely in terms of workmen's tickets established by the 1883 Cheap Trains Act or any other supply variable. The demand to rent or even own new higher quality housing was growing strongly. This implied longer distance journeys within towns, and increasing numbers had sufficient income to pay the railway fares.

Stabilizing technology and accelerating urban growth in the first half of the twentieth century

The final stage in the evolution of public transport in cities was the appearance of motor buses which were 'a major – perhaps *the* major – determinant of London's twentieth century transport map' (Barker and Robbins, 1974a, p. 119). They offered far greater flexibility of routeing, servicing, better time-keeping and the prospect of access from a wider to suburban rail and underground stations, and began operating in small numbers in 1899. As with all the other new methods of transport introduced earlier, high initial operating costs and technical problems retarded their initial impact on urban travel (Barker and Robbins, 1974a, pp. 130–3; also see J. R. Day, 1973). On lightly trafficked suburban or rural routes buses were fairly reliable as feeders to the railway stations but on busy urban routes constant starting and stopping was too punishing. Generally, buses were introduced too quickly and it was only after 1910 that mechanical difficulties and some unrealistic competition amongst operators, were reduced to an acceptable level. 200 buses were operating in London in 1905 but the figure quickly reached 3,000 by 1913; such was the potential that companies such as British Electric Traction set up subsidiaries to develop motor traction services (Barker and Robbins, 1974b, p. 141). Provincial bus services started rather later from 1911 onwards in Birmingham, but by the end of 1914 most major provincial cities were using this new form of urban transport on at least some of their routes, particularly those not covered by the electric tramways. The latter were reaching the peak of their activities around this time (Wilson, 1971; Bett and Gillham, 1957) along with several new electric underground lines and the electrification of existing steam underground railways in London, Liverpool and Glasgow (Ellis, 1959, pp. 110–16).

Throughout the years up to 1914 intra-urban movement increased considerably, but only as far as public transport networks permitted. Services operated along set routes and did not provide the kind of door-to-door travel which was later to be one of the most valuable assets of the motor car. In some ways it also discouraged long-distance travel and helped retain a relatively compact urban form because speed limits and poor protection from the weather did not make bus or tram travel the most comfortable experience. Whatever the difficulties, however, it is likely that few urban passenger road-transport routes were operated by non-motorized transport by 1914 although the movement of goods remained the domain of horse-drawn vehicles. After the First World War the bus consolidated its position; between 1919 and 1921 the number of buses and coaches in Great Britain almost doubled and helped to fuel the pattern of suburban expansion illustrated in figure 1.1. During the 1930s and through into the immediate post-war years some urban bus routes were extended as they gradually replaced trams and trolley buses. London's Green Line services in the outer suburbs and beyond flourished during this period but competition from the private car was already challenging the role of public transport as the dominant element in urban growth. Improvements in bus capacity, comfort or reliability were an inadequate basis for reversing this trend. Personal private transport had really yet to emerge as the most important factor affecting the mobility of urban residents. The bicycle was certainly a form of personal transport which was first used in 1869 but it was only suitable for short-distance movements and was impracticable for the longer journeys made necessary by the expansion of towns and cities during the railway era (Sekon, 1938). Only one of the pre-1910 tube lines extended beyond the edge of the main built-up area of north London, to Golders Green, but during the next two decades the shape of the inner urban tube network was transformed by a flurry of activity, orientated towards the creation of an extensive above-ground system by extending many of the existing lines into open country around London, particularly on the north bank of the Thames. Most of these lines eventually extended up to twelve miles from the centre and gave commuters a maximum journey time of forty-five minutes excluding time taken to reach stations, which could be placed at frequent intervals, 0.3–0.5 miles, along the route, thus providing a catchment area large enough to make the lines economic (Pick, 1926–7, pp. 163–9; Simmons, 1966, pp. 234–40). The rolling stock, which has changed little in basic design, had a high capacity including a large number of standing passengers at the rush hour and, provided that passengers could be attracted to use the suburban stations in sufficiently large number, viability would not be a problem. The development of the motor bus was crucial in this respect and enabled suburban transport services to be properly coordinated so as to extend the catchment areas of the underground extensions to a degree which would make such development worthwhile. These two developments in metropolitan transport were instrumental to the acceleration of infilling by residential and other activities between the urban developments stimulated by the earlier long-distance railways (Dyos, 1955–6, p. 14).

Improvement was the main theme in the development of urban and suburban main-line rail services during the years after 1919 up to the Second World War. The most significant aspect of improvement during this period was the rapid advance of electrification which permitted a larger number of trains, with smaller headways and shorter station spacing, to tap the underdeveloped settlements in and around the edges of the main built-up areas of the larger cities, particularly London, Newcastle, Glasgow and Liverpool. Electrification 'was the means of satisfying a rapidly developing social trend towards better and less congested housing' (White, 1962, p. 185) and its effects are apparent in the estate-location policies of both local authorities and private developers. Instrumental in this respect was the electrification of the Southern Railway, which began running electric trains on the Coulsdon North and Sutton lines in 1925 (Barker and Robbins, 1947b, p. 259; see also Denby Marshall, 1963, vol. 2, pp. 407–21). There was an urgent need to electrify the Southern's network because a large proportion of its revenue came from passengers rather than freight, many travelling to and from London as commuters or going short distances between the provincial centres on the network. The latter traffic grew substantially in the 1930s when electrification, largely confined to a ten–twelve-mile radius from central London, was extended to long-distance routes such as London to Brighton (1933) (Hall, 1964, p. 74).

The influence of the Southern on the suburban expansion of London south of the Thames was almost exclusive because the tube railways, which had also been electrifying from 1903 onwards, agreed not to extend their routes south of the river (with the exception of the Northern Line to Morden in 1926). Main-line electrification north of the river was less extensive and the overland extensions of the electrified underground lines performed the role adopted by the Southern – the Northern Line extension to Edgware (1917), the Piccadilly Line extension to Cockfosters and Hounslow (1932–3) (Hall, 1964, p. 71). Many of these lines opened ahead of suburban development and depended on electrification of the tramways and efficient operation of motor bus services to ensure economic viability. British Railways began the electrification of its Great Eastern lines from Liverpool Street station in London in the 1950s and this simply reinforced suburban development already well-established by the very efficient steam-operated rail service which had coped with commuter demand during the inter-war years. The Southern Region's Kent coast electrification beyond Gillingham was completed in 1959 and beyond Sevenoaks in 1962 and the London-to-Bournemouth line beyond Working was not electrified until 1967. Later still, London Midland Region electrified its line from Euston to Northampton (1966) and Rugby and revolutionized the volume of commuting on the line. Finally, the Kings Cross to Royston (Herts) line will also provide electric suburban services in the near future and will no doubt attract additional travellers and more development along its route. There was also a major electrification of suburban rail services around Glasgow during the late 1950s.

The first motor car to run on Britain's roads in 1888 was an imported Benz but the initial impact of the car was impeded by speed restrictions which applied to steam and motorized transport. These were eventually lifted in 1896, subsequent to which, progress in car development and highway improvement was comparatively rapid. By the 1920s the mass car market had arrived and prices had become sufficiently low for the car to be potentially available to all social groups. As far as urban development was concerned, few areas within cities or within easy reach of them were sacrosanct from the accessibility provided by private cars and other motor vehicles. For the first time urban transport users were able to make their own decisions about frequency, purpose, length and direction of their movements. Within the constraints of income this allowed a wider range of residential choice as well as greater flexibility in the locational behaviour of industry and other land uses in cities. Instrumental in these changes were the motor van and lorry which revolutionized the intra-urban carriage of goods. Both could be adapted to carry a wide range of specialized goods which the railways could not handle, and had the added advantage of providing door-to-door service over any distance. Industry and warehousing therefore became far less dependent on obtaining access to sites near main-line termini or along main-line rail routes.

Not surprisingly, the extended mobility provided by private transport allowed explosive urban expansion which far exceeded anything which took place up to the First World War. Infilling, ribbon development and sporadic building were the hallmarks of the inter-war years (figure 1.1.). Extensive 'semi-detached' estates on green-field sites around the fringes of built-up areas reflected improved housing standards and the ability of a larger proportion of the population to purchase their own homes. This was facilitated by improvements in real income and the related expansion of the building societies which provided the mortgages for the growing band of suburban residents. Decentralization was not confined to residential land uses – there were already signs that industry was beginning to follow, especially in London. The density of development was also much lower than before and, increasingly, it was assumed that residents or companies would make use of motor transport to satisfy their travel requirements. The private car and frequent bus services gave rise to extensive ribbon development along major routes where 'houses, bungalows, shacks, shops and petrol filling stations sprang up in motley disarray; verges, hedges and trees disappeared; views and outlooks were utterly destroyed. The free flow of traffic became impeded by standing vehicles and crossing pedestrians and these ribbon-developed roads became death traps in the full sense of the term' (Buchanan, 1958, p. 60). Sporadic building in open countryside also reflected the increased range of travel permitted by the private car. The worst affected area was the Home Counties, where development often occurred with inadequate services and poor access to retail, educational and other facilities. Since 1945 the private car has dominated expansion and change in urban form. Measures to increase road capacity, to speed the flow of traffic, or to provide adequate parking facilities have made their own distinctive contribution to urban morphology.

Since the last war the construction of new roads and motorways, both inter- and intra-city, has further promoted low-density residential expansion as well as decentralization of industry and commerce from central cities. Motorways have been provided on the assumption that they will help to relieve the congestion caused by vehicular transport, particularly in city centres – such gains are at best short-term and certainly illusory in the longer term because the improvement in journey times created by the motorway helps to generate more traffic until more road space becomes necessary. In cities where planning controls such as the Green Belts effectively limit further expansion at the edges of the continuous built-up area, the motorways have permitted commuters to travel from towns which are forty to fifty miles from their workplaces in the city in sixty to ninety minutes. Meanwhile the identity of the 'motorway city' becomes more and more elusive. At the same time, the dependence of the dispersed city on all forms of transport remains greater than ever, especially the dependence on the private car, and the ease with which movement can take place within the urban fabric is related to whether individuals own a car or whether they happen to live near suitable public transport facilities. Just as the train became an important mode of transport for recreation journeys in the nineteenth century, the car has increasingly adopted this role during the twentieth. In theory at least, the restraints imposed on choice of excursion by public transport networks have been relaxed and every family which owns a car has becomes master of its own urban and extra-urban movements. In practice, every family does not, and cannot, own a car and this, along with other issues to be discussed later, has caused the approach to urban transport provision and its impact on urban development to be drastically revised in recent years.

Transport technology in the last century in general promoted nodality of cities and their activities. In this century this has been replaced by greater dispersal of cities and a concomitant increase in the total volume of movement of goods and people. Road travel almost totally dominates urban movement – between eighty and ninety per cent of all trips in larger cities such as London and over ninety-five per cent of all trips in smaller cities such as Leicester or Sheffield.

This has been a very brief sketch of the major changes in urban transport over a period of almost 200 years, but it goes some of the way towards an understanding of the present structure of cities and their transport networks and problems. The evolution of both public and private transport modes has not ceased, changes will continue to take place in response to contemporary demands for movement and conflicts between transport and other facets of urban life. But changes in transport technology do not in themselves completely explain the changes in the pattern and volume of urban travel and the associated problems which have arisen during the present century. Some of the latter will be discussed at length later but it is worth reminding outselves that all the technological developments discussed in this chapter still leave us with a crucial difficulty to which Owen (1956, p. 1) has referred as follows:

> The capacity of the transport system and the low cost and dependability of transport services has enabled an increasing number of people to seek the economic, social and cultural opportunities that urban living ideally provides. But paradoxically, metropolitan cities have now grown to the point where they threaten to strangle the transportation that made them possible. . . . With the technical ability to solve its transportation problems well in hand, the modern city is confronted by a transportation problem more complex than ever before. Despite all the methods of movement, the problem in cities is how to move.

2
The activities of urban populations and their relationship to urban movement

As a first step in describing and analysing the complexity of movement in cities, it is useful to isolate for consideration the purposes of journeys. Even this small part of the total picture is complex, for there are substantial differences between the reasons why people move and the reasons for moving goods. Moreover, in the last resort each individual and each commodity will display a different pattern of movements motivated by a unique combination of purposes. Clearly, any comprehensive and balanced analysis of urban movement will have to find a reasonable level of generalization to reveal simultaneously the great variety of journeys and their determinants and the regularities of behaviour, spatial pattern and timing of movements between individuals, groups, areas and towns that undoubtedly exist. This can be approached by organizing systematically our impressions and more solid knowledge of the motivations for movement within towns.

The first and more substantial section of this chapter will concentrate on personal movement, by examining the principal influences upon the relative frequency of movements made for particular purposes. This will be done partly through a discussion of the activities, life-styles and journeys of contrasting groups of the urban population. It is hoped that these comments will illustrate at least some of the manifold influences upon individual travel behaviour, but there is a danger that it will lead to an over-simplified view of the position. To correct any such possibility, and in order to evaluate such generalizations, references will be made to findings from empirical studies of personal travel within towns. These will point towards further subdivisions, influences and variations, many of which the reader will be able to construct for himself, and which can be followed by further reference to the sources of this information. A large part of Chapter 4 will deal in similar manner with the movement of goods within cities. Although the published information about intra-urban freight movements is considerably less than that for personal movement, sufficient research on this subject has been carried out recently to enable at least some preliminary comments.

Classifications of personal movement

A majority of the population of most western countries, and more than eighty per cent of the British population live in urban areas, within which the great majority of the urban population's movements are contained. (The alternative methods of defining functional urban areas are not dwelt upon here. As many as 95.7 per cent of the British population in 1971 lived in Metropolitan Economic Areas, as defined in Drewett *et al.*, 1975; 79.3 per cent lived in the more tightly defined Standard Metropolitan Labour Areas.) A book which seeks to deal with movement in cities is therefore examining a high proportion of all personal journeys in a country like Britain. We have already implicitly excluded consideration of movement within and across rural areas because of its different scale and character, and for similar reasons we will not include any examination of the shortest movements that people make within towns, namely those within individual properties, whether homes, shops or places of employment and entertainment. These movements, while numerically great, do not involve considerable distances and, of more relevance, are different in kind from others. By and large, personal movement within a building or property has no external consequences for other individuals, companies or institutions. No public facilities have to be provided and no external economic or social costs are implied by such movement, except in the rare circumstances when considerations of safety, pollution or public order are raised. Another important distinction is one of scale. Our interest is in the geographical patterns and interrelationships with land use and other geographical elements of urban movement. In practice the geographer has not been concerned with the smallest scale of distributions, and likewise we exclude examination of the shortest movements. The plans and layout of buildings, and the patterns of movement within them have rightly been the concern not of geographers but of such specialists as ergonomists, factory managers and architects. Following from this, our definition of a movement of geographical significance and interest is one that involves the crossing of a property boundary and which produces some effect, however slight or indirect, upon a third party.

During the last decade, increasing study of the motives for journeying in towns has taken place. The social scientists and transport planners who have participated in this work have in part been reacting to the lack of attention to the causes of urban movement in much of the previous huge research effort for urban transport planning purposes. According to the argument of recent investigators, only through the understanding of what prompts, encourages and allows people to move will a balanced assessment of the functions and limitations of a transport or movement system be possible, or any sensible prediction of the future scale and pattern of movement be forthcoming.

Correspondingly, the absence of this element in earlier transport studies seriously reduced their value for planning purposes. While an emphasis on studying the travel behaviour of individuals is relatively recent and the results produced to date are widely dispersed and occasionally confusing, many useful individual contributions are easily accessible. They form the background to the following section, which the interested reader is recommended to widen by a closer reading of selected papers. (Kansky, 1967, was an early example of the shift in emphasis towards less aggregated urban travel studies. Interestingly, his approach was to analyse and structure the movement pattern, and then to identify the population groups that were associated with the components of the pattern. He concluded, however, that 'greater attention should be directed toward detailed enquiries about the travel patterns of urban residents and the spatial structure of commercial traffic'. The more recent papers include: Le Boulanger, 1971; Hemmens, 1970; Hurst, 1967; Kofoed, 1970; Kutter, 1973; Stutz, 1973; Wheeler, 1972; and Wiseman, 1975.)

We can each begin to compile a list of the principal purposes of personal movement within towns, by classifying the movements recently made by ourselves or by our friends and relatives. If among this circle of acquaintances we include people of different ages, including schoolchildren, people in work, pensioners and housewives, the circle will approach a cross-section of the total population and a more balanced picture will emerge. This is one approach to establishing the purposes and other characteristics of urban movement which, in the refined form of systematically designed and conducted questionnaire surveys of representative samples of travellers or households, has been extensively employed in urban transport research.

Whether we attempt to discuss motivations from our personal observations or from the results of large and well-organized surveys, we will inevitably have to adopt a classification of journey purposes. This in turn will force upon us a degree of simplification, and in the last resort a measure of arbitrariness in the allocation of journeys to particular purposes or groups of purposes. Classification always implies these methodological difficulties of the definition of classes and the allocation to them of particular cases. However, so long as the adopted procedures are clear and explicit, the gains from reducing information to manageable proportions in a carefully constructed classification will outweigh the disadvantages resulting from the loss of some detail.

Relatively few journeys are undertaken for their own sake. In the language of the economist, trips are usually intermediate goods, which is to say that they are jointly demanded with the benefits that arise at the destination. Generally, the journey forms an essential preliminary to a more important activity, such as shopping, attending school, or earning a living. These more important or 'final' activities may then be used for classifying the

Table 2.1 A classification of purposes of urban personal travel

Activity	Journey Classification	Remarks
I ECONOMIC		
a Earning a living	1. To and from work 2. In course of work	Few employed people work at home, but only 46 per cent of the 1971 population in the UK is in work (Lawton, 1973, tables 2.24, 2.26). Journeys in the course of work may be further divided into (i) peripatetic occupations, (ii) associated with movement of goods, (iii) to and from meetings and conferences.
b Acquiring goods and services	3. To and from shops and outlets for personal services 4. In course of shopping or personal business	Entertainment and recreational services are classified separately, but medical, legal and welfare services are included. Visits to catering establishments motivated by the wish for refreshment would be included here.
II SOCIAL Forming, developing and maintaining personal relations	5. To and from homes of friends and relatives 6. To and from non-home rendezvous	Much of this activity, e.g. within a family, does not generate journeys. 6 is frequently combined with leisure objectives.
III EDUCATIONAL	7. To and from schools, colleges and evening institutes	Affecting at least the great majority of the 5–16 age group, comprising over 16 per cent of the population.
IV RECREATIONAL AND LEISURE	8. To and from places of recreation and entertainment 9. In course of recreation: walks, rides	The distinction between adult education and this category is not always clear. Visits to pubs and restaurants often would be placed here, although they also involve 'acquisition' and social activities. The distinction between recreation and leisure is one of the degree of participation, as in playing or watching sport, and is often unclear.

V CULTURAL
10. To and from places of worship
11. To and from places of non-leisure group activity, including cultural societies and political meetings

The distinction between this category and leisure or social activities is not always clear. The distinction between gainful employment and other activities is not always clear, as here with (local) politicians, artistic performers, or the officers of societies and churches.

trip purposes. Many journeys are clearly and directly associated with a single activity or purpose, such as journeys to and from work, school, shops, places of entertainment or friends and relatives. But there are other journeys which have more than one purpose, or a different purpose or purposes in execution than in intention. Sometimes complex multi-purpose journeys are planned, but at other times a journey in progress is altered spontaneously to include, for example, visits to shops, cafés, restaurants or friends. Depending on the precise use of terms, it can also be suggested that some journeys are undertaken for their own sake, for example when an individual decides to go for a walk, or when a family decides to go for a ride in their car. Yet such journeys can nearly always be linked with a recreational purpose, especially if recreation is defined rather loosely to include passive forms of mental relaxation or stimulation.

Given that the great majority of journeys have only an intermediate purpose, an obvious approach towards their classification is to identify groupings of final purposes. In essence this means identifying the principal activities which the population pursues outside of its homes, and which therefore cause people to make journeys. This is the procedure which has been used to identify eleven principal categories of urban personal travel in table 2.1. As the remarks in the right-hand column of the table reveal, an attempt to simplify all journeys into so few categories is ambitious, and many that are undertaken are not easily placed among them. How, for example, would a journey made to meet a friend, to shop, and to have a midday meal at a restaurant in a local department store be classified? It would not be difficult to think of even more intricate journeys from the point of view of classifying their purpose or purposes. Nonetheless, it is believed that the table represents a useful division of the purposes of journeys and can easily be applied to the majority of movements.

The classification implies that the major human activities associated with personal movement may be grouped under five broad categories; economic, social, educational, recreational, and cultural. Of course, many common activities involve a mixture of these elements, especially in the case of the social, recreational and cultural dimensions of our lives. Broadly, each activity group can generate journeys to and from the places in which they are pursued, but also less frequent journeys in the course of pursuing these activities. There are, for example, peripatetic gainful occupations, of which a majority are undoubtedly involved in transport and distribution itself. As another example, field study trips will probably be a familiar form of peripatetic educational activity, although again these are not uncommonly mixed with social and recreational functions. It is probably the case, though could not be proved, that every individual at some stage participates in the full range of classified journeys. For example, at any one time, nearly fifty per cent of the population is gainfully employed, and nearly all men and women work at some stage of their lives. Cultural activities, including worship, are the only category that possibly does not involve the entire population at one time or other.

Establishing the relative importance of different activities in terms of the time spent upon them, and the regularity and frequency with which activities are pursued has recently become a popular ingredient of disparate research topics. Empirical 'time budget' studies are being pursued by a wide range of social scientists, and there are also theoretical contributions on the allocation

Table 2.2 Average time budgets of adults in Reading, 1973

	Hours spent on activity		
Activity	Men	Working women	Non-working women
1. Activities normally stimulating travel			
Work	4.3 (5.1)	3.6 (4.0)	0.1 (1.3)
Travel in course of work	0.5 (2.3)	0.0 (0.4)	0.0 (0.1)
Shopping, personal business	0.3 (0.3)	0.5 (0.5)	0.6 (0.7)
Drinking in pubs, etc.	0.3 (0.6)	0.1 (0.4)	0.1 (0.2)
Organized leisure	0.3 (0.7)	0.3 (0.6)	0.3 (0.9)
Full-time education	0.3 (2.6)	0.1 (1.6)	– –
Sub-total	6.0	4.6	1.1
2. In travel	1.5 (1.6)	1.4 (1.4)	1.0 (1.0)
3. Activities occasionally stimulating travel			
Eating, Casual social, Miscellaneous	3.1 (3.4)	2.9 (3.2)	3.4 (3.8)
4. Activities normally pursued at home			
T.V., personal hygiene, Domestic, child care, Private leisure	5.2	6.5	9.7
Sleep	8.4	8.6	8.9
TOTAL	24.2	24.0	24.1

Note: The figures in brackets are the mean hours spent by those actually pursuing the activity in question.
Source: Bullock *et al.* (1974), table 1, p. 58.

of time, notably by economists (see, for example, Chapin and Logan, 1969; Szalai, 1972; Evans, 1972). One study which produced interesting comparisons between male and female adults aged 16–70 years was conducted in Reading in early 1973 (Bullock *et al.*, 1974). 450 usable comprehensive diaries of activities for one week were collected, and a sample of the results is reproduced in table 2.2. Unfortunately for our purposes the reported figures are averages over the seven days of the week, and only distinguish men, working women and non-working women. Nevertheless they provide support for the classification of activities and journey purposes which has been adopted, as well as confirming expected differences between men and women. The classification of activities is clearly explained in the Reading study and, although it was not designed with journey purposes in mind, we can reliably distinguish between those activities which will normally generate movement, and those which are normally pursued at home.

On average, about 38 per cent of men's waking time (6 hours out of 16) is spent on activities which will generate journeys, and nearly 75 per cent of this (4.3 hours) is spent actually working. It should be noted that this figure does not include time spent travelling to and from work, or eating and drinking at work, and is a seven-day average computed for both employed and inactive males. Those actually working average 5.1 hours at work, or

Table 2.3 The relative duration of movement-stimulating activities: Reading, 1973

	Percentage of time spent		
Activity	Men	Working women	Non-working women
Work	72.7	78.2	9.1
Travel in course of work	7.3	0.0	0.0
Sub-total	80.0	78.2	9.1
Shopping, personal business	5.0	10.9	54.5
Drinking in pubs, etc.	5.0	2.2	9.1
Organized leisure	5.0	6.5	27.3
Full-time education	5.0	2.2	–
Total	100.0	100.0	100.0
Total hours spent	6.0	4.6	1.1
Travel time as percentage of activity duration	25.0	30.4	90.9

Source: Recalculated from Bullock *et al.* (1974), table 1.

36 hours during the week. Working women spend 25 per cent less time (4.6 hours) on activities stimulating journeys, and non-working women only 1.1 hours per day. Travel not in the course of work accounts for 1.5 hours of men's daily time budget, and even for 1 hour of non-working women's time.

The relative importance of the various activities which are taken to stimulate movement is roughly indicated in table 2.3, in which the contrast between economically active and inactive persons is more apparent. Many features of these tables are unsurprising, but worth noting is the duration of organized leisure activities (cinema, theatre, concerts, etc.; club and society activities; outdoor leisure including watching and playing sport; charitable work and religious activities; and evening classes), particularly relative to all non-home activities for non-working women. The consumption of time actually spent travelling relative to the time spent pursuing activities is another feature of the non-working woman's time budget which is distinctive.

Selected evidence on the purposes of personal movement

The relative frequency of the journeys made for different purposes may be discovered by reference to survey findings. Before reviewing some of these statistics, it must be stressed that while some of the differences among the results of different studies reflect actual variations between towns or dates, other discrepancies arise as a result of differences in methodology. Some studies include walking journeys, others exclude them. There is inconsistency in the coverage of different forms of public transport (trains, taxis, buses), and some studies survey only car drivers while others also examine passengers in cars. Unfortunately, therefore, it is always necessary to examine closely the design of surveys if it is wished to use, and particularly to compare, their results. Only if the methodologies are similar in relevant respects can comparisons be made. A final word of caution unfortunately needs to be said. Publication of survey findings in an impressive format is sometimes accompanied by lack of explanation of the methodology of a survey, and occasionally disguises the fact that a survey has been incompetently designed or executed. It is important for every student to develop a knowledge of good survey practice in order to evaluate the reliability and utility of survey data.

One further point about the availability and quality of movement data should be appreciated. There is no accumulation of comprehensive detailed statistics of movement, as there is with, say, statistics of population. (Transportation studies have a much longer history, and have been more numerous in the United States, enabling remarkably detailed comparative studies as early as the 1950s; see Curran and Stegmaier, 1958. By 1967 more than 200 large-scale studies had been completed, and at least twenty-nine urban areas had had two surveys; see Kain, 1967. Many fewer studies and, therefore, secondary analyses have been carried out in Britain, although one exception is White, 1974.) The history of population censuses in Britain covers 170 years of progressive improvement, refinement, greater precision and expansion of coverage, whereas the length of time over which movement statistics have been collected is to all intents and purposes no more than twenty-five years. By and large, movement statistics have been collected *ad hoc*, or for specialist purposes in individual ways. Even today there is little standardization between the major transport surveys, and the national travel surveys are of a very limited number of people over a restricted range of topics. A specific effect of these variations on any discussion of journey purposes is that different surveys classify and analyse journeys and journey purposes in different ways.

Despite these limitations, it is possible, from the wide variety of sources which are available, to put together a consistent picture of the purposes of travel within British towns and even to begin to form a view of the way in which journey purposes vary among towns of different size, regional location and economic structure. The traffic studies and their successors, land use/transportation studies, are the principal sources of information on individual British towns, and a small selection of their results forms the basis for the following section. We begin by reviewing some of the findings from an early survey of traffic and travel in London.

The London Travel Survey of 1954 was an early attempt to approach comprehensive coverage of travel in the Metropolis. Undertaken by the London Transport Executive (LTE), it took the form of a home-based questionnaire survey in the Central or 'red-bus' area, which was a little more extensive than the present Greater London Council territory. The sample consisted of 4,000 households in this area, but in addition 1,500 households in six surrounding towns also served by the LTE were included. All journeys within one week made by LTE bus and underground services, British Railways, cars, motorcycles and bicycles were included, but those made by people living in hotels and hostels were not represented as they were not part of the sample-frame (LTE, 1956; for a detailed description of the survey, including copies of the questionnaire, see Department of Scientific and Industrial Research, 1965). The relative frequency of journeys in London as revealed by this survey are displayed in table 2.4.

The classification used in this survey obviously relects the operations of the LTE in the mid-1950s, and the value of the findings for other interests is reduced by the high proportion of all journeys in the miscellaneous 'other' category. With hindsight, it is also regrettable that no use was made of general 'social' and 'leisure/recreational' categories, and that journeys during the course of work were not differentiated from journeys to and from

Table 2.4 The purpose of movements in London, 1954

Purpose	Number	Percentage
Work	72,082	58.00
School	8,375	6.74
Shopping	10,074	8.11
Theatre/cinema	4,479	3.58
Sport	2,161	1.74
Other	27,128	21.83
Total	124,269	100.00

Note: Journeys by foot are not included.
Source: Recalculated from LTE (1956), fig. 13, p. 12, and table 7, p. 48.

work. The 58 per cent of all journeys recorded for work purposes is in contrast to characteristic figures of around 40 per cent of all mechanically-assisted journeys, and around 30 per cent of all journeys found in more recent studies. This may reflect an increase in journeys for other purposes in more recent years, and the decline both of weekend working and of the habit of returning home at midday, but it is also possible that the survey was less successful in recording irregular evening and weekend travel than repeated movements. This might well have occurred because individuals were requested to complete their own forms, and it is easier to remember habitual journeys than the occasional and sometimes spontaneous trips we make for social and leisure purposes. Later surveys of London travel suggest a reduced importance for work-travel and higher relative frequencies for nearly all other categories, but particularly for educational, social and leisure purposes.

Similar evidence for other British towns is rare for dates earlier than 1965. Various special-purpose surveys contain some pointers to the position, but for no city to the writers' knowledge were walking trips surveyed. The nature of the information that is available may be illustrated by reference to a 1964 survey of bus travel in Newcastle-upon-Tyne (Burns, 1967). This surveyed journeys only over a twelve-hour period, from 07.30 to 19.30 hours, and obtained the results reproduced in table 2.5. Few details can be retrieved from the published results of the way in which journeys were classified.

The two rows of figures demonstrate that the most useful information on journey purposes from such surveys is obtained by focusing not upon the destination of a journey but upon the activity which has stimulated the

Table 2.5 Purposes of bus journeys in Newcastle-upon-Tyne, 1964

	\multicolumn{7}{c}{Purpose at destination}						
	Work	Business	Education	Shopping	Recreation	Other	Return home	Total
12-hour total	22.8	4.7	3.7	10.9	15.4	4.2	38.3	100.0
Excluding return journeys	36.9	7.8	5.8	17.6	25.1	6.8		100.0

Source: Burns (1967), table 9, appendix, p. xii.

movement. Tabulating destinations resulted in 38 per cent of the journeys being classified as 'return home', when it would be more pertinent to know what activity people were returning from. The Newcastle survey produced the usual dominance of work journeys over all others, but an unusually high percentage (25.1 excluding return home) associated with recreation. Even the simple classification used in this early survey raises problems of interpretation, and emphasizes the importance of a full specification of the classification procedure. It is not clear, for example, whether business journeys are in the nature of personal business and therefore akin to shopping trips, or are journeys undertaken in the course of work.

It is to the more comprehensive land use and transportation studies, set up after the recommendations contained in a joint circular from the Housing and Transport Ministries in 1964, that we must look for more comparable and general results. The early individual study by Taylor (1968), under the auspices of the Road Research Laboratory, of travel in Gloucester, Northampton and Reading exemplified the methodology of at least the survey elements of these studies, and provides some valuable early results. Among the methods used were representative household surveys which were designed to produce information about a normal weekday. They were conducted in Northampton and Reading in September and October 1962, and elicited information about journeys made on Wednesdays and Thursdays respectively. In Gloucester the survey was collected on and refers to Friday, 31 August, which 'proved to be far from a typical day but in many respects the data obtained can be compared with those from Northampton and Reading' (Taylor, 1968, p. 28). Other studies with similar results were the 'Norwich Area Transportation Survey', which collected information about journeys on a normal weekday between the hours of 7.00 and 19.00 by means principally of a 1-in-9 household survey, and the 'Cambridge Transportation Study' which collected information about both weekday and Saturday travel (City of Norwich, 1969; Morgan, 1969). The figures from

Table 2.6 The purposes of travel in selected British towns

Distribution of internal person trips by purpose at destination

	Work	In course of work	Education	Shopping	Social and recreation	Personal business	Other	Home	Total %	Total no.
Weekday										
Gloucester 1962	20.44	1.58	–	16.75	10.10	1.94	4.84	44.33	100.00	198,126
Northampton 1962	24.43	1.72	10.16	9.48	4.17	4.73	1.73	43.58	100.00	314,044
Reading 1962	15.08	3.29	8.84	9.38	6.13	14.96	5.90	36.41	99.99	521,860
Norwich 1967	17.97	3.74	7.75	11.81	3.28	11.35	–	44.10	100.00	460,755
Cambridge 1967	15.00	3.00	10.00	12.00	14.00	9.00	–	37.00	100.00	496,739
Saturday										
Cambridge 1967	6.00	1.00	4.00	21.00	23.00	7.00	–	38.00	100.00	476,161

Notes: The Gloucester study did not include school trips, and had a residual 'other' category with 4.28 per cent of all trips. 'Other' in this table also includes trips to 'serve passenger' or 'change travel mode'. In the Reading study, trips home to eat were not coded as home-bound trips. Other journeys to eat meals are included under personal business in all towns.
The Norwich study did not retain the 'serve passenger' and 'change mode' categories, but, by linking trips and (presumably) classifying the former as course of work trips, allocated them elsewhere.

Sources: Taylor (1968), fig. 44, p. 102; City of Norwich (1969), table 4.1, p. 38.

these various surveys have been rearranged and in some cases recalculated to make them as comparable as possible, but some differences cannot be removed by these tactics. Their findings on the principal purposes of journeys are arrayed in table 2.6, which is accompanied by notes about the important differences between the surveys in individual towns. While there are considerable differences between the purposes of journeys in these towns, there are similar rank-orders of the frequency of the principal purposes. Journeys-to-work are consistently the most numerous of the home-based movements, commonly accounting for in the order of one-fifth of all movements and one-third of non-home journeys. Journeys to home commonly account for about two-fifths of the total, with movements to shops and to educational establishments each accounting for about one-fifth.

One great advantage of these more recent studies is that a more detailed breakdown of trips is given, where a trip is defined as a journey, however complex, in terms of multiple transport modes between one purpose and another. This involves the painstaking reconstruction of such journeys, and the minimization of journeys purely for transport purposes, such as those inelegantly known as 'change mode'. Home is included as a proxy for the range of purposes which bring people back to their homes, and as an origin or starting point for many of the trips. This more detailed classification of journeys produces the broadly comparable figures for Northampton and Norwich presented in table 2.7. In the table, a dash represents the absence of journeys from home to home with no intervening purpose, an '0' indicates that there were no journeys in other cells of the matrix, and '0.00' is produced by rounding from a very low count of journeys.

Inspection of journey purposes in Northampton and Norwich reveals striking similarities. The most common destination of journeys is home, which in both towns accounts for about 44 per cent of all journeys. Trips to work follow in relative frequency, with 24 per cent of the total in Northampton and 18 per cent in Norwich. In comparison to the higher proportions found when studying only mechanically-assisted journeys, it can be seen that trips directly to or from home and work account in Norwich for only 31.7 per cent of the total, and in Northampton for 44.2 per cent. The major difference between the towns is that in Norwich the proportion of all journeys to places of personal business is nearly triple that in Northampton. The questionnaires used in both towns incorporated sections that pre-classified purposes and included a column headed 'Personal business', but as far as can be seen no further guidance to interviewer or interviewee was available on this point. It is difficult to believe that the residents of Norwich consistently interpreted this description more widely than those of Northampton. Little can be made of the discrepancy, which again points to the immense importance of carefully defining any terms or class-descriptions used in a survey. There are other

Table 2.7 Matrices of journey purposes in Northampton, 1962, and Norwich urban area, 1967

Origin or activity at origin	Work	In course of work	Education	Shopping	Social	Personal business	Home	Total
I *Northampton*								
Work	0.11	1.02	0.01	0.61	0.15	0.69	21.74	24.34
In course of work	0.84	0.83	0	0.07	0.08	0.15	1.37	3.32
Education	0.01	0.00	0.02	0.10	0.06	0.12	9.81	10.15
Shopping	0.44	0.17	0.02	3.53	0.19	0.45	4.65	9.47
Social or recreation	0.05	0.09	0.03	0.12	0.28	0.12	3.48	4.16
Personal business	0.54	0.16	0.07	0.63	0.14	0.62	2.57	4.73
Home	22.44	1.17	10.01	4.41	3.23	2.57	–	43.83
Total %	24.43	3.45	10.16	9.48	4.17	4.72	43.58	100.00
II *Norwich*								
Work	0.05	1.04	0.05	1.32	0.26	1.88	16.22	20.81
In course of work	0.78	1.33	0	0.07	0.02	0.19	1.31	3.69
Education	0.04	0.01	0.09	0.16	0.21	0.43	6.91	7.84
Shopping	0.76	0.04	0.06	1.18	0.16	1.26	8.49	11.95
Social or recreation	0.08	0.00	0.05	0.07	0.05	0.10	1.56	1.92
Personal business	0.81	0.10	0.11	1.07	2.24	1.57	9.61	13.49
Home	15.44	1.22	7.41	7.94	2.37	5.92	–	40.30
Total %	17.97	3.74	7.75	11.81	3.28	11.35	44.10	100.00

Notes: The Northampton study distinguished journeys to 'eat meal'; these have been amalgamated with personal business. Similarly 'serve passenger', and 'change mode' have been put with 'in course of work'.
Sources: Taylor (1968), table 32, p. 101; City of Norwich (1969), table 4.1, p. 38.

differences: for example, journeys per head for social purposes appear to be about one-half as frequent in Norwich as in Northampton, but this may be accounted for by the restricted coverage in the former town of journeys only between 07.00 and 19.00 hours. However, it is the similarities between the two towns which are most striking. It appears that when we examine the aggregate pattern of movements in a town, the disciplined timetable of work and school and the habitual patterns of rest and refreshment impose a broadly consistent pattern to the journeys made in a town. Two further examples of similar journey behaviour in Northampton and Norwich are suggestive but need corroboration from other towns before we can claim them to be true more widely. Trips from home to places of education are more common than the direct returns: in Norwich they form 95.6 per cent of all trips to places of education but home-bound journeys form only 88.1 per cent of all those originating from places of education. Conversely, people are more likely to travel from shopping to home than they are to leave home for shopping purposes directly. In Norwich again, 71.0 per cent of all trips from shops are to home, and they form 8.5 per cent of all trips for any purpose, while only 67.2 per cent of trips to shops are from home, and they form but 7.9 per cent of the whole range of trips.

Both studies will of course underestimate shopping, recreational and social trips and overestimate work trips because they represent 'normal' weekday travel. Similarly, because the surveys were carried out during school term, on an annual basis they will somewhat overestimate educational and, to a lesser extent, work journeys. But they reveal enough about aggregate patterns to give us a clear guide to relative frequency of journeys for different purposes.

Factors influencing individual movement

So far, the average or 'normal' weekday purposes of the aggregate pattern of intra-urban journeys have been examined. As already shown by the differences between weekday and Saturday movements in Cambridge, the movements in progress at any moment will depend on the time of day, the day of the week, and the season of the year being examined. The rhythms of movements associated with these time periods are considerable and of great practical importance, for one consequence of individuals structuring their activities and journeys is that congestion and peak-loads at particular times and in specific locations are produced in the urban transport system. In this chapter, however, we will not enquire further into these fluctuations, nor into longer-term changes in travel habits, but will seek to disaggregate the average picture. The total movement pattern is not simply the product of many identical movements by undifferentiated members of the urban population. Large numbers of people do have similar movement patterns, but it is apparent that the population can be divided into several groups, each of which is characterized by a more homogeneous range of activities, and therefore purposes for movement. While it is useful to remember that in the last analysis, no two individuals have exactly similar movements, if efficient groupings are selected strong similarities among the members of a group and pronounced differences between groups are produced.

It has already been argued that an individual's journeys are closely related to his or her activities. It is therefore necessary to identify the principal personal characteristics which bear some relation to an individual's activities. The list of these characteristics can readily become very long. Age, sex, employment status, income, health and personal preferences all influence the frequency with which individuals engage in journeys for different purposes. Some of these characteristics will reinforce the effect of others, e.g. there is a positive relationship between age and income, and both these characteristics may be related to the frequency of journeys made in the course of work. In the absence of considerable research into these particular questions, there are many difficulties in deciding which are the significant characteristics, and what are the most suitable groups. Aside from personal characteristics, a rather different influence upon and individual's journey purposes is his or her location within the city. Spatial patterns of residential segregation not only produce correlations between urban location and personal characteristics, but location also acts independently through its relationship with the accessibility of various attractions. For example, inner-city locations are characteristically accessible to small 'corner' shops whereas locations on the peripheries of towns, because of their more recent development and lower densities, are less so. This purely geographical or environmental distinction is likely to produce contrasting shopping-journey structures between individuals who have similar personal characteristics but who live in different locations. The same effect is produced by the contrasting availability, cost, and convenience of the alternative transport modes in different locations. Public transport services are generally superior in inner-city areas, and this leads to the substution of bus for walking or even car journeys compared to the outer suburbs, and in turn to a reaction upon the frequency of journeys for different purposes.

The influences upon personal movement, then, are manifold and are far from simply structured. Their nature and variation is not well-understood, for while many studies have examined the influence of one, or a larger but restricted number, of personal or environmental characteristics, attempts to carry out a comprehensive study are rare. Research with this aim would be ambitious, but the data and analytical techniques which are now available mean that such enquiries can be readily undertaken. It could well be that such research would be immensely productive. This view is prompted by the achievements of research with similarly comprehensive aims which has recently been pursued in the study of urban residential patterns. So many studies have been completed and published to produce something rare in urban geography, a substantial body of empirical findings with strong consistencies (for concise reviews, see Herbert, 1972, pp. 153–86; or Robson, 1975, pp. 13–28). It has been found that in advanced, western nations, a small number of distinctive spatial patterns contribute to the total urban residential pattern. Each element takes on a characteristic concentric or sectoral (or joint) distribution about the city centre at an ecological or small-area level of analysis, and each is associated with a recurring set of the areas' population and housing characteristics.

Social rank, or the socio-economic status of an area's population, is the most important and consistently occurring basis or dimension of residential segregation. It is highly correlated with occupation, educational levels, income, housing quality and car ownership, and tends to assume a sectoral distribution within the city, at least in North American cities where the dimension has been most clearly identified. The second commonly identified basis for segregation, or dimension underlying the residential pattern, is known variously as familism or urbanism, and it is related to the age, position in the life cycle, type of family and household structure (including number of children and whether adult women are in paid employment) in which a household head is placed. This dimension produces an identifiable spatial pattern of residential segregation because the housing needs and preferences of a household varies with these characteristics. For example, adolescents and young adults who are single or in childless married couples tend to minimize housing expenditure by occupying very small units in the inner city. Their housing often serves little more than a dormitory function, space standards are low, rates of sharing basic amenities are high, and the location has the advantage of good accessibility to places of work, education and entertainment. On the other hand, a young married couple with very young children have clearly different housing needs and preferences. The wife is much less likely to be working, and most likely to be fully occupied child-caring. She is therefore in the home for a great deal longer, prefers more space, finds a garden of immense value, and has greater demands for the exclusive use of basic amenities. The family as a whole is to a greater extent tied to the house, even for leisure and entertainment purposes, and makes in general fewer journeys for entertainment and, with only one earner, for work. The familism or urbanism dimension appears most strongly in American studies, and tends to be supplanted in British studies by a factor relating to housing quality and tenure. Levels of residential mobility in Britain are approximately only one-half of those in North America and consequently individuals do not as readily adjust their housing as they progress through the life cycle. Publicly owned housing is also much more important in Britain, and there is relatively little transfer between the public and private sectors (Morgan, 1976). There tends to be, therefore, a dichotomy in the housing market which severely disrupts the expected spatial patterns of various life-cycle groups, although this does not alter the progression of housing needs, or most important from our point of view, the activity patterns of different groups.

Other dimensions or factors have in various studies and locations been recognized as contributing to the complexity of housing distributions. A commonly occurring and important factor in American studies, for example, relates to ethnic or racial groups. Ethnic segregation, whether from a voluntary or imposed cause, leads to distinctive locations for identifiable groups and is commonly associated with characteristic movement patterns. Several studies have shown, as one example, that blacks commonly face unusually long journeys to work from their areas of concentration, with an unusual number of contra-flow or centrifugal morning trips from their inner-city residential concentrations to suburban manufacturing sites (Wheeler, 1968).

The labelling and interpretation of these ecological dimensions has commonly been couched in terms of behavioural hypotheses, if not conclusions. Although indirect, the pervasive import of the evidence is that there is a tendency for the housing location of individuals to be related to the positions on the social rank, familism, housing tenure, ethnic or other dimensions of the household head or joint heads. The extent to which individuals are able to fulfil their preferences depends on the ease with which people can move between different house types and locations, or on the degree of residential mobility which, in turn, depends upon a wide range of financial, cultural and institutional circumstances. But if even only a small proportion of households actually adjust their housing according to their needs and preferences, then the expected patterns of segregation will occur. Particular areas of towns will acquire an over-representation of people with a certain socio-economic status, life-cycle stage, or racial characteristics. As these characteristics are also related to personal movements the areas will therefore also acquire over-representations of households and individuals with distinctive travel purposes, timetables and means. Although it is certainly not argued that exactly the same structures of influences or of geographical distributions characterize the pattern of movement groups as of housing groups, a reasonable deduction is that strong similarities and parallels do occur. This very uncertainty is the strength of the case for more comprehensive studies of the factors underlying and structuring personal movement to be undertaken. It will be extrememly useful to discover a great deal more about the variations among individuals in their needs and preferences for movements, in their economic power to fulfil the same, and in their susceptibility to the discriminatory or eligibility constraints produced by the administrative and legislative environment. In the meantime it is possible to examine the topic in a less integrated fashion, by reviewing some of the studies which have had more specific aims. There is, for example, a great deal of evidence about differences in car ownership levels between occupational or income groups. Many transportation studies have further revealed the fact that household car ownership is positively related to the numbers of mechanically-assisted journeys generated by a household, although it has not always been appreciated that many of the 'additional' journey made by car-owning households are carried out on foot by those in less well-endowed households (Wootton and Pick, 1967). Nonetheless, there is some evidence about the relationship between socio-economic status and travel habits, and recently several studies have explored the influence of the life cycle on individuals' movements. As the latter have been very productive of new insights, and seem capable of development into more comprehensive enquiries, it is proposed to dwell upon their principal arguments and conclusions.

The life cycle and personal movement

As most commonly referred to in geographical and sociological studies, the concept of stages in the life cycle applies to entire households with differing housing needs, as with Abu-lughod and Foley's (1960, pp. 95–133) categories listed in table 2.8. As journeys are independently undertaken by

Table 2.8 Simplified representation of Abu-lughod and Foley's family life-cycle model

Stage	Access to urban facilities	Space priority	Locational preference
Pre-child	Important	Not important	Centre city
Child-bearing	Less important	Increasingly important	Middle and outer rings of centre city
Child-rearing	Not important	Important	Periphery of city or suburbs
Child-launching	Not important	Very important	Suburbs
Post-child	Not important	Not important	
Later life	Not important	Not important	Widow leaves home to live with grown child

Source: Abu-lughod and Foley (1960); Morgan (1976), table 1, p. 86.

individual members of a household, and not only by the household as a unit, some modification of the scheme for use in movement studies is necessary. The typology needs to be broadened to describe individuals in terms both of their age, sex and dominant activities, and in terms of the family group to which they belong. Once this is done, common observation backed by innumerable findings of transport studies enable us to specify some of the characteristics of both the purpose and other characteristics of personal movement. Further, but more speculatively, it is possible to suggest changes in movement characteristics which are likely to occur in the next twenty years or so. Underlying any predictive statement are a host of assumptions, and in this case the most important is of further increases in median real disposable income at a rate approaching that of the average post-war performance in Britain. This underlies continuing slow improvements in housing, and therefore space standards, and is also assumed to be sufficient to reduce the real costs of car ownership and use. Finally, yet another established trend is expected to continue, that is the faster growth of wage costs compared to total transport operators' costs, so making more difficult the economics of public transport operation.

Basically, therefore, the fundamental trends affecting transport and movement demands are assumed not to change in nature or direction, although recent experience has not only shown that the progress of such changes is often staccato, but also that confident predictions of the short or long-term rate of change are impossible. With these caveats, a summary of probable changes is presented in tabular form in table 2.9, which is arranged by the life-cycle positions believed to be relevant to personal movement.

The table distinguishes eight life-cycle groups which are believed to have both different combinations and frequencies of journey purposes, and distinctive distances, modes and geographical patterns of travel. The classification is loosely based on a wide range of tables from transportation studies (notably Kutter, 1973; Hillman *et al.*, 1973), but also a number of more explicit investigations of the travel habits of different groups. More detailed references to this empirical evidence will be made later in the chapter. The final column of the table, referring to likely trends in journeys are based mainly but not exclusively on the present modal characteristics of each group's journeys, and indicate the likelihood of change in these characteristics. To a very large extent, the comments refer to the access to cars which each group presently has, and whether higher real incomes will enable increased car ownership. For some groups, income is clearly the restraint upon car ownership, as with young single adults or with wives or other second members of households. For others, such as children and the aged infirm, it is not, and physical disqualifications are the constraint. With some groups, mainly established working heads of households, car ownership is already very high, and the scope for further increases as a result

Table 2.9 Life-cycle stage and the characteristics and likely trends in personal movement

Stage in the life cycle	Purposes of movement		Other movement characteristics	Probable trends in personal movement
	Well-represented	Poorly represented		
I Pre-school children with unemployed mothers of young children	Shopping, social, personal business including medical, to and from school accompanying older children	Work, in course of work	Children are completely dependent upon adults, particularly wives with low car ownership. Many short-distance trips predominantly on foot, or with pushchairs, etc., partly because of difficulty of using public transport.	Considerable potential for higher rates of second-car ownership, particularly in young families, but the expected slow rates of economic growth will result in only moderate growth among such families. Slow change in modal pattern.
	Low overall mobility			
II Schoolchildren	School, recreational, social	Work, in course of work	With increasing age, there is a growing proportion of independent journeys, which are exclusively on foot and by bicycle and public transport (mainly bus). Distances of all trips short, and very short with infant school children.	Some increase in accompanied car travel can be expected, particularly for younger children or in areas of deteriorating bus services, but otherwise little reason to expect change.
III Adolescents, young single adults, employed or students, and young childless	Work (or further education), social, recreational	School, shopping	Mainly independent journeys. Considerable dependence on public transport in teenage, although own or others' car use becomes increasingly	The 'affluent' young person is still relatively poor compared to the older working population and car ownership rates remain low, although

married men or women			they may rise quickly. Against this, even higher proportions going on to further education will depress such growth. Some continuing shift towards car travel, but a substantial need and demand for public transport will remain.
		important after age 18 for both work and social-recreational trips. Much longer median journeys than for younger age groups, and much greater diversity of purpose, destination and mode of journeys.	
	High overall mobility		
IV Young married men with young children	Work	Although income is relatively low at this stage of the lifecycle, and there are substantial outgoings, many have sufficient income to support car ownership. Most trips by car, except in largest cities where work trip often by public transport. Lower income groups' heavy reliance for most trips on walking and buses.	Increasing affluence, and continuing improvements in housing standards associated with decentralization of urban populations, likely to increase proportion of car journeys as well as the distance of trips.
	Educational, Shopping		
	High overall mobility with very well-structured patterns		
	For young married women with young children, see category I		
V Older married men with older or adult children at home	Work, in course of work	As in more recent marriages, but average income higher, and car ownership rates higher. Little use of public transport except in large cities for journeys to and from work.	Scope for further transfers from public transport use very limited. Increasing movement demands will derive more from changing of homes and activities than from further increases in income.
	Educational, shopping		
	High overall mobility with very well-structured patterns		

Table 2.9 Life-cycle stage and the characteristics and likely trends in personal movement (cont.)

Stage in the life cycle	Purposes of movement		Other movement characteristics	Probable trends in personal movement
	Well-represented	Poorly represented		
VI Married women at work	Work, shopping	Educational	Rates of second-car ownership still low, and consequently there is a great variety of travel distances and modes by working wives. Where independent access to car is available, nearly all trips by this mode. Where no car available, relatively high use of public transport and walking. Dependence upon car varies with the characteristics of the residential area and with access to public transport.	Scope for increased second-car ownership very great and can be expected to increase with income. Tendency for wives to work increasing, with fewer large families, and more effective chronological planning of children. Population decentralization will be another stimulus of higher second-car ownership.
VII Childless married and single adults aged over 25	Work, shopping, recreational, social High overall mobility	Educational	Rates of independent car ownership probably high, but closely related to income and individual decisions about residential location. In large cities, especially London, many prefer to live in inner districts and forgo car	Increases of income likely to lead to higher rates of car ownership. Improvements in public transport for long-distance commuting can stimulate suburbanization, and oddly, therefore, car ownership.

VIII			
Retired men and women	Shopping, personal business	Work, educational	

High overall mobility

owning. Second-car ownership closely related to income. Relative frequency of social/recreational trips is a spur to car ownership, and proportion of income which can be spent on travel is high (no family, no dependent members of household).

Even owner-occupiers and those with occupational pensions find income reduced, producing lower rates of car ownership than in working ages. Infirmity, difficulties of own maintenance, and reduced trip-making have same result. There is less need for each spouse to have independent access to a car, so rates of 2+ car ownership are very low. Most pensioners have very low incomes and car ownership is impossible. But public transport becomes increasingly inconvenient (distance to bus stop/staion, waiting). Unusually high dependence on short-distance walking trips, or on use of other's (relation's/neighbour's) car.

Likely improvements in real value of pensions which will enable a higher proportion to retain their own cars. But because an income differential will always remain, and because of infirmity and reduced demand for travel, car-ownership rates will always be lower than in working population. Dependence on public transport will therefore remain high, but use will reflect real cost (fares) and ease and reliability of use.

Low overall mobility

of rises in disposable income are not high. The dependence of different groups upon cars, or upon a combination of walking and public transport journeys, is therefore seen principally as a function of physical and mental capacities, and of income, through the intermediary of car ownership.

But, as has been argued previously, geographical factors also have some part to play in the likely changes of travel patterns. As the locations of homes and of the most important activities stimulating movement change, so there will be some interaction with movement patterns and the composition of urban movements. Shopping is one obvious example, but the geography of places of work, education and recreation are also constantly changing and generating movement change. Almost every urban activity and land use is affected by the tendency for its locations to decentralize, and for the concentration of activities at fewer, larger sites. This has the direct result of increasing the average length of journey necessary to carry out an activity, and in general would tend to both stimulate car use at the expense of walking and public transport, and to decrease the frequency of journeys made for particular purposes, by, for example, the reduction of lunchtime trips to and from home, or the frequency of shopping trips.

At this stage it is of interest to consider the principal characteristics and results of Kutter's study of activities, life-cycle groups and personal movement in randomly selected suburbs of three West German towns. Kutter almost overstates the influence of stage in the life cycle. He says that 'individual behaviour within the main groups of society is mostly determined by [the] age and sex of a person. Individual behaviour also depends on certain stages in [the] life-cycle of the family in which the person lives' (Kutter, 1973, p. 239). Given these views, it is not surprising the Kutter's *a priori* grouping of the urban population strongly resembles the classification already presented. He similarly attempted to classify people by age, educational and employment status, but found it most convenient at certain points to use family rather than individual characteristics. He did not include very young children (under six years of age), but otherwise the 24 categories he identifies in the body of table 2.10 produce a more detailed classification consistent with that used in table 2.9.

Using these categories, Kutter collected information about the activities and journeys of a random sample of 2,536 people living in his areas. (Unfortunately, the clarity of Kutter's paper has suffered in translation. Table 2, p. 240, indicates a sample of 2,536, but table 6, p. 247, a sample of 2,788. The former is used here.) This information provides rich evidence about the size of each group, and the distinctiveness or similarity of travel behaviour of each group. In outline, Kutter's procedure was to calculate an 'activity index' for each of the 24 groups, which was defined as the proportion of the group out of home during a given period of the day. The variation of the activity index during the day, which was divided into six

Table 2.10 Kutter's classification of the urban population according to individual and family criteria

		Sex	Age	Employment status	Employment of wife	Percentage of sample
A	Single persons	1. m/f	16–65	Students	–	0.8
		2. m/f	16–65	Employed	–	1.7
		3. m/f	65+	Retired	–	2.3
B	Married persons, no children	4. m	16–65	Employed	Employed	3.4
		5. f	16–65	Employed	Employed	3.7
		6. m	16–65	Employed	Unemployed	2.8
		7. m	65+	Retired	Unemployed	2.9
		8. f	16–65	Unemployed	Unemployed	3.5
		9. f	65+	Retired	Unemployed	2.2
C	Married persons, children aged 1–5 years	10. m	6–15	Pupils	Employed	1.7
		11. m	16–65	Employed	Employed	2.8
		12. f	16–65	Employed	Employed	3.1
		13. m	6–15	Pupils	Unemployed	7.1
		14. m	16–65	Employed	Unemployed	12.9
		15. f	16–65	Unemployed	Unemployed	12.9
D	Married persons, children aged 6–15 years	16. m	6–15	Pupils	Employed	4.1
		17. m	16–65	Employed	Employed	3.1
		18. f	16–65	Employed	Employed	3.3
		19. m	6–15	Pupils	Unemployed	6.2
		20. m	16–65	Employed	Unemployed	6.0
		21. f	16–65	Unemployed	Unemployed	6.5
E	Young adult persons, living in family	22. m	16–65	Students	–	1.8
		23. m	16–65	Employed	–	2.9
		24. f	16–65	Employed	–	2.4

Notes: Each group is of individuals with different characteristics and living in the defined types of household.
Source: Kutter (1973), Calculated from table 2, p. 240.

Table 2.11 Progressive classification of population groups on the basis of activities and journeys

Progressive classification of groups				Activity characteristics of group		
Eleven groups on basis of correlations among 'Activity profiles'	Six groups: 'Activities and social ties'	Three groups: 'Activities'	Determining activity	Duration of determining activity	Location of determining activity
Ia1 Pupils, aged 6–15	Ia (19.1) Pupils 6–15	I (21.6) Pupils and students	School	Medium	Dwelling areas
Ib1 Pupils, over 15	Ib (2.6) Pupils, students, 15				
Ib2 Students, over 15					
IIa1 Unemployed housewives (16–65) in families without children	IIa (23.0) Housewives	II (30.4) Housewives and retired	Shopping and personal business	Short	Dwelling areas and commercial centres
IIa2 Unemployed housewives (16–65) in families with children aged 1–5					
IIa3 Unemployed housewives (16–65) in families with children aged 1–5					
IIb1 Retired persons living alone	IIb (7.5) Retired				
IIb2 Male retired living with adults					
IIb3 Female retired living with adults					
IIIa Employed women	IIIa (12.5) Employed women	III (48.0) Employed	Work	Long	Urban area
IIIb Employed men	IIIb (35.6) Employed men				

Note: The figures in parentheses refer to the proportion of the sampled population in a category. For explanation see text.
Source: Kutter (1973), table 5, p. 224; table 6, p. 247; text, p. 246.

THE ACTIVITIES OF URBAN POPULATIONS 45

periods, was termed the 'activity profile'. This can be split into the profiles for individual activities, and it was found that 93 per cent of the total duration of activities was accounted for by work, education, shopping, personal business and social activities plus recreation. Subsequent analysis was therefore confined to the time spent in the pursuit of, and travel associated with, these five main activities. Activity profiles of the main activities were created for each of the 24 groups of people and compared with each other using correlation coefficients. From the complete matrix of correlation coefficients, it was apparent that some of the original groups had very similar activity and travel profiles. For example, among the eight categories of employed men, there is no correlation between two groups of less than 0.97, and 36 of the possible correlations have values as high as 0.99. When there is this degree of similarity, there is very little loss of information from amalgamating all employed men into one class, whether married or single, and whether childless or with children of different ages. If all groups which show a clear similarity to others are put together (indicated by a correlation greater than 0.95), Kutter's original 24 categories can be reduced to 11 as in table 2.11.

Several of the eleven groups were made up of very small proportions of the total population. For some purposes it will be more convenient to use an even less detailed division of the population, and this may be achieved by further amalgamation of the categories with the most similar activity profiles. The most likely candidates were the three classes of housewives, all pairs of which had correlation coefficients of 0.75 or greater, and the three classes of retired people, which yielded correlations of 0.72 or greater. In fact, Kutter employs factor analysis to produce further economy of description of the full correlation matrix, and his interpretation of the results leads to two further steps in grouping, with respectively six and three categories. His analysis suggests that an urban population can usefully be divided into no more than three groups with contrasting activities and therefore journey purposes, namely: pupils and students, housewives and retired people, and employed people. If this degree of simplification remains useful to a study or for another purpose, family characteristics and gender are seen to be of secondary importance compared to the age and employment status of individuals, at least with respect to their travel behaviour. However, on the basis of small differences in their activity profile and of social ties, Kutter also suggests that the separation of the three groups into two major components, thereby producing six classes, is justified. The details of all three classifications are presented in table 2.11, which also displays his comments concerning the duration and the location of the principal or 'determining' activity of each group.

The activity profiles of the 24 original groups of people are arranged into the six combined classes in figure 2.1. Here can be seen, for example, the

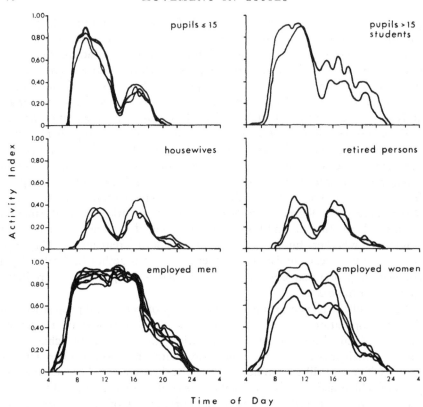

Figure 2.1 *Activity profiles of six population groups in three West German towns* (Source: *Kutter, 1973, fig. 8, p. 247*)

great similarity among the various classes of employed men, and the wider range of differences among employed women. Although these diagrams contain more information about the timing of journeys than their purposes, differences in the activity profiles are clearly related to differences in purpose. Among the six graphs, the contrasts reading from the top to bottom row are greater than those reading across the columns, and this allows us to appreciate visually the justification for a further grouping of the columns to form three basic types.

Let us first examine the principal differences between the two columns. Pupils under fifteen are less active overall, which means that they are less likely to be out of their homes pursuing one purpose or another, than older pupils and students. They are also more likely to return home at midday and less likely to be active in the evenings and, to a lesser extent, during the afternoons. This can confidently be interpreted as evidence of their different school timetable and of the fewer journeys they make for social and recreational purposes in the evening. Retired people are more likely to be active at midday than housewives, and there is a similar but less pronounced difference in the early evening. Housewives undoubtedly are more tied to the home at mealtimes, but on the other hand they show a slight tendency to be more active in the late evening. These differences are minor, however, and the similarity of housewives' and retired peoples' activity profiles is striking. It may be that the profiles disguise substantial contrasts in the activities pursued by the two groups, with perhaps retired people spending more time on social, recreational and cultural activities and housewives more time on shopping and personal business, but this cannot be stated for certain. Turning to the last row, we see that employed men, in contrast to employed women, show only a neglible tendency for reduced activity at midday and also have higher activity profiles in the evening. Clearly women are more likely to travel to and from home at midday, probably for family purposes, and we can suggest that they are also less likely to make work, social and recreational trips in the evening.

Variation among the three rows consists of both volume and distributional contrasts. The employed are by far the most active group, with around 90 per cent of employed men and some groups of employed women active from 08.00 to about 17.00 hours. Pupils and students follow in their volume of activity, but this is more pronounced in the mornings than in the afternoons. Moreover, although their activities begin only a little later than those of employed persons, they have virtually ceased by 20.00 hours, four hours earlier than those of the employed. Housewives and retired people are in total less active than other groups, and have a pronounced bi-modal or double-peaked distribution of activity, which is at its greatest at mid-morning and late afternoon. Their activities begin appreciably later than those of other groups, but they cease only a little before those of employed persons.

There is no obvious reason why a very similar structure of activities and movements should not characterize the population of British towns. Indeed, the more limited evidence of the Reading time-budget research and of the transportation studies points to very similar structures. It should be remembered, however, that Kutter's information refers to West German suburbs, and that the details of activity profiles will be affected by national, and indeed local, practices. For example, work and school begin at an earlier hour in Germany than in Britain, and for younger pupils at least, school finishes at midday. In Britain, the habit of returning home for a midday meal is still common only in the smaller British towns, but it is a more marked feature of family life and travel patterns in other countries, such as France. Such differences are only marginal, however, when compared to the broad patterns, which are produced by the universal influences of employment, of education and, as will be discussed more fully in the following chapter on the timing of movements, of the physiological demands of sleeping and eating.

The purposes of movement of schoolchildren and the elderly

These two groups have already been identified as having distinctive movement patterns, largely because they do not normally engage in the journey to work. There are three fundamental characteristics about schoolchildren's movements: they are very numerous, they are dominated by short-distance travel, and they are remarkably homogeneous in purpose with journeys to and from school forming a large and common component. In 1975 there were 11.14 million children of all ages at school in the United Kingdom, including 18 per cent of the 16-year-olds, and 20 per cent of the 17-year-olds (Central Statistical Office, *Social Trends No. 7*, 1976, table 3.5, p. 88). Nearly all of these pupils travel twice a day for educational purposes, and large numbers also return home at midday for lunch. In addition to these movements, there are many other journeys for social and recreational purposes, particularly by older children.

One of the first attempts to provide evidence of children's independent journeys, as well as much more about the mobility characteristics of different groups, was carried out by Hillman and colleagues for Political and Economic Planning (P.E.P.), some results of which have been published in two Broadsheets (1973, 1976). They contain selected findings of social surveys carried out in five contrasting small areas, four of which are urban in character. Separate surveys of adults or electors, teenagers and junior school children were executed, although in the latter two cases the samples of about a hundred in each location were fairly small. The urban areas chosen reflected a cross-section of town environments, but it was not claimed that in combination they would be representative of British towns or their population. The areas were 'Smallish', an electoral ward of about 10,000 people in a free-standing and mature small provincial town of about 30,000 people. Second was 'Newton', a ward built in the 1950s, of about 13,000 people in a post-war New Town of about 60,000 people (almost certainly Stevenage). The third area, known as 'Suburbury', was a ward of 21,000 people on the outskirts of a provincial town of about 300,000 people; and the last, termed 'Lonborough', an inner London ward of 14,000 people three miles to the north-west of the West End. While the findings from these surveys are heavily drawn upon in the following few pages, information from a number of other sources has also been added (Levin and Bruce, 1968 – their survey covered all modes of travel of 9,000 children at 34 schools in Steveage and St Albans; See also Hillman, 1970; brayson, *1975*; Masterson, 1977).

The P.E.P. study found a surprisingly high participation by even junior school children in independent non-school journeys, as is demonstrated by the figures reproduced in table 2.12. Shops were the most commonly visited facility on unaccompanied trips, followed by parks, playgrounds and the

THE ACTIVITIES OF URBAN POPULATIONS

homes of friends. A more quantitative guide to the volume of movement by children was yielded by a question concerning trips that were both accompanied and unaccompanied by adults during the weekend previous to the survey. The results are summarized in table 2.13, which shows, for the

Table 2.12 Percentage of junior school children who visited facilities unaccompanied by adults

Journey purpose	'Smallish'	'Newton'	'Suburbury'	'Lonborough'	Average of four areas
Shops	88	96	99	97	95
Park/playground	84	90	94	89	89
Visiting friends	85	82	92	78	84
Other leisure	66	66	75	66	68
Entertainment	45	68	77	71	65
Average over five purposes	74	80	87	80	80

Source: Hillman (1973), table 3.21, p. 68.

Table 2.13 Percentage of children travelling to activities in one weekend

Area and car ownership	No travel at weekend	One or more accompanied trips	One or more unaccompanied trips
'Smallish'			
no car	7	67	93
one car	2	90	84
two cars	4	84	88
'Newton'			
no car	0	62	93
one car	1	77	92
'Suburbury'			
no car	0	77	96
one car	0	83	97
'Lonborough'			
no car	0	80	92
one car	2	95	93

Source: Hillman (1973), table 3.22, p. 69.

four urban areas and by the number of cars available in the household, the proportion of those who did not go out, those who were taken out and those who went out unaccompanied by adults.

Nearly all children in all four areas make independent journeys to shops, nine out of ten travel independently to parks and playgrounds, and more than four out of five will take themselves to visit friends. The differences in behaviour among the contrasting urban areas are not considerable, but using a crude average of the percentage participation in unaccompanied trips over the five stated purposes, children's mobility is below average in 'Smallish', the small provincial town, and above average in 'Suburbury', the suburb of the medium-sized provincial town. Some of the variations in the incidence of particular journey purposes will be related to the accessibility of opportunities, shown most clearly by the low participation (45 per cent) of journeys for entertainment purposes in the small town. The low figure (78 per cent) for visiting friends in the inner London area in comparison to that (92 per cent) in 'Suburbury' could be explained by a combination of differences in house types, such as more flats and fewer houses with gardens, and a lower encouragement for children to play at home. The important information revealed by table 2.13 is that the possession of a car by a household has surprisingly little effect upon the propensity of children to make unaccompanied trips, although it does increase the approximate percentage making accompanied trips from 70 to 85.

Unaccompanied trips are by definition journeys that children make additional to adult trips. While some of these, particularly shopping trips, may substitute adult movements, it is clear that children generate an immense number of journeys in their own right. It is symptomatic of the lack of attention their travel habits have received that there is no known evidence about the distances, modes or spatial patterns of this movement, but it would be shortsighted to dismiss this ignorance with the argument that children's mobility is unimportant. It is true that the minority of their journeys which do have an impact on urban transport facilities will be traced by existing studies and monitoring procedures, either because the journeys are accompanied by adults in private cars, or because they are made by public transport. If, however, children's travel is examined from other points of view and not those of the transport operator and planner, there are very good reasons for advocating more concern and investigations. Mobility and travel is a necessary part of children's personal and social development and their environmental education, and therefore should be encouraged and facilitated. Yet children are unusually vulnerable to road accidents, and are particularly constrained in their movements by their lack of knowledge, confidence and money. It can hardly be argued that their needs and preferences are sufficiently understood or taken into account in discussions or urban management and planning. Children cannot be expected to form themselves into pressure groups to advance their own interests, and their parents have only given voice in support of two aspects, improved road safety and subsidized journeys to and from school.

The piecemeal evidence that is available on the travel and mobility of the elderly in British towns is broadly consistent with some detailed findings from a survey in New York in 1963 (Markovitz, 1971; see also Carp, 1972, and Neilson and Fowler, 1972). It was found that whereas the total population (except very young children) made 1.78 trips per day, the elderly made only 0.79 trips. Examining non-home trips only, work trips were still surprisingly important, accounting for over a quarter of all trips even though only 16 per cent of the sample were employed. Shopping, personal business and social trips were all relatively more numerous among the elderly's journeys, yet only six out of every hundred elderly persons made a trip for social purposes each day, at least by mechanically assisted transport.

In Britain also the elderly have only low mobility; indeed, many are housebound. One survey showed that 10 per cent of the pensionable population of the London Borough of Greenwich (1972, p. 5) were bedfast or housebound, and that another 7 per cent were only able to get out with difficulty. A more recently national survey found that 12.1 per cent of the respondents were bedfast, housebound or needed help to go out (Age Concern, 1978). One-quarter of a sample of elderly people in Exmouth, Devon, could not walk beyond a quarter of a mile, and four out of ten not beyond half a mile (Glyn-Jones, 1975, pp. 47–62).

Among the purposes of trips, shopping and personal business, including many trips to doctors or other health facilities, are relatively more important than for younger age groups, although it must be remembered that there are enormous variations among the elderly population. For example, no doubt there are many active retired men who hardly ever make a shopping trip, while some others shop almost every day. In Greenwich, one-quarter of the elderly were found to visit their doctor at least once a month, but on the other hand thirty per cent attended less than once a year. Given these ranges of travel habits, averages, such as the figure of six visits by elderly people in Greenwich to their general practitioners per year, are not very informative.

Very recently the difficulties faced by elderly people in reaching facilities and their relatives and friends have belatedly received considerable attention (Age Concern, 1978; Norman, 1977). It is certainly the case that the increasing tendency for the elderly population to live in independent small households, combined with the concentration of retail and medical facilities into fewer locations and the decline of public transport facilities, strongly suggests that it is becoming more difficult for large numbers of elderly people to maintain independent lives. The implications of these questions range far beyond transport policies and planning, and concern particularly physical planning, and health and social service polices and expenditure, but those who have been advocating that greater attention should be paid to the mobility of specific population groups can draw some encouragement from a statement made late in 1978 by the Minister of Transport: 'With regard to allocating financial resources, Mr Rodgers felt that a national policy of concessionary fares for the retired and disabled should be a priority.'

The variability of travel purposes

More detailed attention has been given to the activities and travel of two broad groups of the urban population who frequently have been relegated to a status of residual unimportance in aggregate travel studies. Other groups also have strong claims for a heightened awareness of their activity structures and their movement needs and behaviour. Hillman, *et al.* (1976) for example, have discussed the particular requirements of teenagers and collected evidence on their strongly independent patterns of activities and movements and their low access to cars.

Housewives are another group with distinctive travel patterns. Their trips, although not as time-consuming as the journey to work, are characteristically numerous but short-distance, are not made by car, and are therefore unusually demanding. This is partly because so many of their activities are structured around serving the needs of other members of their family. They are generally the principal shoppers of a household, and many are obliged to accompany young children to and from schools, medical facilities, and social and recreational activities. The tendency for wives to take up employment is further complicating and stimulating their day-to-day travel. There is no doubt that several trends in the structure and facilities of urban areas are increasing the difficulties of their journeys, e.g. the movement towards group practices of doctors and dentists; and re-organizations of publicly provided services such as hospitals and schools.

At the same time, housewives have generally low access to cars and the quality of public transport services is deteriorating. In 1971 only seven per cent of households possessed two or more cars in the United Kingdom. In its constituent urban areas, the comparable figures are much less. An unknown number of these supplementary private cars are in any case owned by adult sons and daughters, or by non-household heads in non-married couple households. In 1965, only ten per cent of women in Great Britain were driving-licence holders. Greatest access to cars by women is probably in the 30–39 age group, when eighteen per cent are licence holders. It is true that at the younger, more car-oriented population ages, a greater percentage of women will become licence-holders even if there are no further increases in living standards. Nevertheless, the available evidence firmly points to the fact that for many decades to come, only a minority of households will possess more than one car and that only a minority of women will have independent access to a car. It would be quite inappropriate to plan our towns on the assumption that cars will cater for the population's entire mobility needs in the foreseeable future, even if it were realistic to predict that virtually all working households will possess at least one car.

Excepting educational journeys for the young, it is not the case that the separately identified sub-groups of the population travel for purposes individual to their group. Rather, it has been shown that the different groups exhibit contrasting frequencies of journeys for particular purposes. Even the elderly practise many work-trips, and employed men do a great deal of shopping. Contrasting activity structures are not the entire explanation for contrasting travel profiles. The spread of car ownership has facilitated increased aggregate mobility, but it has also created inequalities in access to travel related to an individual's status in a household. In many respects, the differences between car-owners, licence-holders in car-owning households, and between the fully-active, the immature and senescent in non-car-owning households may be more important than the mobility differentials stemming from persisting income differences.

Those individuals who face the greatest difficulties in travelling are most likely to undertake no more than the minimum of necessary journeys. They may therefore spend less time on social, recreational or even medical visits than they would otherwise wish. In other words, it may be deduced that different travel patterns are not solely a reflection of contrasting preferred activity patterns, but also that the activities pursued by each individual and group are considerably affected by their access to, and ease of, travel. The geographical patterns of a group's journeys will also reveal its mobility constraints, as shopping trips may be shorter and less varied, or at a different frequency, than those that would be undertaken if cars were available or public transport better. At the other extreme of the range of personal mobility, although the costs of journeying are an inevitable restraint on mobility even for the richest and most privileged inhabitants of cities, many employed people presently earn sufficient to support car ownership and to enable a level of usage giving considerable choice in the number and destination of movements. It follows that when we examine the present patterns of personal movements within towns, we are dealing with the aggregate product of innumerable decisions to travel made by very diverse individuals.

3
The aspatial characteristics of movement

This chapter is concerned with the non-spatial characteristics of journeys, including their timing, duration, frequency and method or – in the specialist term – mode. As well as describing these characteristics as they occur in British cities, some aspects of the decision process of trip-making will be reviewed. Information will later be given about the geographical characteristics of journeys, reflecting their particular interest to the aims of this volume. The aspatial characteristics of urban journeys are particularly important for the design and efficiency of an urban transport system, for it is the concentration of movements at particular times which leads to the highest or peak demands on each modal system. Consequently those concerned with the management and planning of transport systems have been very active in monitoring and predicting volumes of movement, and when, where, how and how often movements occur. Numerous surveys of transport and travel have been carried out in Britain in recent decades, and although many may be faulted on the grounds of their selective coverage and their over-simplicity, they do provide a very wide and useful body of information which can be drawn upon to illustrate this section.

The timing of movements

There are easily recognizable, regular and predictable temporal patterns of movement in all towns and cities, although the timing and duration of the peaks and lulls of movement vary with the size of towns, their occupational and industrial character, and their location. In decreasing order of their amplitude, and also of their practical importance, the principal rhythms of travel are diurnal, weekly and seasonal. Each of these periodicities is related to temporal rhythms of the population's activities.

The diurnal pattern

The diurnal pattern is largely shaped by the astronomical clock and the requirements of human physiology for rest, sustenance and maintenance during each twenty-four hours. The physically determined pattern is much reinforced, however, by social and economic conventions about, for example, the normal hours of work, shop-opening and school, and the timing of public entertainments and social events. The good sense and convenience of these conventions is in most cases unquestionable: many institutions and enterprises such as schools and factories cannot operate successfully unless all participants are present at approximately the same time. Frequent physical and other contacts between and among commercial and industrial enterprises also mean that there are considerable benefits if most economic activity is taking place at the same time. A major disadvantage of this synchronization of activity across many enterprises is, however, the unusually high demand for passenger transport facilities at the beginning and end of the working day. The problems of the journey to work, both from the point of view of the suppliers of facilities and capacity and from that of the comfort and convenience of the passenger, are especially severe and will be considered in some detail in this section. Fortunately goods movements in towns are more evenly distributed through the waking hours, and tend to a maximum during mid-morning.

Details of the hourly variation of trips may be obtained from the local transportation studies. These have often collected the hourly distribution of trips for different purposes, and by different modes, although it will be no surprise that walking and cycle trips have often been ignored in these surveys. Some examples of characteristic results are given in figure 3.1. The graph has been redrawn from a composite figure in Taylor's study (1968), and gives the half-hourly distribution of all trips, by six categories of purposes on an 'average' weekday in Northampton in 1964. These data cover journeys made on foot as well as by mechanically assisted means. As is often the case, in the interests of economy, less than the full 24-hour period was surveyed, but this is of little consequence because very few trips are made during the period from 22.00 to 06.00 hours which was not surveyed.

A clear triple-peak pattern of movements is shown. These can be approximately defined as a morning peak between 07.00 and 09.00 hours, when 19.5 per cent of all movements occurred; an extended midday peak between 11.30 and 14.30 hours, during which 32.1 per cent of all movements took place; and an evening peak between 16.00 and 18.30 hours containing 24.1 per cent of all movements. The 7.5 hours making up these three peaks therefore contained 75.7 per cent of all journeys in but 46.9 per cent of the 18-hour survey day. It would be reasonable to suggest that well over 70 per cent of the movements made during the whole 24 hours occurred during these peaks.

Some more detailed features of the distribution of journeys through the day are also revealed. The morning and midday peaks were themselves bimodal, although of different form. The morning surge of journeys

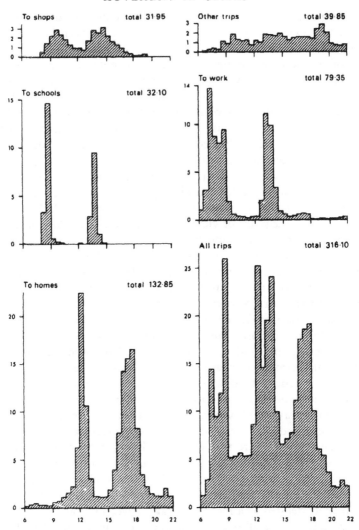

Figure 3.1 *The half-hourly distribution of internal person trips in Northampton, 1964* (Source: Taylor, 1968, fig. 45, p. 103)

comprised a minor peak in the half-hour following 07.00 hours, 95 per cent of which were journeys to work, followed immediately by a slight slackening of journeys, but the half-hour after 08.30 hours comprised a very much higher volume of movement, of which work journeys made up only 37 per cent but school journeys as many as 56 per cent. The midday bimodal peak was more symmetrical, although the earlier mode after noon was largely the result of journeys to home (89 per cent), and the second mode after 13.30 hours of the return journeys to schools (40 per cent) and workplaces (41 per cent), as well as some shopping trips (7 per cent). There

THE ASPATIAL CHARACTERISTICS OF MOVEMENT 57

was a steady decline of journeying after 18.30 hours, and only a slight increase of journeys in the late evening. The final feature of note is that journeys in the mid-morning were about 20 per cent lower than in the mid-afternoon. The single busiest half-hour was that between 08.30 and 09.00 hours when 25,800 journeys were made.

When we examine the diurnal distribution of journeys for different purposes, then clear and by-and-large expected differences emerge. Journeys to schools had the most pronounced peaking, 98.6 per cent occurring in the

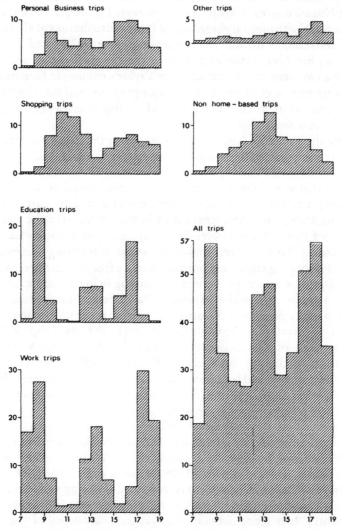

Figure 3.2 *The hourly distribution of trips in Norwich, 1967*
Source: *City of Norwich, 1969*

three hours 08.00–09.30 hours and 13.00–14.30 hours. Journeys to work and to homes were also strongly peaked, although both occurred at all periods of the day, and the peaks understandably were quite different. Of the remaining categories, journeys to shops had a surprisingly symmetrical, although broad, bi-modal distribution with the greatest frequencies occurring in mid-morning and mid-afternoon, while the residual category, which would include a large element of social and recreational trips, was most well-distributed, but with a minor maximum in the early evening.

Very similar features of the diurnal distribution of journeys were revealed by the Norwich Area Transportation Survey. Figure 3.2 shows, in similar form to the Northampton diagram, the hourly distribution of all movements and of movements for different purposes between 07.00 and 19.00 hours. While, as has been mentioned before, the Norwich study also included pedestrian movements, it differed from Taylor's enquiry in Northampton by classifying trips according to their major purpose, and not simply by their destination. Thus in the Norwich study a trip from work to home was classified as a work trip.

While the use of an hour as the time-unit in the Norwich study gives a much coarser view, the similarities between the Norwich and Northampton results are striking. The triple-peaked pattern of all movements is repeated, although that at midday was relatively less pronounced in Norwich than in Northampton. Education trips were most-confined to particular hours. Shopping trips, with the exception of 'other trips', were the most evenly distributed through the twelve-hour day, but had a noticeable bi-modal distribution. There were, however, some interesting differences: in Norwich shopping trips were noticeably more frequent in the morning than in the afternoon, journeys in the mid-morning being about twenty per cent greater than in the mid-afternoon. The different methodology of the Norwich study permitted the observation that non-home based trips were unusual in having a distribution through the day approaching the characteristics of the theoretical normal distribution, with the peak of movements occurring between 13.00 and 14.00 hours. The temporal distribution of personal business trips is seen most closely to resemble that of shopping trips, although a distinction between the two modes was that the former had a higher frequency in the afternoon.

To what extent the temporal distributions of movement in Norwich and Northampton represent patterns repeated in most British towns is questionable. The clearest features are certainly repeated: almost every town sees a peak of movements around 08.00–09.00 hours and 17.00–18.00 hours and relatively low daytime movements during mid-morning and mid-afternoon. Journeys to and from schools will have similar timings in almost all parts of the country. It is more difficult to assess the consistency of the degree of peaking, or whether the relationship between the midday and other peaks is subject to fluctuation. It is clear that in the largest cities and conurbations, the relative importance of the morning and evening peaks is greater, and of the midday peak considerably less, even when allowing for the fact that survey results from larger areas are more likely to ignore walking or short-distance (intra-zonal) trips. For example, in the West Midlands in 1964, only 188,200 or 10.2 per cent of the 1,839,000 mechanically-assisted journeys made between 07.00 hours and 00.00 hours were carried out during the two hours after noon (Freeman Fox et al., 1968, vol. 1, table 4.4, p. 42). Further contrasts for a number of British towns are shown in table 3.1.

Detailed evidence concerning the relationships between the diurnal pattern of trip-making and age, sex, income, occupation and location in the city has been collected in a number of studies. Kutter's work was extensively referred to in the previous chapter, and his activity studies of different groups (figure 2.1) reveal a clear contrast between employed and unemployed persons in their daily profiles – the former have a high level of activity throughout the working day, whereas the latter's activities decline (when they return home) at midday. Pensioners and housewives showed quite symmetrical patterns of activity around the midday low, while children and students tended to be more active in the mornings than in the afternoons (Kutter, 1973). Similar variations have been shown to exist in British towns and cities, as by Hillman et al. (1976) in their presentation of some aspects of the mobility of housewives, teenagers and pensioners, and as will be described in greater detail in Chapter 5, in a detailed study of shopping

Table 3.1 The concentration of personal movement into peaks: the percentage of daytime (07.00 to 19.00 hours) movements carried out at midday (noon to 14.00 hours) and in the evening peak (16.00 to 18.00 hours)

Town	Year	Midday peak: noon—14.00 hours	Evening peak
Cambridge	1967	16.6	17.5
Northampton	1964	30.9	22.3
Reading	1962	19.7	22.8
Reading	1971	17.5	20.7
Norwich	1967	20.4	20.4
West Midlands	1964	13.3	25.6

Note: The West Midlands data excludes walking trips.
Sources: Taylor (1968), fig. 45, p. 103, fig. 42, p. 97; Downes and Wroot, (1974) fig. 17a; Morgan (1969), table 16.01, p. 76; City of Norwich (1969); Freeman Fox et al. (1968).

habits in Watford (Daws and McCullogh 1974). Among other variations, this revealed that pensioners were unusually likely to make their shopping trips in the morning.

The weekly pattern

The second major rhythm of urban movement is that associated with the division of each week into working and leisure days, at least for the majority of families. In broad terms the pattern of activities and movements of both people and goods is very similar on each of the five working days, although it is common for the frequency of shopping trips and of evening social and recreational journeys to increase towards the end of the week. The evening journey from work tends to occur somewhat earlier on Fridays, and overtime working is much less common towards the weekend, which may cause a greater peaking of homeward movements on this day. As with movements of people, there is little variation between the working days of goods movements in towns.

For both classes of journey, then, it is principally the strong contrast of Saturday and Sunday movements with each other and with weekday movements which produces the weekly rhythm of movements. In the postwar period in Britain, Saturday morning working has declined to very low levels except in the retail and personal service industries. In 1955, normal basic hours of work were slightly over 44, and nearly 49 were normally worked on average, but by 1975, normal basic hours of work were just under 40 and 44 hours were normally worked (Central Statistical Office, *Social Trends No. 7*, 1976, Chart 10.1, p. 173). Most of the considerable personal travel on Saturdays is for shopping, personal business, social and recreational purposes. Journeys are well spread through the day, but there are noticeable lulls of movement around lunchtime and midway through the evening, perhaps especially in smaller towns. Goods movements are at a lower level on Saturdays than weekdays, particularly in the afternoon and evening. Personal movements on Sundays are less frequent than on any other day, and are principally for social and recreational purposes, while goods movements are extremely few. Characteristics of the temporal distribution of journeys on Sundays are the exceptionally low numbers of journeys before noon, and the variation of the frequency of leisure trips with weather conditions.

Among the most comprehensive data available on the weekly distribution of journeys is that collected by the earlier National Travel Surveys. The first of these surveys, commissioned by the Ministry of Transport, was carried out in 1964 and was largely experimental. Its emphasis was upon car ownership and public transport use. Improved surveys were organized in 1965 and 1966 (Ministry of Transport, 1967h, p.iii of each part). A multi-

Table 3.2 Variation in modal split according to the inclusion of short walk data

	Percentage of journeys by mode					
	car	Motocycle	Bus	Rail	Bicycle	Walk
Excluding	60.0	1.4	21.9	2.6	3.8	10.2
Including	40.6	0.9	13.2	1.6	2.8	40.8

Note: Short walk data covered journeys on foot of less than one mile.
Source: National Travel Survey 1972/3; Mitchell and Town (1977), table 3, p. 13.

stage random stratified sample was taken of the British population of sufficient size (134,000 individuals in 13,000 households) to allow analyses to be made at the regional level. The survey was designed to establish overall national travel patterns and did not therefore confine itself to movements within towns, nor did the published tables give any information for specific cities except for some details about the patterns within Greater London. The great majority of the respondents (over 80 per cent) were drawn from urban areas. One limitation was that walking journeys of less than one mile were excluded from consideration, although all other trips were surveyed. This makes for difficulties in making comparisons either between regions, for in some facilities may be within shorter walking distances, or between individuals with and without cars.

The 1972/3 National Travel Survey (see Department of the Environment, 1975, 1976c, 1976d) included data on short walk trips (of less than 1.6 km = 1 mile) for one day of the seven-day travel diary which was administered to 20,242 persons. On other days of the recording period such trips were excluded. The published data from the survey exclude short walk trips, but several researchers have published their own analyses of the full data (Goodwin, 1975; Mitchell and Town, 1977; Viekerman, 1972). The difference the short walk count makes to the analysis of the modal split is shown in table 3.2. Walk trips were 41 per cent of all trips on the seventh day, as compared to 10 per cent for the other six days. As the purposes, timing, duration and frequency of journeys on foot are very different from those made by other modes of transport, the inclusion of these short journeys will radically change our impression of the aspatial characteristics of movement.

Published tables from the earlier surveys permit the disaggregation by trip purpose of the weekly rhythm of personal movements (figure 3.3). Personal trip making increased from Monday to Friday, when it was 112 per cent of the daily average, but receded slightly to 110 per cent of the daily average on Saturday, and more substantially to only 65 per cent of the daily average on Sunday. Only journeys to and from work showed a very consistent number

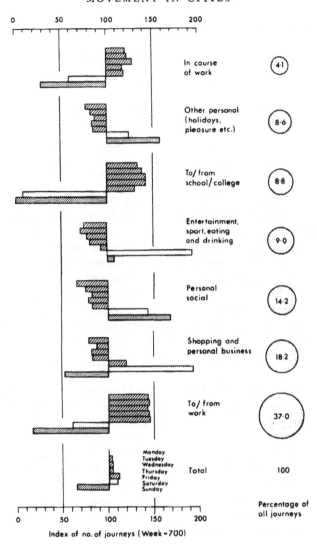

Figure 3.3 *The variation from day to day of personal trip making for different purposes: Great Britain, 1964* (Source: *Ministry of Transport, 1967, table 7, p. 9*)

from Mondays to Fridays, for both educational trips and journeys made in the course of work tended to slight maxima in the middle of the week. On the other hand, shopping and the various categories of personal trips, which tended to be most prevalent at weekends, during the week showed a tendency for minimum rates in the middle of the week. Not all of the differences in detail between the weekly patterns for the different purposes were self-evident. Personal, social and pleasure travel peaked on Sundays at around 160–170 per cent of its daily average volume; shopping and personal business trips were above average on Fridays in contrast to other weekdays, but peaked at 194 per cent of their average daily volume on Saturdays.

The seasonal pattern

As well as clear daily and weekly rhythms of movement, there is a less pronounced annual or seasonal fluctuation in journey habits. This fluctuation owes something even today to the physical influences of the length of daylight hours, weather conditions, and the progress of the growing season, but of greater contemporary importance in Western countries is the preservation of an annual rhythm to our activities by the continuing observation of annual Christian festivals, and the somewhat rigid conventional arrangements for holiday periods, particularly for school children and their families. The last two decades have witnessed substantial increases in holiday entitlement, as well as additions and alternations to the annual sequence of statutory holidays. This is giving the British population greater flexibility in their holidaymaking and day-tripping, which may be leading to a stronger correlation between weather conditions and leisure travel than used to be the case (table 3.3).

Substantial studies of the annual rhythms of movement are rare because of their demanding data requirements, but automatic traffic counts permit such analysis, as has been done in surprising detail by the Road Research Laboratory. In the early 1960s monthly traffic flows increased steadily from 76 per cent of the average figure in January to 124 per cent in August. In the remaining half of the year, there was an equally smooth decline to the low

Table 3.3 Manual workers' paid holiday entitlement, United Kingdom

Annual holiday entitlement	Percentage of manufacturing manual workers			
	1960	1966	1970	1975
Two weeks or less	97	63	28	1
Two to three weeks	1	33	5	1
Three weeks	2	4	63	17
Three to four weeks	–	–	4	51
Four weeks and over	–	–	–	30

Source: Central Statistical Office, *Social Trends No. 7* (1976), table 10.3, p. 174.

Table 3.4 Extremes of the monthly variation of traffic flows

Type of road	January minimum	August maximum
All roads	76	124
Rural roads	71	136
Urban roads:	81	112
Trunk	77	113
Class I	81	110
Class II	81	130
Class III	89	106 (July)
Monday–Friday	83	109
Saturday	77	115
Sunday	72	120 (June)

Note: Except for the figures referring to different days of the week which relate to vehicle flows, the data give vehicle miles per standard week in each month averaged over 1959 to 1962, expressed as a percentage of the average over all months.

Source: Department of Scientific and Industrial Research (1965), table 2.12 and fig. 2.7, p. 39.

winter flow. Seasonal variations were less on urban roads than on rural roads (table 3.4), and within urban areas least pronounced on the most minor roads. Further analyses showed that the seasonal fluctuation on urban roads was greatest for weekend and particularly Sunday traffic. At weekends, traffic flows remained at an unusually high level throughout the months from June to September, while weekday traffic tended to a narrower peak in August. These differences can be readily accounted for by considering the likely locations and routes of both the least and the most variable purposes of traffic, such as shopping trips on the one hand and leisure trips as a category which predominates in the summer (Department of Scientific and Industrial Research, 1965, pp. 37–42).

Other patterns

There are other temporal rhythms which influence individuals' behaviour and travel. Chapin added those of the life cycle, which, as we examined in the last chapter, patterns the journeys an individual makes at any particular age. In broad terms, however, the hour, day or month is not cyclically related to the proportion of people at any particular stage of the life cycle. There are relevant secular trends, for as has been mentioned the proportion of the population in the retired age groups has been increasing during this century, and there are marked fluctuations in annual birth rates which lead to a redistribution of the age structure. The travel effects of such demographic trends are to a large extent swamped by endogenous transport changes.

The temporal variation of weather conditions may be associated with the generation of social, recreational and leisure journeys. These variations are assuming greater importance for the transport networks and in particular the road networks in certain rural areas and regions such as the South West and Lake District, to the extent in some cases that peak demand for these travel purposes exceeds that of peak workday demand. The degree to which these partly unpredictable and relatively infrequent demands should be catered for by the expensive enlargement of capacity currently exercises road planners, but by and large this is more a problem of highly attractive rural and coastal areas than towns and cities. However, even in a large city, an exceptionally fine weekend will generate outward and inward flows which exceed peak rush-hour flows on the major arterial roads, and can still lead to overcrowding on certain public-transport routes, such as the railways extending to proximate coastal resorts. Other less important periodicities of movement can be identified. For example, there is the temporal variation associated with business or economic cycles. These are particularly related to the generation of freight trips as part of the manufacturing and first-distribution processes, as for example with coal and steel consignments. In at least a small way, the level of unemployment has an effect on the number of work trips – lower purchasing power depresses retail sales, and possibly also shopping and leisure trips. Since the Second World War, however, while economic fluctuations have been very much in evidence, economic management has been sufficiently successful to prevent fluctuations in unemployment and personal real income great enough to have other than marginal effects on travel behaviour.

From time to time it is suggested that the present timing of work, school and other activities are unnecessarily rigid, and that changes can be made that will save resources such as those required for peak-hour public transport capacity. Certainly more variable hours of work as between places of employment would spread the peak loads upon the public transport and road systems, and enable individuals to enjoy less crowded and in some cases cheaper (off-peak rail) travel. Resistance to such changes comes not only from employers faced with the problems of organizing daily work programmes and of coordinating with other firms and organizations, but also from the employees whose activity patterns are intricately meshed with those of their families. Schoolchildren require to be despatched, collected, fed and cared for; adult children or spouses are driven, met or accompanied.

It cannot be doubted that the 'nine to five' core of activity time is a precondition of the dovetailing and interweaving of individual activity patterns. Only a social cost and benefit appraisal of unusual scope could determine who would gain and lose from alterations to the present conventions and customs of our daily lives and travel.

The number, duration, frequency and distance of journeys

The number of trips undertaken by an individual varies according to his or her activities, age, income, access to different types of vehicle, and according to facets of the local environment, such as accessibility to various facilities and the quality of local roads and transport services. These interrelated variations have been the subject of innumerable studies over a long period, with a popular focus being upon the role of car ownership on trip generation. Recently attention has been given more readily to the relationship between trip making and an individual's family and household position, so giving a different perspective on the inter-correlations between age, household circumstances, income, activity patterns and trip-making habits (see, for example, DoE, 1976d; Hillman *et al.*, 1976). Hillman and colleagues, for instance, have concentrated upon the contrasting journey requirements and the great differences in income and access constraints of schoolchildren, teenagers, housewives with young children and pensioners. This disaggregated approach has been a useful antidote to a former view that car ownership is the most influential variable on trip-making levels, based on an inappropriate longitudinal interpretation of cross-sectional data on the trips made by car-owning and car-less house holds.

On average, according to the 1975/6 National Travel Survey, the number of journeys made by the population ranges from 3.0 to 3.5 per day depending upon the area of residence. In other words, people make of the order of two return trips from home, or one four-stage return trip from home per day. The lower trip rate is found in rural areas and in urban areas of the lowest residential densities, while the highest trip rates are found in urban areas of the highest density (figure 3.4). Goodwin's analysis of these data also showed a clear and strong inverse relationship between the trip rate and the distance of journeys. Whereas in the densest urban areas the average daily distance travelled was approximately 12 kilometres, in the least densely inhabited areas average daily travel distances were of the order of 20–25 kilometres. These findings are consistent with those from the earlier 1972/3 National Travel Survey (which excluded short walk data), which found that people made approximately 1.5 round trips per day, ranging from 1.2 in the least densely settled areas, to over 1.7 in urban areas with 40 persons per acre. Webster's analysis (1977, fig. 8, p. 44; fig. 10, p. 45) of this earlier data also revealed that on average the population spent approximately 55 min-

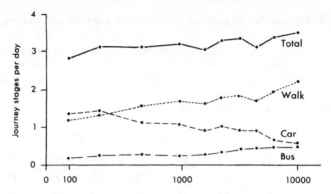

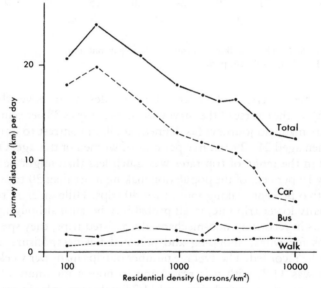

Figure 3.4 *The variation between residential densities and trip rates and distances: Great Britain, 1972/73* (Source: *Mitchell and Town, 1977, fig. 5, p. 22)*

utes on travel each day and that this figure varied little according to the nature of the residential area.

There are marked variations in the number of trips made, and in their distance, duration and cost, according to the age and sex of the individual. Data from the 1972/3 National Travel Survey exaggerates the contrast in trip rates by excluding short walking journeys, but even so it cannot be doubted that it is younger adult males who have the highest trip rates, and

Table 3.5 The number of personal journeys per week by age and sex, 1972/3

	Percentage distribution of/number of journeys							
	0–6		7–12		13–20		21+	
Age	m	fm	m	fm	m	fm	m	fm
0–16	46		26		21		7	
16–20	12	17	17	22	38	39	33	23
21–29	8	24	15	26	33	28	44	21
30–59	11	29	17	28	33	28	38	15
60–64	16	44	24	30	33	20	26	6
65+	51	70	24	20	15	9	9	1
All	32		23		26		19	

Note: These data exclude walking journeys of less than one mile.
Source: DoE (1975), table 14, p. 5.

children, pensioners, and older adult females who have the lowest (table 3.5). In the course of the survey week, as many as 33 per cent of female pensioners made no journeys (as defined) at all, in contrast to only two per cent of men aged 21–29 and five per cent of women of this age. Overall the variation in the range of trip rates was much less than might be expected, with only 19 per cent of the population making more than 20 trips per week and but two percent making more than 40 trips. Hillman *et al.* recorded a substantially lower trip rate, in all probability because although they were scrupulous in recording non-mechanically assisted trips, they specified that only trips for work, shopping and personal business, leisure and social purposes be recorded. The average number of trips made per week for these purposes was 11.3, with it is reassuring to note a minimum of 8.6 trips recorded for a village population, and 11.7 for their samples in a small town, a new town and an inner London suburb (Hillman *et al.*, 1976, table 14, p. 6). Their interpretation of the most important characteristics of the intercorrelations between trip rates and socio-economic variables was that:

> People in households with cars reported making a greater number of work trips than those in households without a car, no doubt partly owing to the high proportions of retired people without a car. Trips for social and especially for leisure purposes are also more frequently made by those in households with cars, though they make fewer shopping trips – in view of the greater ease with which large quantities of goods can be carried by car. Licence-holders in the car-owning households reported more leisure

and work trips and fewer shopping trips than those without licences. While this reflects the differences in licence-holding and trip purposes between men and women, the survey indicates that similar differences occur in the travel of women with and without licences. (pp. 4–5)

As has already been noted, there is an inverse relationship between an area's trip rate and the average distance of its journeys. There are also very clear and widespread differences in the distances of journeys for different purposes. Education and shopping trips tend to be the shortest, reflecting the large proportion which are carried out on foot, while work journeys tend to have the longest mean distance. These differences can be detailed by reference to one more set of data from the 1972/3 National Travel Survey, which also makes a comparison with the situation in 1965 (table 3.6). It is evident from these two distance distributions that there had been little change in this characteristic of personal trip making during the intervening seven years, although the pervasive tendency was for modest increases in journey distances. Fifty-two per cent of work journeys in 1972 were less than ten miles, a fall of four percentage points from the earlier date: similarly, the percentage of educational journeys in this distance range fell during the period by five percentage points to 69.

There is no simple relationship between journey distance and duration, for the latter depends not only upon the mode of the journey but also upon the local transport environment. It was noted above that in aggregate there is little variation between areas in the amount of time spent on all types of travel. This is simply explained by the compensating facts that although journeys made by residents of rural and low density areas are unusually long, they also have relatively low trip rates and a high proportion of their journeys are made by car at relatively high average speeds. Webster's analysis of National Travel Survey data showed for example that in rural areas the average personal car travel time was approximately 33 minutes compared to about 16 minutes in the densest urban areas. On the other hand, the comparative figures for bus travel were approximately 6 and 11 minutes, and for walking 12 and 25 minutes (Webster, 1977, fig. 10, p. 45).

The mode of journeys

Modal choice has been the subject of extensive empirical and theoretical research, the latter because of the subject's importance as one of the four stages of urban transport modelling. This aspect will be more fully discussed in Chapters 6 and 7, but the rationale, simply put, is that future urban transport demands will be related to the household and economic characteristics of the urban population. The important intermediary factor between the outcome and its underlying causes has been regarded as car ownership. It has been widely argued that if the population becomes more affluent, or to a

Table 3.6 The distance of personal journeys by purpose: Great Britain, 1965 and 1972

Distance (miles)	To/from work		Education		Shopping and personal business		Entertainment sport, drink		Personal social	
	1965	1972	1965	1972	1965	1972	1965	1972	1965	1972
0–3	13	12	25	29	19	20	11	12	7	8
3–10	43	40	49	40	40	39	35	30	28	27
10–15	15	17	11	13	11	13	13	13	11	12
15–30	16	19	10	12	14	12	19	18	18	18
30–100	12	11	4	4	13	11	18	22	27	25
100 or more	1	2	–	2	4	5	4	5	9	10
Total (000s)	458	324	58	46	211	210	134	128	292	265

Note: These data do not include walking journeys of less than one mile.
Source: DoE. (1976c), table 18, p.33.

lesser extent if there is a trend towards smaller households, then car-ownership rates will increase and this in turn will increase trip generation rates. The present patterns of the spatial distribution of movement within the city can then be used, together with information concerning approved or possible additions to routes and capacity, to allocate or assign the increased number of trips to the available capacity or routes. (The innumerable papers on modal split include: Demetsky and Hoel, 1972; Notess, 1973; Royal Institute of Public Administration, 1973; Quarmby, 1967.)

It can be argued that this approach is based upon a simplistic view of the causal relationship between car ownership and trip generation. Such a view is encouraged by recent research into the modal characteristics of movement, which reveals that a large proportion of urban personal journeys continue to be made on foot, and that these trips represent a vast potential demand for more expensive modes of transport. This demand can be readily and unintentionally actuated both by the relaxation of income and time constraints upon mechanically assisted travel, and by planned or spontaneous environmental changes which reduce the accessibility of facilities to the population. Urban transport policies based on projections of present conditions including modal split patterns, therefore need to be flexible and adjustable to their own results, for, as Jones (1978) has put it, 'latent suppressed demand is apparently [even] more policy sensitive than modal choice'.

In this section the emphasis will be upon establishing the present patterns of modal split in British towns. It is in fact only recently that surveys have taken a comprehensive view of urban personal movement, and that we have moved away from the study of modal choice in imaginary towns where no trips are made on foot. The latest National Travel Survey shows that between 36 and 40 per cent of trips in urban areas were made on foot (table 3.7). In urban areas about 43 per cent of trips were made by car or van, and about 12 per cent by bus. The presentation of the data from the 1975/6 Survey for the major conurbations and for five size categories of other urban areas permits many more details to be studied. The clearest relationship between urban population size and modal split in Britain in the mid-1970s was that car use was inversely related to urban scale, with 47 per cent of all journeys being by this mode in towns of less than 25,000 people, but only 39–40 per cent in the four largest conurbations and in London (figure 3.5). To complement this relationship, bus use increased from 7 per cent of all journeys in the smallest towns to 18 per cent in the large conurbations. However, the relationship was not linearly projected for London, in which metropolitan area bus use declined to 13 per cent, but there was an even greater reliance on public transport by the exceptional use made of railways, which in London attracted 6.5 per cent of all trips.

Similar relationships are found within cities and metropolitan areas, in

Table 3.7 Mode of travel by area; Great Britain 1975-6

Type of area	Rail	Bus	Bicycle	Walk	Car/van	Other	Total trip sample
London built-up area	6.5	13.2	2.5	36.5	39.7	1.6	7,660
Birmingham built-up area	0.7	16.0	1.7	35.8	44.2	1.4	2,849
Manchester built-up area	0.7	16.5	2.2	39.8	37.9	2.9	3,494
Glasgow built-up area	2.7	24.3	1.5	36.8	32.0	2.7	952
Liverpool built-up area	1.4	22.3	1.7	34.7	36.9	2.9	1,000
Population							
Urban 250,000–1,000,000	0.6	16.1	2.7	37.6	40.6	2.4	8,982
Urban 100,000–250,000	0.9	14.1	4.1	35.9	42.0	2.9	9,738
Urban 50,000–100,000	1.9	10.3	3.8	38.3	43.1	2.5	5,145
Urban 25,000–50,000	1.4	8.7	3.8	37.7	45.5	3.0	6,529
Urban 3,000–25,000	1.0	7.0	3.8	37.8	47.0	3.5	13,321
Total urban	1.8	12.3	3.2	37.3	42.7	2.7	59,673
Rural < 3000 population	0.6	5.9	3.6	27.1	57.4	5.4	10,946
Total	1.2	11.3	3.3	35.7	45.0	3.2	70,619

Note: Data for short walks for one day only; other data based upon one-day average of seven-day records.
Source: 1975/6 National Travel Survey, cited by Potter (1978).

THE ASPATIAL CHARACTERISTICS OF MOVEMENT

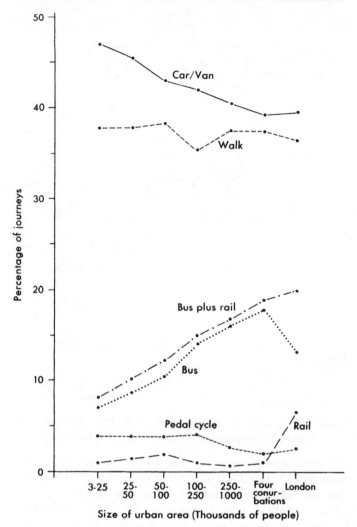

Figure 3.5 *Mode of travel by area: Great Britain, 1975/6* (Source: Potter, 1978, table 1, p. 4)

part reflecting transport supply factors and in part the varying characteristics of people and their activities. Inner-city areas tend to have lower rates of car ownership, both as a result of the low median income of their inhabitants and as a rational response to the superior public transport facilities of densely developed urban areas. In the West Midlands in 1964, for example, whereas 40 per cent of the residents of central Birmingham walked or cycled to work, and over 45 per cent of the residents of the inner districts of Smethwick, Acocks Green, Erdington and Handsworth used public

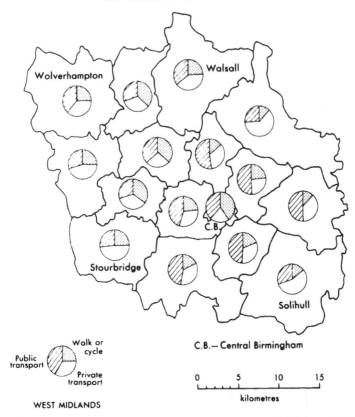

Figure 3.6 *The mode of travel to work in the West Midlands, 1964*
(Source: *Freeman Fox* et al., *1968, table 4.11, p. 52)*

transport for the journey to work, only 14 per cent of the residents of the affluent suburbs of Solihull and of Sutton Coldfield walked to work, and less than 30 per cent used public transport. Instead they made exceptionally high use of private transport, with 55 per cent of Solihull's and 62 per cent of Sutton Coldfield's employed populations travelling by car or other vehicles (figure 3.6).

More general results have been produced by Goodwin's analysis of National Travel Survey data (figure 3.4). This shows that the number of walking and bus stages per day increased as the density of an individual's residential area increases, but that the number of car stages falls. It is also recorded that whereas the distance of walk stages varies little with residential density, the distance of bus and car stages increases as the density is reduced; it is these two modes, then, which are responsible for the inverse aggregate relationship that has already been noted (Goodwin, 1975; reproduced by Mitchell and Town, 1977, fig. 5).

THE ASPATIAL CHARACTERISTICS OF MOVEMENT

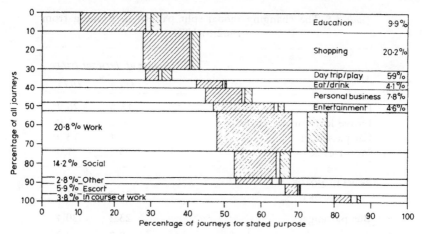

Figure 3.7 *The mode of personal travel according to trip purpose: Great Britain, 1972/73* (Source: *Mitchell and Town, 1977, table 4, p. 13*)
Note: *The diagram represents the distribution of journeys by purpose (horizontal rows), and the mode or means of travel (vertical divisions). Reading from left to right in each row, the modes represented are: car and motor-cycle; bus; train and tube; bicycle; and walk.*

The variations by residential area in the use of different modes of transport are largely accounted for by obvious contrasts in transport and accessibility environments, with the socio-economic differences in area populations apparently taking a more minor role. But within each area of the country, considerable individual variations in modal characteristics of trip-making do exist, principally because of the strong relationships between trip-purpose, trip-distance and trip-mode. Rigby has analysed the short walk National Travel Survey data by trip-purpose, and found very considerable differences in the various categories' modal splits (figure 3.7). While, in the aggregate, three modes – car, bus and on foot – account for 94 per cent of all personal trips, the relative importance of walking and of motorized trips varies markedly with the trip-purpose. Over 65 per cent of education trips were on foot, as compared to less than 15 per cent of the journey stages made in the course of work. Only about 42 per cent of shopping trips were mechanically assisted, as opposed to nearly 70 per cent of work trips and nearly 65 per cent of social trips. The relative importance of motorized travel is clearly related both to the distance distribution of the trip-purpose and to the access to motor-cars of the participants. Trips to schools and shops are dominated by walking both because these facilities are very widely scattered within residential areas and because of the restricted access to cars of children and women.

During recent years there has been a well-known trend for the proportion of journeys made by bus, rail and bicycle to fall and for the motorized

Table 3.8 The changing modal split pattern in Reading from 1962 to 1971

Mode	Percentage of personal trips	
	1962	1971
Drove car	15.3	26.9
Drove lorry or passenger	0.1	1.0
Car passenger	6.4	13.0
Motor-cycle	3.2	1.0
PRIVATE MOTOR TRANSPORT	25.0	41.9
Bus passenger	23.2	20.4
Taxi	0.2	0.2
Train	0.0	0.1
PUBLIC TRANSPORT	23.4	20.7
Walked	38.9	32.5
Bicycle	12.7	5.0
NON-MOTORIZED TRANSPORT	51.6	37.5

Note: Only internal person trips were surveyed, totalling 521,860 in 1962 and 589,750 in 1971.
Source: Downes and Wroot (1974), fig. 15.

proportion to increase. Longitudinal data on walking trips is scarce, but the evidence from Reading is that between 1962 and 1971, walking journeys also declined in relative importance (table 3.8). The nine years witnessed a fall of 6.5 percentage points, to 32.5, in the representation of journeys on foot. In other words, 38,334 additional motorized personal journeys were being made during the week in the town because of the change in the modal split pattern. The fall in the relative importance of cycling journeys from 12.7 to 5.0 per cent generated a further 45,411 additional mechanically-assisted trips (Downes and Wroot, 1974, fig. 15). These changes produce not only the very well-known difficulties for public transport operators of declining use, but clearly bring about considerable community costs in the provision of road capacity and vehicles for the larger number of car trips (Webster, 1977; Mitchell, 1977).

4
Goods movements within towns

While most of this book is concerned with personal or passenger movements within towns, in this chapter an attempt is made to bring together a coherent account of commodity movements within the city. Although a great deal of data has been collected by many transportation studies on at least the road-vehicle movement of goods, there have been few secondary analyses or appraisals of this information. The spatial patterns of freight movement and the situation in British towns are particularly badly served. This information gap has arisen for many reasons, but among the most important are the assumptions made in the early transportation studies that most internal goods movements are handled by road vehicles, that these vehicles form a small proportion of road traffic and that their movements are distributed more evenly through the day than passenger vehicle movements and therefore are of little importance in creating the peak flows which at that time were seen as the dominant urban transport problem. (This point is well documented in quotations from early transport studies in Meyburg *et al.*, 1974; for example, a 1959 report of the Chicago Area Transportation Study stated that 'generally the patterns of automobile trips and truck trips are sufficiently alike to suggest no special treatment of 'necks will be warranted'.) The more minor role of municipal or central government authorities in goods as opposed to passenger movement, together with the fiercely competitive structure and often anti-planning attitudes of the road haulage industry can also be related to the subject's neglect. As a result more reliance will be placed in this chapter on material from other countries.

Several exceptions to the general neglect of freight studies have provided a foundation for this account. Among the earliest studies was Chinitz's monograph (1960) on the impact of America's transport revolutions on the New York region, which was produced as part of the independent Regional Plan Association's broader study of this metropolitan region. The book focused on two kinds of freight: that involved in international trade mainly through the port, and that which served and was generated by manufacturing industry within the region, Although Chinitz's books and the survey commissioned for it were mainly concerned with international and inter-regional movements, they provided a substantial introduction to goods movements in a large city. More recently in the United States, there has been a surge of urban goods movement studies led by some of the transportation study teams, such as the Tri-State Transportation Commission, and most energetically and successfully by the Chicago Area Transportation Study (Wood, 1967, 1970, 1971; Blaze and Raasch, 1970). Indeed, sufficient interest had been generated by 1972 to have enabled at least three substantial bibliographies to be published (Smith, 1971; Watson, 1972; Goeller, 1972; see also Chappell and Smith, 1971). In Britain, interest lags far behind, and it is only during the last five years that the Transport and Road Research Laboratory (TRRL) has undertaken specific studies on a limited scale of urban freight movements (Hitchcock *et al.*, 1974). The Freight Division of TRRL has completed twenty reports on various aspects of commercial vehicle operations in towns, and studies of movements in Hull and Lewisham are in progress. The latest publications concerning the Swindon study are: Purcell *et al.*, 1977; Christie *et al.*, 1977; the first Hull report is W. Smith and Associates, 1977. Both studies are also described in TRRL, 1977. Urban and transport geographers on both sides of the Atlantic have paid very little attention to the subject – most of the few studies of freight transport have concentrated on inter-regional and external movements (Chisholm and O'Sullivan, 1973; Helvig, 1964; Perle, 1964; Ullman, 1957). This emphasis has persisted despite the surveys carried out by the transportation studies, and the productive and interesting research by Starkie (1976b, 1970) on the generation of road vehicle movements at manufacturing plants within the Medway Towns in 1964. As is reflected in several paragraphs of the latest government statements of its transport policy, recent trends in public opinion are forcing more attention upon the environmental nuisances associated with urban good-vehicle movements, and stimulating research aimed at reducing or diverting lorry flows through residential and commercial areas (Department of Transport, 1977, paras 192–203). These welcome developments have not yet led, however, to anything approaching a satisfactory study of the geography of intra-urban commodity movement.

The scale and types of urban goods movement

Freight movements are not as concentrated in urban areas as are personal movements, for considerable tonnages of commodities flow on extra-urban and inter-urban journeys. For example, 14 per cent of national commodity tonne-kilometres are handled by coastal shipping, and only a small amount of this total (e.g. in East London and between Liverpool Bay and Manchester Docks) are intra-urban movements. Nationally, freight movement is handled by a greater variety of transport modes than personal movement, for in addition to the shared systems of road, rail and air transport, commodities are also moved by pipeline, conveyor, coastal shipping and inland waterways. Even walking has its goods equivalents in the use of messengers and barrows, the latter being sufficiently important in some trades and areas to have attracted at least one 'transportation survey' (Kahan et al., 1976). Another point to remember, especially when studying the development of urban goods movements through time, is that just as some personal movements are substituted by other forms of communication when social or business exchanges are performed on the telephone, some commodity movements can be substituted by other forms of transmission. Formerly, considerable volumes of coal movements to business and domestic premises have in effect been replaced by the 'non-transport' distribution of electricity and the pipeline movement of gas for heating and other purposes (reduced use of coal is cited as one factor in the closure during the 1940s of a 62-mile underground freight railway in Chicago which once carried annually 600,000 tons of merchandise, including 57,000 tons of coal and large volumes of unders; see Fruin, 1972). Looking to the future, it seems likely that at least some letter and packet movements between commercial premises will be substituted by the telecommunication of documents (Post Office, 1972, cited by D. Clark, 1976).

The dominant mode of transport for freight in the United Kingdom is by road, which accounted for 66 per cent of total tonne-kilometres in 1975 not including pipeline movements, except of oil and petroleum (Central Statistical Office, *Annual Abstract of Statistics 1976*; additional statistics on goods movements appear in Government Statistical Service, 1977). This percentage has been increasing steadily (it was 36 in 1953 and 57 in 1965), largely at the expense of rail (down from 41 through 21 to 18 per cent) and coastal shipping (down from 23 through 21 to 14 per cent). Road movements, it should be noted, form a higher proportion of intra-urban than national goods movement. There is little doubt that, overall, road vehicles are taking a larger share of an increasing volume of freight movement in Britain. It has been estimated that in 1965 there were 120.6 thousand million tonne-kilometres of commodity movements, and that by 1975 this figure had increased to 133.5. This is widely related to the economic growth of the country, and indeed for the purposes of forecasting freight transport growth, a figure of between 0.65 and 0.85 is posited for the ratio between the annual percentage increases in road plus rail freight traffic and the annual percentage increase in gross domestic product (Ministry of Transport, 1963c, para. 20 and table 5, cited by Chisholm, 1970, p. 432; also see O'Sullivan, 1970, pp. 443–50). This aggregate growth disguises, however, contemporary decreases in the volume of bulk commodity (coal, iron and steel materials and products) movements, and a faster growth rate of general merchandise traffic. It seems reasonable to infer from this changing composition of commodity movements that intra-urban goods movement is increasing faster than national freight movement. Because of concurrent changes in vehicle capacities, and possible changes in the mean size of a consignment, this does not necessarily mean, however, that there have been proportionate changes in goods vehicle trips.

The importance of goods movements relative to passenger travel can be measured in several ways: by the expenditure incurred, the employment it creates, the traffic generated, or by its social and environmental costs and benefits. It has been estimated that in the United States of America, approximately 45 per cent of transport expenditure is for freight movement (including all expenditure on international transport) (Fresko et al., 1972). The freight bill accounted for approximately 10 per cent of its gross national product (Hille, 1972). The comparative distribution by modes of passenger and freight expenditure reveals that in 1967 over 75 per cent of internal freight movement expenditure in the U.S.A. was on road transport, and that rail movement accounted for nearly 15 per cent of the total bill (table 4.1).

In terms of its percentage share of urban road traffic, goods vehicle movements are more important than many studies and authorities have assumed. Surveys carried out in twelve United States cities during 1953–9 showed that 'trucks' accounted for 16 per cent of the total number of urban

Table 4.1 United States transport expenditure, 1967

Mode	Personal movements $m.	Personal movements %	Freight movements[4] $m.	Freight movements[4] %
Rail	520	0.6	9,950	14.8
Road[1]	77,732	90.7	51,406	76.6
Water[2]	12	0.01	1,246	1.9
Oil pipelines			1,147	1.7
Air[3]	7,471	8.7	1,062	1.6
Miscellaneous			2,302	3.4
Total	85,735	100.0	67,113	100.0

Notes: 1. Includes private cars, taxis, buses, urban *rail* transit and trucks.
2. Coastal, inland and Great Lakes domestic movements, but not international movement expenditure.
3. Includes private planes and expenditure on international movement.
4. Includes costs of freight forwarders, loading and unloading traffic operation and of the shipping department.

Source: Fresko et al. (1972), p. 2.

Table 4.2 Predominance of truck trips in four U.S. cities, 1960–3

		San Diego 1963	Salt Lake City 1960	Nashville 1961	Cleveland 1963
Truck trips in 14-hour day	no.	23,080	277,294	16,964	282,626
	%	14.0	18.0	15.5	8.8
Truck trips in morning peak	no.	6,000	5,000	2,904	6,504
	%	19.4	16.7	16.4	11.6
Truck trips in evening peak	no.	6,000	6,000	2,421	5,718
	%	14.6	16.7	11.6	9.0

Notes: Alternate rows express truck trips as a percentage of all trips. The peak-period figures refer to crossings of each survey's screenline.
Source: Fresko *et al.* (1972), table 2, p. 4, in which references to the original sources are given.

vehicle trips, with a range of results for the individual cities from 12.0 to 22.2 per cent and a mode of 14 per cent (Smith and Associates, 1961, p. 67, cited in Starkie, 1976b, p. 18). Later United States transportation studies report similar findings. Particularly unexpected is the finding that truck trips generally form a higher percentage by a small margin of all peak-period trips than they do of trips throughout the day (table 4.2). So in a country with much higher rates of car ownership and use than Britain, and even at a time when higher-capacity goods vehicles were operating on American than British roads, characteristically between one-tenth and one-sixth of intra-urban mechanically assisted trips were freight movements.

A very similar relative incidence of freight movement is indicated by most sources for British towns. Evidence from the West Midlands suggests that urban goods vehicle mileage increased substantially during the late 1950s and early 1960s (table 4.3) but because private car mileage increased even more rapidly, goods movements declined from a third to a fifth of the total (Freeman Fox *et al.*, 1968a, vol. 1, p. 144). By 1971, 16 per cent of vehicle mileage on urban trunk and classified roads in Great Britain was accounted for by goods vehicles (Central Statistical Office, *Highway Statistics 1971*, cited by Warner, 1973, table 1, p. 235). The percentage representation of goods vehicles had continued to fall since 1966 as the mileage travelled by cars and taxis had increased 30 per cent but also because the mileage of goods vehicles had actually decreased by one per cent. Among goods vehicles, light-van mileage increased its share relative to all other types of goods vehicles. These trends continued until 1976, but more from the stronger decline in goods vehicle mileage since the growth of private car mileage after 1971 was less than 5 per cent. The continuing growth of mileage by light vans (of less than 30 cwt) is notable in comparison to the 8-per cent decline in goods vehicle mileage since 1960. Much of the fall in

Table 4.3 Goods vehicle mileage on urban roads relative to all vehicle mileage, West Midlands 1954–64 and Great Britain 1966–75

Vehicle Type	West Midlands Weekday million vehicle miles			Great Britain Annual million vehicle miles			
	1954 number	1954 per cent	1965 Index (1954 = 100)	1971 number	1971 per cent	1971 Index (1966 = 100)	1975
Cars and taxis	1.90	54.3	252	41,265	80	130	136
Motorcycles	0.25	7.1	120	972	2	59	74
Goods vehicles	1.10	31.5	136	8,114	16	99	92
(Light vans)				(4,204)	(8)	102	105
(Other)				(3,910)	(8)	95	77
Buses and coaches	0.25	7.1	92	1,116	2	90	80
Total	3.50	100.0	195	51,467	100	120	124

Sources: Freeman Fox et al. (1968a), vol. 1, table 9.3, p. 144; Central Statistical Office, Highway Statistics 1971; Government Statistical Service (1977), table 25, p. 41.

urban goods vehicle mileage can be attributed to the increasing size of vehicles and the reduction of retail delivery services.

More specific surveys of the relative incidence of goods vehicle trips have been conducted at innumerable sites. A small selection of the results, principally from London and Liverpool, is considered here. Evidence from various parts of London in 1968–9 resulted from the screenline studies of Munt (1970). Light and heavy goods vehicles comprised over 20 per cent of all traffic. Each survey site, illustrated in figure 4.1, returned similar figures (table 4.4), but commercial traffic was most prevalent across the River Lea screenline, in London's north-eastern suburbs, and across the central area cordon. In a commentary on these figures, Moulder has suggested on the basis of limited studies in Hammersmith and Wembley that the concentration of goods vehicles in town centres during business hours is much higher, for in studies of the traffic flow during a single day along the main streets of Hammersmith and Wembley town centres, goods vehicles represented 38 and 32 per cent of all vehicles (Metra Ltd., 1970, cited by Moulder, 1971). Later studies in three other towns by Christie and others (1973a, b) have irritatingly not separately distinguished light good vehicles (generally defined as these with less than 30 cwt unladen weight), and do not produce exactly comparable results, but provide further indications of the local importance of commercial vehicle movements. In general the results from the Merseyside Transport Study suggest a higher representation of

Table 4.4 Prevalence of goods vehicles in traffic at selected sites in Greater London, on Merseyside, and in shopping streets in south-east England.

Site and date	Total vehicle flow (000s)	Composition (%)		
		Light goods	Heavy goods	All goods
1. Greater London				
Central area cordon 1968	672	11.5	11.6	23.1
Inner area cordon 1969	900	11.9	9.0	20.9
Outer composite cordon 1969	907	10.3	9.1	19.4
River Thames screenline 1968	445	11.0	10.6	21.6
Kent/Surrey screenline 1969	56	11.1	7.2	18.3
North-west screenline 1969	140	10.4	9.7	20.1
River Lea screenline 1969	60	10.7	13.3	24.0
All cordons and screenlines	3616	11.1	9.0	21.0
2. Merseyside, 1966				
River Mersey screenline	36			31.7
Mid-Wirral screenline	41			17.4
Birkenhead Docks screenline	29			30.3
Liverpool ring screenline	207			27.5
Seaforth–Kirkby screenline	70			26.6
Aigburth–Huyton screenline	104			23.6
All screenlines	487			25.9
3. High Streets, Towns in S.E. England				
East St, Hammersmith, westbound 1970	7.23	10.6*	27.5*	38.1
Wembley High Road, two-way 1970	11.87	11.3*	21.0*	32.3
Northbrook St, Newbury 1972	10.43	9.6*	0.4*	10.0*
High St, Camberley 1972	4.76	9.6*	0.3*	9.9*
High St, Putney 1972	17.24	12.3*	1.7*	14.0*

Notes: The cordon and screenline data refer to a sample weekday of 24 hours. The Hammersmith and Wembley data refer to sample Mondays from 08.00 to 16.00 hours. The Newbury, Camberley and Putney data refer to sample weekdays from 08.00 to 17.30 hours.
* Buses are included as commercial vehicles, except in the Newbury, Camberley and Putney data. In Hammersmith and Wembley light goods vehicles were estate cars and small vans. In the three towns studied, light goods vehicles are not separately tabulated from cars, and the second column is used here to represent goods vehicles with two axles only. The third column indicates vehicles with more than two axles.

Sources: Munt, (1970), cited by Moulder (1971), table 3–5, p. 11; Christie *et al.* (1973b) table, p. 3; Traffic Research Corporation (1969c), plate 4, p. 12.

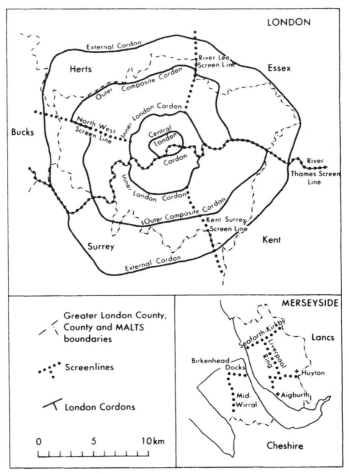

Figure 4.1 *Screenlines and cordons used in commercial traffic surveys of London and Liverpool* (Sources: *Moulder, 1971, fig. 1, p. 12; Traffic Research Corporation, 1969c, plate 61, p. 102*)

goods vehicles in the conurbation's traffic than in London. Just over a quarter of all surveyed vehicles were commercial (Traffic Research Corporation, 1969c, plate 4, p. 12).

The functions of urban goods movements

There have been many different approaches to the structuring or classification of urban goods movement, although because of the diversity of these movements and the contrasting objectives of different studies little by way of an agreed or most useful approach has yet emerged. It is possible to compile a long list of characteristics by which to disaggregate goods movements. These can be grouped into those concerned respectively with the functions or social and economic role of goods movement, and the trip characteristics (particularly frequency, mode and the spatial pattern of origins and destinations) of different classes of movement. Several suggested classifications mix functional and trip characteristics, but more systematic understanding of the patterns and determinants of urban goods movements is likely to result from initially keeping separate functional and trip characteristics.

Commodity movements are involved in all types of economic activity, but their number and character varies considerably. In general terms the ordinal classification of economic activity from primary to quaternary is correlated with the demand for goods movements. The primary production of food or mineral raw materials generates bulky and high-weight movements, while most tertiary activities require only small quantities of tangible raw materials and produce only small volume and low weight outputs. There is, however, great variation within each class of activity in the relationships between the volume of goods movements and the invested capital, level of employment and value of output. This is amply demonstrated by contrasting the transport requirements of cereal farming or milk production as opposed to soft-fruit farming; of iron-ore and diamond mining; of furniture stores and jewellers; of laundries and opticians; and even in the quarternary sphere of activity, of pharmaceutical and aerospace research. As the industrial and occupational structure of urban areas is typified by a low proportion of primary activity and a high proportion of tertiary activity, it is the variations in goods movement generation within the manufacturing and service industries that most concern us here.

The transport intensiveness of different industries has recently been calculated by Mackie and Urquhart by combining a number of sources (Mackie and Urquhart, 1974, table 3, p. 7, which table was compiled from 1966 Sample Census of England and Wales *Economic Activity Tables*, and for expenditure on transport from the *1968 Census of Production*; see also Tulpule, 1971, pp. 1–12, and Edwards, 1969). Their results, which are reproduced in table 4.5, are consistent with Starkie's findings on the numbers of commercial vehicle trips per day at various manufacturing plants in the Medway Towns. He conducted site traffic counts on a single and different day at each of 77 plants employing more than ten people, so there are few observations for each industrial category. Thirty-seven engineering and allied trades plants (Standard Industrial Classification Orders 5, 6, 8 and 9) were isolated as a standard for the relationship between plant size and traffic generation, and a satisfactory and significant log-linear regression equation calculated. The regression line was plotted on graphs together with the lines representing plus and minus two standard errors of the estimate of y

Table 4.5 Transport expenditure by industry, England and Wales 1966; and commercial vehicle traffic generation in the Medway Towns, 1964

Standard industrial class		Total transport expenditure as a percentage of net output	Road transport expenditure as a percentage of total	Road transport expenditure as a percentage of net output	Medway Towns plants, mean traffic generation in S.E. (y)
No.	Description				
20	Distributive trades	21.6	83.0	17.9	
13	Bricks, pottery, glass, cement, etc.	13.9	90.3	12.6	$+5.3$
3	Food, drink and tobacco	12.6	95.1	12.0	$+3.0$
2	Mining and quarrying	9.8	79.3	7.8	
14	Timber, furniture, etc.	7.2	95.7	6.9	-0.3
17	Construction	5.3	99.4	5.3	
4	Chemicals and allied industries	6.8	72.4	4.9	0.0
15	Paper, printing and publishing	5.6	77.8	4.4	-0.1*
5	Metal manufacture	5.8	73.2	4.3	0
16	Other manufacturing industries	4.5	89.6	4.0	-2.6
9	Other metal goods	4.0	88.7	3.6	0†
11	Leather, leather goods, fur	3.3	87.9	2.9	
10	Textiles	2.9	90.5	2.6	
8	Vehicles	2.8	93.4	2.6	0†
18	Gas, electricity and water	4.6	51.5	2.4	
12	Clothing and footwear	2.2	80.2	1.8	-3.3
6	Engineering and electrical	1.9	85.3	1.6	0†
7	Shipbuilding etc.	1.5	76.6	1.2	

Notes: * Refers to ten plants under Minimum List Headings 481–3, 486–9.
† These industrial orders were combined and taken as the standard for traffic generation (see text).
Sources: Starkie (1967b), figs. 5.4 and 5.5, p. 37; Mackie and Urquhart (1974), table 3, p. 7.

and the observations for the remaining manufacturing plants. The position of these observations in relation to the engineering traffic generation standard has been measured from Starkie's diagram in units of the standard error of the estimate, and the unweighted mean values are entered on table 4.5 (Starkie, 1967b, chapter 5; see figure 7.2, below, and accompanying section).

Transport expenditure as a percentage of net output varies from 1.5 in the case of shipbuilding and marine engineering to 21.6 for the distributive trades. The latter, together with the only two manufacturing orders spending more than 10 per cent of the value of their net output on transport (building materials, etc., and food products, etc.) are classed by Mackie and Urquhart as 'high transport intensive' industries. All orders except the utilities (gas, electricity and water) spend at least 72 per cent of their transport expenditure on road transport. Construction is most dependent on road movement, with 99.4 per cent of all transport expenditure devoted to this mode. This is further evidence that, except for the pipeline distribution of gas and water and non-vehicular movements, there are very few urban commodity movements that do not involve road trips. It has been found on Merseyside that only one per cent of the tonnage carried on internal trips by road and rail was transported by the latter, whereas 23 per cent of the commodity movement to or from the urban area was handled by rail (Traffic Research Corporation, 1969 c, plate 37, p. 75). It can also be seen from table 4.5 that the number of trips taking place in a town will be related to the industrial structure of the area and to the size-distribution of its manufacturing and other plants. Interpolating values from his engineering standard regression line, Starkie found that a plant with 50 employees generated about 16 commercial vehicle trips per day, one with 100 employees 24 trips, one with 500 employees 49 trips, and a plant with 1000 employees 70 trips. The scatter of the observations he made of other manufacturing sectors suggests that these figures approximate the average trip-generation rates for all manufacturing industry in the Medway Towns.

There is also considerable variation in the volume of goods movement stimulated by different branches of the retail trade, as has been shown most clearly by Masson's study (1970) of two French towns, Aix-en-Provence and Metz-Thionville. The former is primarily a residential town of 90,000 people, while Metz-Thionville is an industrial conurbation of 390,000 people. To facilitate comparison between the two areas, each town's requirements have been adjusted as for a town of 100,000 population, by dividing the recorded tonnages by 3.9 for Metz-Thionville and by 0.9 for Aix-en-Provence (table 4.6). Fuels were also discounted because, for climatic reasons, they are in heavier demand in the northern urban area. With these adjustments, it is seen that similar quantities are delivered to each area. Groceries accounted for approximately 35 per cent of the tonnage delivered

Table 4.6 Goods delivered to the retail trade of Aix-en-Provence and Metz-Thionville

Type of trade	Aix-en-Provence		Metz-Thionville	
	tons	%	tons	%
Groceries	57,778	36.8	59,827	35.5
Garages and service stations	51,133	32.5	50,507	29.9
Bakeries	6,240	4.0	8,867	5.3
Hotels and catering	4,738	3.0	6,933	4.1
Bar rooms	3,120	2.0	4,933	2.9
Furniture	982	0.6	4,773	2.8
Meat dealers	4,160	2.6	3,267	1.9
Hardware and electrical household appliances	1,098	0.7	3,133	1.9
Department stores	3,813	2.4	2,013	1.2
Household utensils and maintenance items	1,560	1.0	1,600	0.9
Clothiers	867	0.6	1,120	0.7
Dairy shops	1,387	0.9	973	0.6
Tobacconists	231	0.1	720	0.4
Books and stationery	1,675	1.1	693	0.4
Flowers and seeds	9,360	6.0	533	0.3
Pharmacies	809	0.5	533	0.3
Pastry and confectionery	347	0.2	493	0.3
Shoes	462	0.3	440	0.3
Miscellaneous	6,587	4.2	17,293	10.3
Total	157,156	99.5*	168,654	100.0
Fuels	24,093	–	42,333	–

Note: *rounding error.
Source: Masson (1970); based on the adaptation by Fresko *et al.* (1972), table 3, pp. 10–11.

to the retail trade, and supplies to garages and service stations for another 30 per cent. Particularly for the latter category, the high tonnage delivered by each trip will reduce the dominance of these bulky or high-weight commodities when measured in terms of traffic generation.

Some details of the variations in commercial vehicle trip generation among retail trades is available from limited postal surveys carried out in the early 1960s of business premises and hauliers in Birmingham. While department stores generated many more trips than other types of retail premise, when their trips are expressed per 1000 square feet of floor space,

Table 4.7 Commercial vehicle trip generation in Birmingham, c.1962

Trade type	n_1	No. of trips	n_2	No. of trips (08.30–17.30) per 1000 sq. ft. floor space
Department stores	8	54.5	5	0.32
Variety chain	–	–	5	0.40
Clothes and furniture	19	5.0	15	0.50
General retail and service	19	5.8	11	2.21
Food and entertainment	16	9.6	–	–
Restaurants and cafés	–	–	8	1.23
Warehouses	12	36.3	–	–
Offices	19	0.8*	–	–

Notes: Two distinct surveys were carried out. n_1 and n_2 refer to the sample of stores for each result.
* 3.8 calls per working week.
Source: Rose (1963).

the general retail and service trades (including food traders) are shown to be the largest users of commercial vehicles, and department stores to have low trip-generation rates (table 4.7) (cited by Pain, 1967, figs. 10 and 11, pp. 71–2).

However, commodity movements directly associated with manufacturing industry and with deliveries to retailers (1973, table 3, p. 235) constitute less than one-half of commercial vehicle trips within towns. Warner has put forward the unusually high estimate that 80.2 per cent of all trips in the West Midlands conurbation in 1964 were to residential areas, and that over 60 per cent were retail deliveries and another 11 per cent refuse collection trips (table 4.8). (The basis of Warner's estimates are not fully explained, and are dissimilar from those of Freeman Fox et al., 1968 a, vol. 1, p. 140, e.g. he estimates that the daily mileage of refuse collection vehicles was 11,700 miles.) The West Midlands Transport Study data excluded local authority refuse-collection trips, but nevertheless found that 78 per cent of the trips and 23 per cent of the mileage covered by commercial vehicles on an average weekday in 1964 was to residential land. The very high trip percentages arise because of the large number of journeys serving numerous destinations, such as those made delivering bread, milk or laundry. Other data from Norwich and Merseyside use some aggregation of these delivery stops and therefore produce much lower estimates of the proportion of trips destined

Table 4.8 Journeys by resident goods vehicles in the West Midlands, 1964

Destination land use	Journeys (000s)	(%)	Mileage (000s)	(%)
Residences	1 100	80.2	290	23.4
Refuse collection	(160)*	(11.6)*	(20)*	(1.6)*
Milk delivery	(500)*	(36.5)*	110	8.9
Other retail	(340)*	(24.8)*		
Shops	90	6.6	200	16.1
Industry and utilities	90	6.6	400	32.3
Offices and commerce	40	2.9	150	12.1
Others	50	3.6	200	16.1

Note: * estimated by Warner (1973).
Sources: Freeman Fox *et al.* (1968a), cited by Warner (1973), table 3, p. 235.

for residences. (The special procedures for recording journeys with numerous short trips, such as milk or laundry deliveries, are described in Freeman Fox *et al.*, 1968 a, vol. 2, chapter 20, p. 19. The Merseyside survey (Traffic Research Corporation, 1969b, pp. 13–14) distinguished 'subsidiary stops' within a street and not more than 700 yards apart, and did not use these for defining trip ends.) Table 4.9 attempts to present the information from the three urban areas in a comparable form. As remarked before, the lack of standardization of definitions and survey practice confounds most attempts to generalize from these sources, but the *relative* magnitude of journeys to different land uses does have many similarities. Even when aggregating short-distance deliveries, about a third of all trips are to residences, and of the order of a sixth of all trips are made both to retailing and to industrial uses. Commerce or warehousing and trading uses are the only other land use category to attract substantial percentages of the total, most other specified uses attracting less than five per cent of the total. In terms of vehicle mileage, a different rank order of land uses is seen in the West Midlands, with trips to industrial destinations accounting for nearly a third of the mileage. Trips destined for residential areas are on average much shorter than others, and account for only 22.5 per cent of the total internal commercial vehicle mileage.

Many other details of urban commodity movements are available from some but not all of the British transportation studies. (Apart from the studies

Table 4.9 Destination land uses of internal commercial vehicle trips: Norwich, 1967; Merseyside, 1966; and the West Midlands conurbation, 1964

Land use at destination	Norwich		Merseyside[2,3]		West Midlands Conurbation[3]		
	No. of trips	%	No. of trips (000s)	%	No. of trips	%	Mileage
Residential	7,145	29.4	83	37.0	945,402	77.9	271,952	22.5
Retail	4,300	17.7	40	17.5	87,267	7.2	195,009	16.1
Industrial	4,195	17.3	27	12.0	83,987	6.9	376,292	31.1
Transport	—	—	16	7.0	17,783	1.5	80,577	6.7
Public utility	2,315	9.5	—	—	4,952	0.4	21,678	1.8
Commerce	1,845[4]	7.6[4]	30	13.0	29,550	2.4	122,656	10.1
Office	705	2.9	4	2.0	7,814	0.6	23,124	1.9
Education, health	625[1]	2.6[1]	3	1.5	7,581	0.6	22,606	1.9
Service, garage	1,320	5.4	—	—	—	—	—	—
Open spaces	—	—	10	4.5	26,078	2.2	87,973	7.3
Docks	—	—	8	4.0	—	—	—	—
Public buildings	—	—	3	1.5	2,506	0.2	8,236	0.7
Others	1,870	7.7	—	—	—	—	—	—
Total	24,315	100.1	224	100.0	1,212,920	99.9	1,210,103	100.1

Notes: 1. Education only.
2. The Merseyside figures are interpolated from a diagram.
3. The Merseyside and West Midlands data exclude journeys by refuse collection vehicles.
4. Warehousing rather than commerce.

Sources: City of Norwich (1969), table 4.5; Traffic Research Corporation (1969a), plate 13, p. 21; Freeman Fox *et al.* (1968a), vol. 1, table 8.8, p. 126.

already cited, valuable information on goods vehicle movements was produced by Traffic Research Corporation, 1968.) The most exhaustive tabulations appear to have been provided by the West Midlands Study. Not only can the trips be analysed by destination land use and mileage, but they are also occasionally classified according to the commodities being carried and the purpose of the trip at destination (deliver to base, pick-up, pick-up and deliver, personal use, other firm's business). Commodity analysis is of interest as it shows the relative importance of goods movements for various urban activities. As table 4.10 indicates for Merseyside and the West

Table 4.10 Commodities carried by commercial vehicle trips: West Midlands, 1964 and Merseyside, 1966

Commodity	West Midlands				Merseyside*	
	No. of journeys 000s	%	Mileage 000s	%	No. of journeys 000s	%
Dairy produce	571	47.0	36	3.0	21	9
Bread, fruit, vegetables and meat	270	22.2	92	7.6	35	15
Beverages, other food and tobacco	34	2.8	59	4.8	14	6
Refuse	5	0.4	18	1.5	2	1
Coal, petrol and fuels	18	1.5	38	3.2	12	5
Other materials	16	1.3	75	6.2	12	5
Machinery, electrical	6	0.5	28	2.3	7	3
Motor vehicles and components	9	0.8	42	3.5		
Other manufactured goods	123	10.1	239	19.7	55	24
Unallocated and mixed	38	3.1	56	4.7	23	10
Empty	102	8.4	434	35.8	51	22
Passengers	22	1.8	93	7.7	–	–
Total	1,213	99.9	1,210	100.0	232	100

Notes: *Interpolated from diagram. The classes are specified as dairy produce, other perishables, other food, drink and tobacco, refuse, fuel, other raw materials, machinery and transport equipment, miscellaneous manufactured goods, mixed commodities, and empty.

Midlands, when trips are defined scrupulously a very large percentage are involved with the carriage of perishable foods, to the extent that of the remaining categories, only 'empty' and 'other manufactured goods' trips exceed 3.1 per cent of the total. In terms of total mileage, however, empty trips accounted for over a third in the West Midlands, and the carriage of manufactured goods was the leading loaded trip commodity with nearly a fifth of all mileage. The Merseyside data, with its looser trip definition, exaggerates the prevalence of trips carrying manufactured commodities.

The same sort of picture emerges from a far from clearly specified survey of goods movements in Tokyo in 1970, which found that

> Consumer goods related to the distribution of information have together come to constitute the greatest percentage of urban freight. [These] 'daily necessities' accounted for 33 per cent of all trips and 20 per cent of total tonnage. 'Food industry products', 'other special items' (i.e. waste, returned materials, etc.) and 'other items' followed in that order. These four groups accounted for 72 per cent of all trips and 58 per cent of total tonnage. (Nakanishi, 1973)

Further details are reproduced in table 4.11, and it is seen again, in a very different setting, that there is on the one hand a very large number of trips associated with daily food and convenience goods: these are characteristically short and involve consignments of small size and low weight. On the

Table 4.11 Goods movements by road in Tokyo on 25 December 1970

Type of freight	Truck trips		Tonnage	
	No.	%	Total	%
Daily necessities	2,972	33.2	6,554	19.8
Other	1,534	17.1	2,502	7.6
Waste, returned materials	887	9.9	5,082	15.3
Food and kindred products	1,027	11.5	4,970	15.0
Farm and mine products	314	3.5	2,095	6.3
Timber and minerals	68	0.8	404	1.2
Metals	401	4.5	2,728	8.2
Machinery	387	4.3	1,686	5.1
Chemical products	520	5.8	2,693	8.1
Other manufacturing products	854	9.5	4,423	13.3
Total	8,963	100.1	33,137	99.9

Note: 'Other manufacturing products' includes cement (0.4% of trips), other ceramic products (2.4%), refined petroleum products (0.3%), paper & pulp (2.3%), textiles (0.9%) and miscellaneous (3.3%)
Source: Japan, Transport Economic Research Center, *Research Surveys on Physical Distribution in Large Cities* (Tokyo, 1971), cited by Nakanishi (1973), table 1, p. 14.

other hand, there are many fewer trips to non-resident uses, but these are characteristically longer and involve heavier and bulkier consignments.

It has been asserted that commodity movements are associated in varying degrees of importance with all types of economic activity. It may clarify matters further if we consider in a little more detail the function of commodity movements for these activities. Broadly, commodity movement takes place either before the principal activity, when it may be described as enabling the assembly of raw materials, components or fuels; during the course of the activity, as when semi-manufactured goods are transferred from one plant or firm to another, or when documents are exchanged between offices; or after the completion of the economic activity, when the commodities are delivered to their final purchasers or consumers. The second category of movement can further be divided into between-firm, and within-firm but between-plant movements. Movements within plants do not normally use the public road system, although as well as employing conveyors and pipelines, road vehicle transfers are often involved. As with 'within-hereditament' personal movements, they take place at a scale not normally considered in geographical studies, and are with some exceptions regarded as being beyond the scope of this volume. The third category of movements, associated with final distribution, are a very important generator of goods movements within towns, and indeed employ so many vehicles and people that 'distribution' is commonly regarded as an industry in itself. An attempt is made in table 4.12 to present an illustrated classification of urban goods movements using the functional basis which has been outlined. During the construction of this table it became apparent that commodity movements are more diverse in function than personal movements, the majority of which were classified in chapter 2 on the basis of the type of activity being pursued at the destination. The home origin of each day's personal trips is so universal a feature that the pursuit of social and family life that they permit is not normally regarded as a function of personal travel. This practice cannot be followed in the case of commodity movements, for a substantial percentage if not a majority of them have no direct link with the maintenance of homes or of daily residential life. An important implication of this difference is that it is far more difficult to conceptualize and model commodity trip generation than passenger trip generation without double-counting journeys. If the residential population or residential land uses are regarded as commodity trip generators, which may be the most convenient procedure when dealing with milk delivery or refuse collection, then direct delivery trips from manufacturing or retail establishments will be double-counted if these land uses are also regarded as commercial-vehicle trip generators. In short, the complexities introduced by multi-stage trips, multi-purpose trips, and various origins, although found among a minority of personal journeys, are much more prevalent among

Table 4.12 A functional classification of urban commodity movements

I FACILITATING ECONOMIC ACTIVITY

A *The assembly of raw materials, components and fuels*

  1 The delivery of fuel and power (including gas and oil), raw materials, and components. Includes materials such as water and food for employees. The frequency of movement varies from the continuous delivery of gas and water, through postal services and other daily deliveries, to the infrequent delivery of capital equipment and materials used in small quantities, e.g. stationery or paint.

B *Movements during the course of economic activity between plants or firms*

  2 Movements to or from supplying and customer firms, dealers, wholesalers and markets, including postal movements.

  3 Movements between establishments of a single firm carrying out the same level of economic activity, including postal movements.

  4 Movements associated with the maintenance of buildings, plant and utilities.

C *The distribution of tangible products*

  5 The disposal of waste products from the economic activity itself or ancillary waste products including sewage. Includes the infrequent disposal of waste capital equipment, hardcore, etc.

  6 The distribution of tangible products to dealers, wholesalers, markets, or directly to the next stage of economic activity (including retailing). Includes transfers between establishments of the same firm carrying out different types of economic activity.

  7 Distribution of products to final consumers, including postal delivery.

II FACILITATING SOCIAL AND RECREATIONAL ACTIVITY

A *Assembly of raw materials, consumer goods and fuels*

  8 The delivery of water, gas, oils, foods and household items (normally) to residences. N.B. Does not include commodities carried by consumers, viz. shopping goods, and is identical to the movements described immediately above (section 7).

B *Movements during the course of social and recreational activity*

  9. Negligible, but certainly exist, e.g. movement of sports teams' equipment such as track bicycles, or of processional floats.

  10 Movements associated with the maintenance of buildings and residential services, including utilities.

C *Commodity movements on 'completion' of social and recreational activity*

  11 Disposal of personal wastes and refuse, chiefly from homes in the forms of refuse collection and sewage.

commodity movements. Even in functional terms, many movements have dual roles. In table 4.12, many movements under category 6 which are distributing the products of one level of economic activity will also occur in category 1 as being part of the assembly of component materials for executing another kind of activity.

The trip characteristics and spatial patterns of commodity movements

The trip characteristics of urban commodity movements are at least as diverse as those of personal movements, for in addition to their own patterns of timing and frequency, they have additional characteristics related to their mode and the spatial patterns of their origins and destinations. Not even the dominant road vehicle mode is a homogeneous category, because there are considerable ranges in the size and capacity of vehicles, and also diverse specialist vehicles and methods of handling consignments. The bulk transport of a homogeneous and 'handleable' material, such as petroleum or grain, contrast sharply with the labour-intensive transport of small consignments, often by common carriers and as part loads, as exemplified by the postal packet service. Among the distinguishing trip characteristics of goods movements is the larger proportion having their origin or their destination (or both) outside the functional urban area. This is another complicating feature for analytical purposes, for the volume, type and distribution of urban commodity movements is to some extent dependent upon extra-urban factors, such as the surrounding distribution of population and economic activity. These comments only serve to emphasize the value of disaggregating total goods movements, and to accentuate the fact that most empirical data on urban goods movements refer only to a selection of all journeys.

Information on the trip characteristics of urban goods movement has been drawn from several British surveys. One of the earliest was carried out in Greater London for the Road Research Laboratory in 1960 (Dawson, 1963). A random sample of 3,407 vehicles was drawn from the Goods Vehicle Index of the Licensing Authority of the Metropolitan Traffic Area. 1,247 of these vehicles were either scrapped, operating outside the London area, or principally passenger vehicles, and a further 919 were rejected by the interviewers for unspecified reasons. 364 vehicles were 'non-contacts' or 'refusals'. One-week travel diaries or 'logs' were therefore available from 864 vehicles, although only 754 of these actually made journeys during the week. In this survey a journey was defined as being from base back to base, except that all journeys were deemed to terminate or commence when they crossed the boundary of the survey area or to terminate at the end of the day. Journeys were divided into stages: a stage being from one business step to the next. Each stage had to be recorded separately, except for journeys in which the vehicles made many stops close together; in these cases the addresses of stopping places half a mile apart, and the total number of stops made in between were recorded. For journeys beyond Greater London only full details of that part of the journey made within the conurbation, and of the origin or destination of the journey outside London, were required. The Merseyside and West Midlands Transport Studies carried out goods vehicle surveys of similar scope. The West Midlands Study also used the Goods Vehicle Index as a sampling frame, but in Merseyside the consultants attempted to overcome the deficiencies and problems of this source by adopting a more complex sampling procedure. As Post Office, local authority, and certain government departmental vehicles do not require carrier's licences, the West Midlands study explains in some detail how these were inventoried and surveyed. Nevertheless, like the Merseyside study, it did not include trips made by local authority refuse vehicles in its main survey on the grounds of their fixed and regular itineraries. The multiple destinations and extremely short average trip length of the travelling dustcart journey obviously presents analytical problems. The West Midlands study explains its method of sampling hire vehicles, and vehicles registered outside the conurbation study area, and makes clear its exclusion of agricultural, emergency services and armed forces vehicles. 14,687 vehicles constituted the West Midlands sample, but on Merseyside only 3,929 vehicles were investigated. In both cases, information was collected on the trips made during an 'average' weekday of 24 hours, and this information was linked with vehicle counts at selected survey points arranged along cordons and screenlines. Completely detailed interviews were obtained for 9,475 vehicles, or 64.5 per cent of the sample, in the West Midlands. (For further details see Freeman Fox *et al.*, 1968a, vol. 2. pp. 19–21; Traffic Corporation, 1969b.)

Journey distances and stages

In the 1960 London survey, 754 vehicles were found to have made journeys which consisted of 7,840 trips and 41,500 stages (only 22,000 of which were recorded in detail). In the course of these they travelled 102,500 miles and performed 82,300 ton-miles. The distribution of the distances of the 41,500 stages and of the vehicle mileage according to the length of the stage is given in table 4.13. The average length of all stages was 2.4 miles and the standard deviation was 3.7 miles. It is clearly established that nearly three-quarters of travel by commercial vehicles in London was in short stages of less than 2.5 miles. Of the journeys that took place entirely within the survey area, two-thirds of the mileage was on journeys with a number of stops, averaging 10 per journey, and one-fifth of the mileage consisted of stopping journeys

Table 4.13 Commercial vehicle movements in London, 1960. Distribution of stages and vehicle miles by length of stage

Percentage distribution of:	Distance in miles of journey stage										
	$0-1\frac{1}{2}$	$1\frac{1}{2}-2\frac{1}{2}$	$2\frac{1}{2}-3\frac{1}{2}$	$3\frac{1}{2}-4\frac{1}{2}$	$4\frac{1}{2}-5\frac{1}{2}$	$5\frac{1}{2}-10\frac{1}{2}$	$10\frac{1}{2}-15\frac{1}{2}$	$15\frac{1}{2}-20\frac{1}{2}$	$20\frac{1}{2}-25\frac{1}{2}$	$25\frac{1}{2}$ plus	All	No.
Journey stages	61	12	7	4	3	8	4	1	0.4	0.1	100	41,500
Vehicle miles	15	10	8	7	6	25	18	7	3	1	100	102,500

Source: Dawson (1963), table, p. 248.

that were all within 3.25 miles of base (Dawson, 1963, p. 248). Results from Merseyside and the West Midland are consistent with these findings, and in addition show that there are positive relationships between the size of vehicle (light, medium or heavy), the average trip and journey length, and the average load by weight. For example, in the West Midlands the average journey length for all commercial vehicles was one mile, but for light vehicles it was only 0.70 miles and for heavy vehicles 4.69 miles (Freeman Fox *et al.*, 1968a, vol. 1, table 8.6, p. 125). On Merseyside the average vehicle load for journeys of up to 10 miles was 1.2 tons, but for journeys between 51 and 100 miles was 5.2 tons (Traffic Research Corporation, 1969a, plate 41, p. 79).

Timing

Turning to the timing of urban commercial vehicle movements, a clear characteristic of both their annual and diurnal flows is the lack of fluctuation about the average state, which is in marked contrast to the summer, morning and evening peaks of passenger car flows. Monthly flows on urban roads in Great Britain are illustrated in figure 4.2., where it can be seen that the incidence of heavy goods vehicle movement in towns is particularly stable. Except for a 10 per cent reduction in August from the monthly mean of 523 million kilometres, in no month do these vehicles register more than a 6.5 per cent departure from the average figure. Movement by light goods vehicles is 6.5 to 10.5 per cent lower than the mean monthly flow of 860 million kilometres throughout December to the end of March, and apart

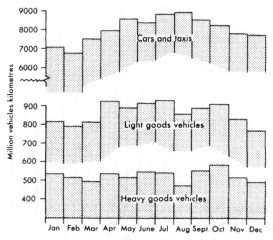

Figure 4.2 *Monthly traffic flows on urban roads in Great Britain (*Source: *Government Statistical Service, 1977, table 26, p. 42)*

from a lull in August, is from 3.5 to 7.8 per cent above the mean throughout April to the end of October. Light vehicle flows do no show the same degree of monthly peaking as passenger car movements, which exhibits a smooth annual cycle from a nadir in February of 16 per cent below the monthly average of 8,036 million kilometres, to a zenith in August of 11.2 per cent above the average. For more specific types of urban road goods movement, different annual rhythms will be found, as shown by Horwood's study (1958) of deliveries to department stores in central Philadelphia in the late 1940s and early 1950s. These peaked during the last quarter of the year, to the extent that the daily volume of goods handled was about twice the daily average throughout the year. Deliveries of domestic heating fuel, of petrol, and of garden supplies are among the other commodities most likely to have pronounced seasonal variations, but these are partly complementary. As these commodities also stimulate far less movement than the incessant requirements of homes, industry and commerce, they do not disturb the temporal stability of aggregate urban commodity movements.

The hourly distribution of trips through a working day has been recorded in many surveys of commercial vehicles. Results from the 1960 London survey are reproduced in figure 4.3. When measured in terms of commercial vehicle miles, and compared with person-miles by car, the main distinguishing feature again is the absence of the peaking so characteristic of personal travel. In London, apart from a decline at lunch time, there was a fairly steady flow from 08.00 to 18.00 hours, during which period 84 per cent of the total commercial vehicle mileage was run. Only 61 per cent of personal mileage occurred during this period, mainly because of its greater incidence

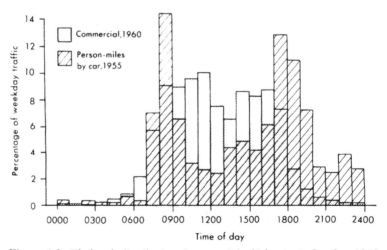

Figure 4.3 *The hourly distribution of commercial vehicle trips in London, 1960* *(Source: Dawson 1963, fig. 2, p. 249)*

Table 4.14 Hourly distribution of commercial vehicle arrivals in the main streets of Camberley, Newbury and Putney

	Percentage distribution by hour of day									
	07.30–07.59	08.00–08.59	09.00–09.59	10.00–10.59	11.00–11.59	12.00–12.59	13.00–13.59	14.00–14.59	15.00–15.59	16.00–16.59	17.00–17.30
Camberley	0.8	8.5	11.2	15.3	16.4	9.4	11.2	11.7	7.6	4.3	1.7
Newbury	3.1	12.3	12.7	12.1	15.5	12.3	11.9	8.4	7.1	4.2	0.4
Putney	6.7	0.2	9.0	18.5	17.0	11.6	15.5	10.8	5.6	3.7	1.5

Note: In Putney the earliest and latest periods refer to the full hour.
Sources: Christie *et al.* (1973a), table 5, p. 10; (1973b), table 10.

during the evenings. A commercial trip count carried out as part of the Norwich transportation study also reported very consistent numbers, in fact between 2,200 and 2,800 every hour between 08.00 and 17.00 except for the lunch hour after 13.00 hours when the flow reduced to 1,300 trips (City of Norwich 1969, fig. 4.10). Similarly, from counts through the Mersey tunnel and across a Liverpool-ring/screenline, the lack of variation in goods vehicle movement through the day is striking (Traffic Research Corporation, 1969a, plate 11, p. 19). A similar spread of movements through the working day is indicated by the more specialized data on the numbers of commercial vehicle visits to traders in the High Streets of Camberley, Newbury, and Putney. As one would expect from the timings of outward journeys, there is a concentration of these arrivals in the morning (table 4.14).

Spatial patterns

The geography of urban commodity flows and of commercial vehicle movements is less easily described than their other trip characteristics. Several different spatial distributions comprise this geography, consisting not only of those pertaining to any combination of the units, volumes and weights of the consignments, commodities and vehicles, but also to the separate geographies of origins, destinations, and of actual movement. While each constituent spatial pattern is related to the others, there are marked differences between them, in that, for example, origins much more than destinations focus on industrial, retail and wholesaling land uses in the inner areas. Destinations tend towards a more dispersed distribution related to the density of homes within the city. Few clear features of the spatial pattern of urban commodity movement have emerged from the studies which have been consulted, but on the other hand there is widespread discussion of the concentration of flows in the central and inner areas, and research is beginning to establish some systematic relationships between urban employment and commercial vehicle trip generation (and attraction). Both aspects are examined in this account, which will go on to assess the likelihood of establishing relationships between land uses and trip generation.

Most British surveys reveal a concentration of movement in the central and inner areas of towns, although some of the more extreme findings result from the exclusion of certain types of commodity movement. The 1960 London survey found that one-third of all commercial vehicle journey stages were wholly or partly within a central area of five square miles centred on Charing Cross, an identical proportion to that of passenger-miles in 1954 (Dawson, 1963, p. 249). This area represented only 4 per cent of the survey area.

A similar degree of central concentration appears to be true of most large modern cities. C. Nakanishi's summary (1973, p. 12) of goods vehicle surveys in Tokyo in 1968 states that there was a high concentration of the city's physical distribution facilities in its five central wards (Chiyoda, Chuo, Minato, Koto and Taito). These wards occupied only 14 per cent of the total area, but contained 57 per cent of all the wholesale establishments in Tokyo, 52 per cent of all truck terminals berths, 78 per cent of all warehouse floor space, 52 per cent of the total tonnage of incoming and outgoing railway freight, and 34 per cent of all incoming and outgoing freight carried by truck.

Returning to British urban areas, the West Midlands Transport Study data are notable for the detailed spatial information that it collated for goods vehicle trips. Its study area was divided into 116 zones and a full origin and destination matrix for the 1.2 million recorded trips has been published (Freeman Fox *et al.*, 1968a, vol. 3, table 40.5). Given the scrupulous recording of even the shortest trips which is a feature of this study, it is not surprising to find that no less than 948,248, or 79.1 per cent, of all trips were completed within one or other of the 116 zones. Only when intra-zonal journeys are discounted does the central area of Birmingham appear as a principal generator and locale for commercial vehicle movements. This is shown in figure 4.4, on which inter-zonal flows of more than 500 vehicles are mapped. Apart from the identification of the Birmingham central business district, these exceptional flows occur in the centres of Wolverhamptom and Walsall, as well as between other contiguous zones that for idiographic reasons generate an unusual degree of interaction.

Some analysis is also provided of the spatial distribution of trips within the West Midlands by land use at the destination. In this case, only 16 sectors are employed (Freeman Fox *et al.*, 1968a, vol. 1, table 8.14, p. 135). The frequency table of land-use destinations is again dominated by the residential category, and to emphasize the varying incidence of trips to other land uses, table 4.15 employs location quotients to express each sector's over- or under-representation of each type when compared to its percentage of all trips internal to the West Midlands. These quotients clearly identify central Birmingham as having the most distinctive goods vehicle movements, with trips to offices, commercial, transport and industrial land uses being overrepresented in that order.

This central sector attracts only 4.3 per cent of all trips, but if residential trips are excluded, over 14 per cent of the total in this extensively defined and multi-nuclear conurbation. Elsewhere within the urban area, the varying composition of goods vehicle movements is clearly related to the economic and residential geography of the conurbation. The sector boundaries are indicated in figure 4.4. Woverhampton (sector 15) repeats to a minor degree many of the features of central Birmingham, with the second

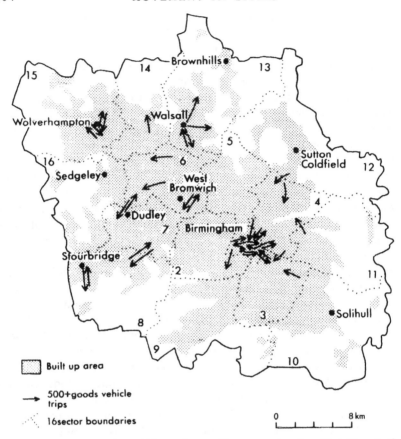

Figure 4.4 *Flows of over 500 trips between Transportation Area Study Zones in the West Midlands, 1964* (Source: Freeman Fox et al. (1968a), vol. 1, fig. 8.2, p. 123; vol. 3, table 40.5)

highest office-trip location quotient, and over-representations of retail, commercial, industrial and transport destinations. Four adjacent Black Country sectors (6, 7, 14 and 16), lying between the suburbs of Birmingham and Wolverhampton, are the only other areas to show an under-representation of residential destinations, and they also have the highest over-representation of trips to industrial destinations. The easternmost sectors, and particularly the peripheral areas of Solihull, Sutton Coldfield and Stechford, have the greatest over-representation of commercial vehicle trips generated by residential land uses, and considerable under-representations of all other trip destinations.

Similar if less detailed information is also available from the Merseyside Area Land Use and Transportation Study. Its study area, which was a little less extensive than the official Merseyside conurbation, is illustrated in

Table 4.15 Destinations of trips by sectors and land uses in the West Midlands, 1964

Destination sector	Number of trips total (000s)	Trips to residences as per cent of all trips to sector	Ratio (location quotient) of sector's percentage of West Midlands trips to stated land use against the sector's percentage of all trips						
			Residential	Offices	Industrial	Commercial	Retail	Open space	Transport
1 Central Birmingham	51	27.9	0.36	8.45	3.72	6.99	1.96	1.53	4.37
2 Smethwick/Harborne	143	85.7	1.09	0.59	0.59	0.54	0.80	0.45	0.59
3 Acocks Green/Kings Health	112	79.6	1.01	0.38	0.80	0.85	1.20	1.23	0.74
4 Erdington	135	80.3	1.02	0.82	1.01	0.94	0.93	0.40	1.29
5 Handsworth	85	86.6	1.10	0.30	0.44	0.51	0.86	0.89	0.28
6 Wednesbury/West Bromwich	71	70.4	0.90	1.49	1.68	0.94	1.27	1.46	1.06
7 Dudley/Rowley Regis	60	68.7	0.87	1.02	1.98	1.19	1.33	1.19	1.17
8 Stourbridge/Halesowen	52	80.1	1.02	0.64	0.96	0.74	1.04	1.12	0.56
9 Rubery/Northfield	84	86.9	1.10	0.19	0.54	0.54	0.66	0.94	0.42
10 Solihull	70	91.4	1.16	0.62	0.26	0.41	0.42	0.59	0.60
11 Stetchford/Coleshill	87	90.6	1.15	0.13	0.24	0.33	0.60	0.63	0.48
12 Sutton Coldfield	48	88.4	1.12	0.33	0.26	0.32	0.67	0.85	0.37
13 Walsall/Brownhills	83	79.4	1.01	0.55	0.99	0.66	0.86	1.33	1.54
14 Willenhall/Wednesfield	22	55.6	0.71	0.75	2.84	1.47	1.82	1.81	2.12
15 Wolverhampton	61	68.5	0.87	1.71	1.45	1.39	1.49	1.59	1.50
16 Sedgeley/Brierley Hill	35	68.8	0.87	1.44	1.23	0.91	1.58	2.64	1.25
All areas	1199	78.6	1.00	1.00	1.00	1.00	1.00	1.00	1.00

Source: Freeman Fox et al. (1968a), vol. 1, table 8.14, p. 135.

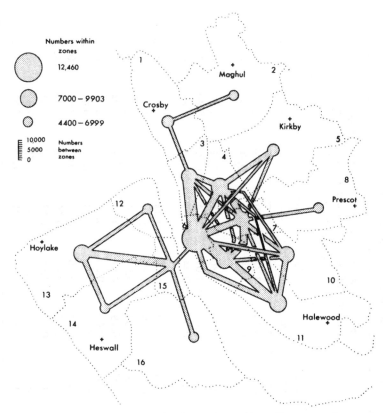

Figure 4.5 *Number of internal goods vehicle trips within and between study zones in Liverpool, 1966.* Note: *only inter-zonal flows of more than 1000 trips are shown.* (Source: *Traffic Research Corporation 1969a, plates 3 and 8, pp. 11, 15*)

figure 4.5. On an average weekday in 1966 some 250,000 goods vehicle trips were made within and across the study area boundaries. As described by the study, in their geographical patterns of origins and destination these trips are of three main types:

1 internal movements (those trips with both ends in the study area) numbering about 215,000 and accounting for 86 per cent of all journeys by commercial vehicles
2 external movements (those trips with one end in the study area) numbering about 35,000 trips
3 through movements (those trips with both ends outside the study area) numbering less than 500 trips. (Traffic Research Corporation, 1969a, p. 10)

On average two in every three commercial vehicles making journeys on the

Table 4.16 Number of commercial vehicle trips per day within and between zones in Greater Liverpool, 1966

Zones	1	2	3	4	5	6	7	8	9	10	11	12	13	14	15	16
1	4,822	553	1,672	443	257	557	202	74	246	192	242	21	27	32	72	13
2		5,549	1,481	670	775	610	425	198	209	121	257	47	32	22	35	31
3			9,752	2,874	843	5,338	1,267	276	1,466	498	814	658	317	73	536	239
4				7,787	2,099	3,439	3,531	788	1,558	1,010	1,116	356	152	95	214	118
5					6,490	1,086	1,370	620	639	334	518	37	119	49	125	32
6						12,460	3,666	743	6,163	1,426	2,746	251	429	211	1,696	375
7							8,638	1,703	3,031	1,984	1,430	38	146	18	259	49
8								4,406	468	975	387	0	34	0	36	0
9									9,239	1,807	3,376	219	64	112	247	102
10										7,054	2,506	78	78	0	51	36
11											8,083	43	66	33	117	61
12												6,891	1,444	437	1,268	191
13													9,903	2,308	3,451	461
14														5,628	1,945	396
15															6,931	1,862
16																6,392

Source: Traffic Research Corporation (1969a), plate 7, p.14.

internal roads were light vehicles, but this type made up only 14 per cent of the vehicles engaged in external trips. The internal legs of external trips accounted for almost 40 per cent of the total goods-vehicle miles on the primary road system.

Merseyside was divided into 16 zones for the origin and destination survey. As can be seen from the results reproduced in table 4.16, and as illustrated in figure 4.5, 75 per cent of all the trips end in the same zone as they originated. Only 14 per cent (29,986) of the internal trips were not intra-zonal or between contiguous zones – (if zones 6 (central Liverpool) and 15 (Birkenhead) are regarded as contiguous). The concentration of movement in and around the city centre is repeated as a feature of the distribution of trips. Almost 13 per cent of all internal trips began or ended in the central area, and this percentage rises to almost half when immediately adjacent zones are included. There was surprisingly little movement across the Mersey, as such trips accounted for only 4 per cent of the total. Also familiar is the characteristic that residential destinations dominated the origin-destination matrix. The Merseyside study, by revealing that of the 80,000 trips to residences 67,000 originated from this land use, and by showing that 70 per cent of the trips to residential land were made by light vehicles, provides further evidence of the diversity of urban goods-vehicle movements and of the complex elements that make up its geography.

In the absence of a more comprehensive analysis of the intra-urban spatial patterns of goods vehicle movements, as has been carried out for other types of urban movement using multivariate summarizing techniques like factor analysis (see, e.g., Wheeler, 1970; Holly and Wheeler, 1972), only a tentative summary description can be formed. The evidence that has been reviewed suggests that the aggregate pattern of movement is composed of three elements:

a A very widely distributed 'base' of short-distance movements related to the residential areas of a town. This comprises of the order of 75 per cent of all commercial vehicle trips, and the density of trips is related to the gross residential density of population.
b A focus of trips upon the retailing, office and commercial uses in the central area, which is enlarged by movements to industrial uses in the immediately surrounding area. These trips account for of the order of 5 to 10 per cent of the total in a large conurbation.
c A residual category of trips, mainly related to non-residential uses, and given its spatial character by the distribution of trips to industrial uses. The geography of this dimension is probably particular to each city, and related to the idiographic influences upon the location of industry in non-inner areas.

Modelling urban commodity movements

These generalizations represent no more than a preliminary statement about the geography of intra-urban commodity movements. The ignorance that has been encountered is reason enough for further studies of the subject, for which the existing inventories of trip movements provide readily available sources. As well as exploiting these opportunities for empirical research, it may be profitable to pursue an obvious theoretical approach to the description of these spatial patterns. This builds upon the models of commodity movement generation and assignment in both inter- and intra-urban studies (see Chisholm, 1970, 1971; Chisholm and O'Sullivan, 1973). The former have concentrated on the utility of different independent variables (such as population, economically active population, employment by industrial categories, and retail turnover) in models of the trips generated by different commodities, and they have also examined the variation in the friction of distance, or the distance exponent in interaction models, for different commodity-trips or by location within study regions. On the other hand, intra-urban studies have characteristically disaggregated urban commodity movements by the location of their origins (or destinations) rather than by the commodities being transported. Following the evidence presented above about the numeracy of light-vehicle short-distance movements to residential areas, these trips are distinguished as a category and commonly modelled as a function of population. Other goods trips are more commonly carried out by heavy vehicles, have a longer mean distance and more often involve external movement, and most transportation studies have found that they are most sucessfully modelled as a function of employment at the destination (see, e.g., Wood, 1971; Freeman Fox et al., 1968a, vol. 2). Approaches vary in detail, but greater accuracy is normally attempted by a further disaggregation of the latter category according to the type of economic activity or land use. The Merseyside study adopts the less common procedure of a threefold classification of movements according to the location of origins and the type of vehicle making up the flow (Traffic Research Corporation, 1969b, pp. 44–64).

Ogden (1977) has adopted the conventional transportation study procedure of disaggregating trips by purpose at destination. This appears to be feasible but less illuminating than a spatial or commodity approach. His research also exemplifies the strengths and weaknesses of simple regression modelling. Sixteen separate regression equations are developed to model the generation and the attraction of eight trip purposes in Melbourne, Australia. Ogden explains that, 'in all cases the regression coefficients were significantly different from zero according to Student's t-test', but the confidence levels are not stated, nor are the standard errors of the estimates given. Many of the variables are not in fact independent but overlapping and undoubtedly highly correlated measures of different categories of employment and of population. The most satisfactory models appear to be the home-base, pick-up, wholesale delivery, personal trips, and 'all trips' models, for in each case high coefficients of determination (R^2) are achieved with three terms. The retail delivery trip-production and industrial delivery trip-attraction models are least satisfactory because of the lower explanation achieved by four terms and their overlapping independent variables. Although the variables are described as land-use measures, it can clearly be seen that they are head-counts requiring very detailed enumerations of the and land-use distributions, rather than being based upon population and employment distributions.

This approach can produce acceptable models of flows for individual cities, but requires not only the mentioned employment and population inventories but also a considerable effort in carrying out and processing trip surveys. The spatial descriptions produced by these studies are detailed but necessarily idiographic, and the urban geographer is also likely to be interested in the progressive development of generalizations for sets of cities. In order to achieve these, it will be necessary to accumulate studies of many more urban areas, but few large-scale and comprehensive urban transport studies are likely to be commissioned in the near future (Atkins, 1977). It would be extremely useful, therefore, if the movement pattern could be indicated from the more commonly available data, of population and land-use distributions, rather than be based upon population and employment distributions.

Although most city planning authorities will have at their disposal very detailed maps of land uses, there has been little spatial analysis of their distribution in widely distributed published literature. Lowenstein (1963) has provided some useful descriptive generalizations for American cities although little detailed analysis of within-city patterns, while Browning (1964) has published a substantial analysis of land-use distributions in Chicago with respect to distance from its city centre. He found that between the centre and nine miles from the centre, over 95 per cent of the available land was developed, but beyond this circumference until 26 miles from the centre, developed land fell steadily to about 15 per cent of the total. Of developed land, the principal features are the increase in residential uses from about 5 per cent of the total in the central area to about 40 per cent eight miles from the centre, and its minor fluctuation about that percentage thereafter until 30 miles from the centre; the decline of commercial land uses from over 20 per cent of the total at the centre to about 9 per cent of developed land 9 miles from the centre, and its much gentler decline thereafter to about 2 per cent of the total 28 miles from the centre; the peaks of manufacturing uses over a mean of above five per cent in the zones from 3 to 6 and 9 to 13 miles from the centre, of transportation uses over a mean of about 11 per cent in the zones within 2 miles and between 14 and 17 miles from the centre, and of public building and street uses in the outer suburbs.

Enough errors have been committed in the past to warn us against generalizing from the case of Chicago, but if Browning's land-use analysis were widely replicated, then some reliable statements describing intra-urban land-use distribution would be arrived at, and one step towards a more direct procedure for predicting urban commodity trip generation would be accomplished. A second necessary step would be to develop commercial trip generation rates with respect to unit-areas of different types of land use, to avoid the need for employment figures by establishments. There are great difficulties and additional sources of error associated with this approach, however, because even within a single employment or land-use category there is not a direct relationship between the floor space or plot area of an establishment and its employment. Morever, as mentioned earlier in the chapter, the relationship between plant employment and commercial trip generation is curvi-linear. It is also likely that non-linear relationships exist between household size or house-plot size and the residential generation of commercial vehicles. As an obvious demonstration of this, milk delivery and refuse collection trips in most circumstances will be invariant with household size, and will certainly not vary linearly with the independent variable. Nonetheless, the potential for estimating the spatial pattern of goods movement flows from the distributions of population (or households) and of land- use should in our view be explored, for the topic is a sizable lacuna in our knowledge of urban spatial organization.

Transport planning and urban freight movements

There is currently a widespread advocacy in the transport-planning profession for a greater number of more exhaustive urban commodity-movement studies. The arguments employed, and the research which has become available already, have proved useful earlier in the chapter, but it is questionable whether the advocated course of study is the most desirable or realistic. The stated advantages from large-scale studies often appear to be confined to the study process itself, and rarely amount to substantial direct practical benefits. For example, Ogden has difficulty in finding an applied justification for his own research field in freight trip-generation modelling. He recognizes that regression equations of cross-sectional data are unsuitable as forecasting models, but in his introduction argues that, 'these models may have application in large-scale strategic transportation planning where approximate estimates of freight and truck requirements are needed' (Ogden, 1977, p. 108). No more specific or direct practical applications are described, which, in a paper exclusively concerned with a major strand of the conventional study procedure, cannot but suggest that few exist. Static modelling exercises certainly advance knowledge and understanding, but perhaps only indirectly through this contribution do they help to alleviate or solve existing problems or to plan for change. More modest and progressively refined studies of specific aspects of urban freight movements may be more appropriate and productive than comprehensive surveys and analyses. A less glamorous planning approach might also avoid the rigidity in study recommendations, policy objectives, and action which have been associated with the need to justify and act upon expensive 'package' research.

Brief reference to the scope of applied research now being recommended is salutary. Meyburg and Stopher (1976, p. 652), as one example, stress the need for a diversity of data-collection tasks, and specifically mention eight subjects: (1) consignment flow patterns, (2) commodity flow patterns, (3) inventory of physical facilities for freight movement, (4) collated statement of the regulations affecting freight movement, (5) institutional constraints, (6) land-use patterns, (7) inventory of technological options for tackling problems of freight movement, and (8) inventory of the characteristics of shippers, receivers and carriers. Such an exhaustive exercise would employ many experienced professional and clerical workers, but its expense would need careful and detailed justification. This does not appear to be a matter of great concern to many advocates of goods-movement studies. Fresko *et al.* (1972, p. 15) concluded strongly on the need for more research:

> A number of areas will need extensive research. These include the development of relationships between goods movement generation and the socioeconomic characteristics of the population, methods for determining the production and attraction zones for goods movement, the determination of cost and travel characteristics of different modes and the prediction of their technological environment, the construction of optimization models and of multimodel conceptual schemes for freight transportation.

Further points are added, but again the justification for all this is insubstantial. 'Proper integration of goods movement in the total transportation process will make network assignment procedures more realistic because the impact of goods vehicles will have been considered in the forecasting process.' Apart from this and other advantages endogenous to the study process, the authors' claims are quite limited, as shown by the final statement in their paper: 'The potential gains appear to be primarily in the areas of peak period highway design projections, urban area truck route studies and in more economically beneficial land use planning.'

Not all transport planners think that the best way to proceed is by setting up large-scale comprehensive studies. Meyburg *et al.* (1974, p. 796) do recognize the importance of tackling immediate and specific problems, which will vary from city to city. They list nine specific options 'as promising candidates for alleviating urban goods movement problems'. They are: (1) urban freight consolidation, (2) standardized shipping units, (3) spatial separation of goods movement and people movement, (4) temporal separation of same, (5) off-street (un)loading facilities, (6) urban fringe terminals, (7) combined passenger and freight movement, (8) automated systems, and (9) automated terminals. Many North American (e.g. Fruin, 1972; Blaze 1972; Lovejoy, 1972; Mayer, 1971) and British studies reflect the popularity of just these proposals. In Britain much of the recent effort of the Transport and Road Research Laboratory has been concerned with issues such as the diversion of through traffic from city centres and congested streets, the reduction of noise pollution from commercial vehicles and lorry-parks, and in Swindon the advantages from an experimental trans-shipment depot to serve the town centre. (For other British studies, see Collins and Pharoah, 1974, pp. 461–7; Robertson, 1971; Moulder, 1971.). Proposals range from modest traffic management measures to ambitious schemes for the development of multi-modal freight corridors and rationalized terminal and break-bulk facilities. The latter have the objective of reducing the volume of commodity movement, and particularly of road vehicle trips, required to serve the city. In this, as in other specific studies, existing spatial patterns are being critically examined, and improved distributions are the objective. Although a beginning has hardly been made, there is in these matters great scope for the application of the spatial techniques which have been developed in recent years within and outside geography.

The preparation of this chapter has uncovered a surprising volume of discussion and data concerning goods movements in towns, particularly in the United States transport planning literature. There is still great disorder in the debate, and knowledge of this field remains partial. As yet there has been little analysis of the spatial aspects of this category of movement. It is believed that many valuable geographical studies could be carried out in this field, which would contribute not only to urban and transport geography but also provide a healthy alternative approach to applied transport studies. The accumulation of research findings on the spatial and economic implications of experimental changes in urban freight facilities and route networks would form a useful reference by which to evaluate projections and proposals from the broad-brush transportation study process. Research interest and activity is not so highly developed in Britain as in America, but the subject is rapidly being established in the British transportation planner's catalogue, and if there is to be an equal debate, academic research in this field must also be expanded.

5
Spatial patterns of urban movement: the organizing principles

So far in this book we have simplified our approach to the analysis of urban personal movements by separately examining particular characteristics. Now we seek to integrate at least some of the variety of characteristics that have been described and to introduce the most interesting aspects of movements to the geographer, namely their spatial pattern. The intention is to produce generalizations about the volume, density, directions, purposes, and the modal characteristics of journeys in and between different parts of cities. Although the emphasis will be on the spatial dimension, we shall not completely lose sight of the temporal dimensions, for the daily and weekly fluctuations of flows in certain parts of cities are a very important aspect of the geography of movement. In seeking generalization, the principal and common features of movement in British towns will necessarily be emphasized, while the effect of unique characteristics and distinctive local geographies will tend to be passed by. In many cases, however, it will be argued that consistent differences can be identified among sets of towns and cities defined in terms of size, function or form, and generalization at this level will frequently be preferred.

This chapter will elucidate the most important factors in the spatial distribution of personal movements within towns. Firstly, the description of movement patterns will be tackled by examining theoretical approaches and generalizations concerning the nature of towns and of urban travel patterns. For example, it is possible to make general statements about the distribution of different types of trip origins. If to this geographical knowledge is added a few simple generalizations about trip characteristics – number, frequency, distance – a broad picture of the geography of urban movement is produced. It will then be possible to evaluate such a general picture, as well as to fill out some details as they occur in particular cities, by turning to other approaches to the problem of description, namely the selection of findings from the travel and transport surveys to give an empirical view of urban movement patterns.

Principles and order in urban personal movement

The number of movements taking place within any urban area is considerable. They could produce a completely chaotic and meaningless picture of agitation. If there were no consistent influences upon the manner in which journeys were undertaken, the result would be a completely random temporal and spatial dispersion of movements. No logic, order or pattern could be discerned. Such an anarchic situation, however, never occurs, for there are always constraints and influences upon individuals' movements, and these are responsible for the clear and predictable patterns present in every urban area. This is not to say that every movement is predictable. Not only does personal idiosyncrasy, impulse, irrationality and spontaneous reaction to transitory circumstances affect each individual's decisions of whether, when and how to journey, but also from the perspective of total urban movement many trips which are not inherently random can conveniently be regarded as such. Each individual's characteristic journeys are on occasion disrupted by a variety of comparatively rare events, such as illness, hospital treatment, unemployment, job-searches and family celebrations, but it remains true that for most of the time, most of the population exhibits spatial behaviour which in its main elements is both repeated and conventional. The point is illustrated by drawing the distinction between our relatively high ability to predict the daily volume of demand or flow on a particular road, bus or train service, or of pedestrians along a High Street under average weather conditions, and our much weaker ability to predict which individuals would constitute that flow.

While the fundamental role of common constraints and influences, or of what can be called the organizing principles of urban movement is apparent, it is an unavoidably subjective exercise to identify, describe and evaluate even the major principles. In table 5.1, six clearly important principles are identified, together with some of their main spatial or geographic effects. The first is the temporal coordination of activities for different purposes, which as described in chapter 3 is practised simply to facilitate co-operative activities, and which is the cause of pronounced rhythms of travel upon the town. This principle alone would, therefore, have a spatial effect, by creating mirror-image movement patterns at the beginning and end of the duration of each activity for which people travel. However the clearest spatial consequences of temporal coordination are produced not in isolation but in association with the principle of spatial coordination, whereby individuals assemble at selected locations to join with each other and with the necessary equipment to pursue an activity. This results in the segregation of different activities and, therefore, land uses within the town. Because journey purposes fluctuate through time, and because different purposes focus journeys on different areas of the town, the combined effect of these two

Table 5.1 Organizing principles of urban personal journeys

Principle	Some geographical effects
1. Temporal coordination of human activity	Creation of parallel daily, weekly and seasonal rhythms of activities and movements for large groups of the population.
2. Spatial coordination of human activity	Spatial agglomeration at different scales of activities and land uses, leading to a focusing and segregation of movements for different purposes and by different groups. Volume of movement to each part of a town related to its attractiveness, or number of activity-opportunities.
3. Minimization of travel effort	Strong distance decay effect among urban movements. Dominance of short-distance movements, creating a relationship between the population and movement distributions. In each purpose category, relationship between accessibility to opportunity and frequency of journeying.
4. Economies of scale in provision of movement facilities	The controls of threshold demands for higher-capacity and higher-speed facilities limit their provision for the dominant medium- and longer-distance journeys. A hierarchy of roads and a mainly radial pattern of public transport routes is produced, leading to a relationship between the distance and mode of journeys, and generating feedback effects on individuals' destination, route and frequency decisions.
5. Structuring of human activity by position in the life cycle or role in the household	Temporal and spatial ordering and grouping of journeys by personal characteristics. Creation of contrasting movement patterns for schoolchildren, housewives, the employed and the retired.
6. Increasing differentiation of activities, land uses and movements with increasing urban size	Economics of urban size determine a greater range and volume of activities in larger places. Geometry determines an increasing diversity of journey distances with increasing scale. Together they produce both greater complexity of the aggregate movement pattern and greater diversity of individuals' patterns with larger town size.

co-ordinating principles is to produce the marked contrasts in movement patterns at different times of the day, week and year. They are also directly responsible for the highest volumes and concentrations of flows, and for the worst congestion in the movement system. These, in turn, stimulate secondary effects which complicate the relationship between the underlying organizing principles and the resulting pattern. For example, as the fourth principle argues, concentrations of movement in time and space encourage the provision of higher-cost and higher-capacity/speed facilities such as urban motorways or rapid transit routes. But the provision of these facilities tends to stimulate the demand for movement along these already most heavily-used corridors, and to a lesser extent, at the most popular times. The net effect is to generate relationships between the time, purpose, distance and route of journeys and their modal split. Also complicating the picture is the fact that the congestion produced by temporal and spatial concentration reacts upon individual decisions as to timing, destination and mode of travel. One general effect is to suppress the numbers attempting to use the most congested facilities, notably roads, and to divert people either on to other routes or on to other forms of movement. This is most clearly shown by the exceptional demands for intra-urban rail facilities in the largest cities.

Efficiency of production, administration, exchange or education is by no means the sole cause of agglomeration, but in some form it usually plays a part in the organization of space. Patterns of residential segregation within the city play an important role in structuring the spatial and temporal patterns of movement, but they are more the result of past decisions by developers and builders, variations in incomes, discriminatory practices, and varying evaluations of the importance and cost of access to employment and the city centre, than of a concern to coordinate and make efficient social activities and information exchange among residential groups. Nevertheless, as Dahya (1974) has shown, a clear and valued attribute of the tight residential clusters of recent overseas immigrants in British towns is the efficiency and intensiveness of employment and housing information exchange, social and cultural activity, and internal welfare support networks that they make possible (see also Boal, 1976). Spatial agglomeration is therefore a feature of some social as well as economic activities. It allows journeys to friends, relatives, cultural centres and places of worship to be more frequent and less burdensome, and therefore stimulates social activity among the occupational and ethnic groups of a heterogeneous urban population. Geographers and others have commonly deprecated residential segregation, but against the reasons for disapproval should be set a wider recognition of its advantages in terms of movement and of community formation. Whatever the causes of residential segregation, the result is to emphasize the distinctive movement geographies of the various social and occupational groups. For in addition to the tendency for different activities and occupations to cluster in contrasting areas of a town is now added the tendency for the various participating groups to be residentially differentiated in space. If a residential area can be distinguished by the age or sex of its population, by its relative number of pupils, students, housewives and retired people, or by its ethnic composition, it is also likely to possess a distinctive pattern of personal movements in time and space. (For a succinct review of geographical research on urban residential patterns, see Robson, 1975.)

The third organizing principle identified in Table 5.1 is the application to journeying of the widespread human principle of least effort. This does not imply that human beings attempt to minimize their activities, or that *homo sapiens* is a naturally sedentary animal, but rather that once a course of action is decided, some attempt is made to follow the minimum-effort path to its achievement. The principle is fundamentally, therefore, a manifestation of the reasoning and information-storage abilities of the human brain. In terms of movement, the effect is that within the limitations of our knowledge of equivalent opportunities and perception of travel routes and facilities, our journey behaviour normally reveals an attempt to minimize the cost and inconvenience of the journeys we choose to make. While we do not always succeed, a failing does not necessarily offend the general principle, for the revision of a set of travel choices requires both mental and physical effort in the gathering of information, evaluation, and possibly experimentation, which would not always be justified by the improvement that could be achieved. The fact that there is a tendency to minimize the inconvenience and therefore distance of journey for each purpose is, accordingly, not just a result of rational economic behaviour or, in less abstract terms, of minimizing petrol consumption, bus fares or time costs. It also derives from the effect of space on our learning and cognitive development, which in turn has an effect on the way in which we organize our spatial behaviour. There is, then, a circular or interactive causal relationship between space or distance and our movement behaviour, because we will tend to travel only to the locations which we already know and along the paths or by the means that have been proven. As a result, our cognizance of the urban area will only slowly expand, and our movements tend not to break out of a degree of spatial restriction established either during childhood or in a matter of weeks after moving to a different area of a town. (There has been a long-standing effort to conceptualize these effects, although little in the way of experimental research has appeared to support the various suggested ideas. For an overview, see Golledge *et al.*, 1976. The minimization of commuting distances is discussed and illustrated for Philadelphia and Milwaukee by Catanese, 1970, and elucidated by comparing white and black populations in Detroit by Deskins, 1972.)

An interesting conceptualization of the way in which we make use of and move within a town has been put forward by Horton and Reynolds (1971). Drawing in part on behavioural studies and conceptualizations of the decision processes involved in intra-urban migration, they distinguish firstly the *objective spatial structure* of a household, which is its situation relative to the set of locations of, and routes to and from, potential activities. Secondly they suggest that each individual has an *action space*, which is the set of locations and paths about which the individual has information. While in theory each location can be assigned a relative attractiveness and accessibility according to objectively measurable characteristics, in practice and individual's action space will be associated with subjectively determined relative utilities which differ from those generated by the objective spatial structure. Thirdly, Horton and Reynolds define an *activity space* within the action space as the set of locations and paths with which the individual has direct contact as a result of day-to-day activities. They suggest that individuals' activity spaces are much more variable than their action spaces and that they are often discontinuous and asymmetric.

Adopting this theoretical framework, an investigation was carried out into the characteristic action spaces of samples of residents from two contrasting but socially homogeneous areas of Cedar Rapids, Ohio. One was a low-income central district, the other, an upper-middle-income peripheral suburb. An individual's action space was operationally defined as the area with which a respondent was familiar, and determined from city maps on which he or she indicated their level of familiarity with 27 sub-areas on an ordinal scale from 0 to 5. The inner city sample's action space was characterized by a pronounced familiarity with the home area and the adjacent CBD, and an increasing unfamiliarity with increasing distance from the home area. There was one exception to this clear expression of the distance-decay principle, in that a north-easterly radial corridor, which contained both a major thoroughfare and the town's principal suburban shopping centre, generated higher levels of familiarity. The peripheral sample had a higher level of overall familiarity, despite its generally shorter length of residence in the area, perhaps reflecting its higher education, income and mobility levels. It also had a much more clearly linear pattern of familiarity, with the best-known areas being the home area, those in the direction of the CBD, the CBD itself, and to a lesser extent, the north-east corridor and shopping plaza. In other words, while an absolute distance-minimization principle could still be said to operate, its spatial expression was heavily modified by the concentration of activity-opportunities at the CBD and the reinforcing effects on this linearity of the road network.

The general success of the minimization principle is revealed by the very strong distance-decay relationship which characterizes most categories of urban movement. The relationship is illustrated for different types of journeys in a variety of British towns in figure 5.1. It produces a situation

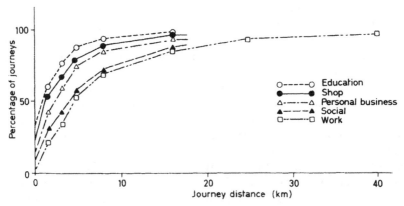

Figure 5.1 *The distance decay of urban personal journeys*

whereby a large proportion of all journeys are short-distance, and are completed locally around people's homes, and to a lesser extent, places of work and education. Consequently, the spatial result is not to focus or concentrate journeys, except in a minor way on the smallest neighbourhood retail and service centres, but to produce a wide scattering of journeys, broadly in relation to the distribution of the population. The minimization principle is given institutional and administrative recognition, and it is thereby strengthened, in the formal definition of catchment areas for schools and, less rigidly, for medical and welfare facilities and in church districts. Sometimes it operates indirectly, as when for reasons related to the efficient running of their business (and the comfort of their juvenile staff), newsagents refuse to arrange delivery outside of an area regarded as convenient. As a result, journeys to pay newspaper bills will probably have a shorter mean distance and a stronger distance-decay characteristic than if determined by the reading public. This example may appear unimportant, but it is hardly exceptional, and could be replaced by many others. Consider, for example, the similar constraints which are sometimes imposed on movements to and from public libraries, rates payment points, polling booths, political constituency meetings, evening classes, or sports and games league matches. Often, it can be concluded, individual impulses to minimize movement for different purposes are reinforced by collective, institutional or entrepreneurial arrangements. In other circumstances, institutional arrangements may operate in the other direction, and produce longer journeys than consumers would wish.

It has already been commented that the principles underlying the urban movement pattern do not act in isolation but in combination to produce a set of intricate relationships between the purpose, timing, destination, routing and mode of journeys. At any single time, the major features of the movement pattern can be understood in terms of the joint influence of movement minimization and the spatial coordination of activities. This combination gives rise to a gravity principle or effect, by which the volume of journeys made from an origin to a destination is functionally related both to the ability of the respective areas to generate or attract movement, and inversely to the friction acting against movement which derives from the space or distance separating the two areas. More specifically, in the case of a movement from a residential area for a particular purpose, the flow to a particular destination area will be related to the number of residents wishing to travel for that purpose (and at that time), and to the number of opportunities for the activity at the destination. The latter, however, will need to be adjusted according to the number of alternative, intervening or competing opportunities in other parts of the city. These prevent the development of any simple relationship between the distance of a movement and its inconvenience or friction, and result in a mainly relative evaluation of the accessibility of alternative destinations. Most people probably adopt a loose ratio scale of evaluation, by which the inconvenience of a journey to a destination is judged against that of the most accessible place known to satisfy a given purpose. This implies that in parts of cities or in towns where the density of opportunities is high, the mean distance of journeys for the purpose in question would be below average, and conversely, that low densities imply unusually long journeys.

Opportunity density would, however, only modify the frictional effect of distance within a limited range. It would not always be true, for example, that a dearth of short-distance opportunities would stimulate long journeys. Less frequent journeys would be a common response, as with the elimination of shopping trips or lunch-time commuting; or journeys may be substituted by other devices, such as home-based social activity or entertainment. A third option is not to participate in an activity which can only be fulfilled by unacceptably difficult journeys, for there is a maximum effort or distance of travel which can be justified by the benefits of the journey. This range effect, familiar to the urban geographer because of its fundamental role in central place theory, comes into play when, for example, the inconvenience and cost of a journey to work exceeds the psychic and income benefits of employment, as is frequently the case with part-time occupations (for a full exposition of central place theory, see Beavon, 1977; for a useful introductory review, Lewis, 1977). Journeys to and from these occupations are, therefore, short-distance, and much more concentrated around the places of employment opportunities than are journeys to full-time employment. In the past, this fact has been partly responsible for the concentration of part-time work in the inner areas of British towns, notably in service occupations in the commercial core and in small manufacturing units (e.g. in clothing manufacture) in the inner suburbs. The suburbanization of retailing and commerce, and the growth of part-time employment in education are now reducing this concentration (Dicken and Lloyd, 1978; Keeble, 1978). At the other extreme, exceptionally high opportunity densities would not necessarily generate volumes and frequencies of journeying above a certain level. Once the purpose or activities underlying movement are satiated, no further movement will be generated. Four home–work or four home–school trips per day would rarely be exceeded; a single movement to, and another from, a football match would satisfy the keenest supporter even if he lived next door to the ground. Overall, it can be summarized that the frictional effect of distance on reducing movement is a complex function involving both the absolute or objective time and inconvenience implications of distance (although not in a simple linear relationship), and the relative or subjective evaluation of each destination and path in relation to likely alternatives. From this it can be appreciated that although in outline the gravity principle is relatively straightforward, in detail its operation is subtle and variable. Indeed, many of the factors modifying the basic principle are elusive and difficult to reproduce in predictive models. These problems, and the use of gravity functions in transportation studies, will be discussed further.

The remaining three organizing principles that are identified in table 5.1 further modify and complicate the more fundamental influences that have just been considered. The fourth principle derives from the fact that movement itself consumes resources, or is an economic activity, which causes several economic principles to bear on the geography of urban movement. For example, the supply of a transport facility, whether a particular grade of road or a public transport service, is dependent on a sufficient or threshold demand. The existence of higher-grade facilities will alter, however, the objective and subjective time and money costs of movement along the route, and will therefore interactively modify demand. The net spatial consequence of all this is invariably intricate and quite often marginal, but these circular relationships do explain observable relationships between journey purposes and modes, and journey distances and modes. Higher-speed and -capacity roads, and all forms of public transport, but especially rail services, are limited to the most popular paths, and so tend to emphasize and in part create the strong radial pattern of urban movement.

The fifth principle arises because each household does not have a standard complement of pupils, workers, the retired and a housewife. Instead, as individuals progress through the life cycle they tend to belong to differently constituted households, and to some extent these tend to focus in different parts of towns. Unmarried adults and elderly people tend to live in small households, the former concentrating in inner parts of the towns, the latter being notably absent from housing developed during the previous two decades. Married couples at the child-rearing stage, it is evident, tend to live in large households, and these are unusually over-represented in recent residential developments and in the outer suburbs. The life cycle therefore adds another dimension to the temporal and spatial ordering of activity and movement. Because different stages of the life cycle can be related to different activities, access to modes of transport, and perception of the frictional effect of distance, movements within and from each residential area are structured according to the life-cycle composition of its population (Bell, 1958; Morgan, 1976).

Finally in this selective list of organizing principles are separately identified the urban-size effects on movement. Both the scale or geometry of town size, and the fact that an urban area is amongst other things a setting for a wide range of activities which are organized according to economic principles, cause pertinent urban features to vary in relation to the size of the settlement. Larger places can obviously contain longer and more varied internal journey distances. The demands of larger populations support higher threshold functions, and a greater number of lower threshold functions. This tends to encourage the specialization of activity and the segregation of land uses. More people also produce greater peaks of movement in time and space. Altogether, the size effects increase the diversity of movement patterns in larger urban areas. A simple illustration is that many small towns have no internal public transport services, aside from taxis, while only the seven largest British metropolitan areas have suburban rail services of any consequence. More generally, movements in small towns have an unusual lack of diversity, for they take on the characteristics of all short journeys, e.g. an unusually high percentage of journeys to and from work are on foot. In the largest towns as well, urban movements reveal atypical features; even car-drivers have high rates of public transport usage during peak hours, school journeys sometimes involve rail trips, and the relative importance of shopping trips to the centre decreases as distances lengthen and the scale and attractiveness of suburban centres rises. By this stage the reader will not expect any principle to operate in a simple or isolated manner, and as a final consideration it must be remembered that the size effects are modified and reduced by the substitution of external or inter-urban journeys for internal movement as urban size decreases.

A spatial typology of urban personal movement

Having discussed some of the most influential factors affecting journeys within towns, it is possible to order and to some extent understand the complexity of urban movement. Personal and goods movements will continue to be dealt with separately, and it will be useful to preserve a

Table 5.2 A typology of characteristic residentially-based intra-urban personal movements

Geographical description			Destinations and modes in order of incidence					
			Daily journeys		Weekly journeys		Infrequent journeys	
'Field' description	Maximum range	Range population	Destination	Mode	Destination	Mode	Destination	Mode
Neighbourhood (small town) journeys	1 km	5,000	Convenience-goods shops, primary schools, relatives, clubs, pubs, work	Walk Car Bike	Food shops, clubs, pubs, places of worship	Walk Car Bike	Clinics and doctors, parks	Walk Car Bike
Suburban (medium town) journeys	5 km	50,000	Secondary schools, work	Bus Car Bike Walk	Food shops, personal business, friends or relatives, sports	Car Bus Bike Walk	Durable goods shops, personal business, friends or relatives, entertainment	Car Bus Bike Walk
Urban (city) journeys	15 km	500,000	Work, higher education	Car Bus Rail			Specialist shops, entertainment	Car Bus Rail
Metropolitan or conurbation journeys	75 km	10,000,000	Work	Rail Car			Specialist shops, entertainment	Rail Car

distinction between personal journeys on the basis of the presence or absence of a link with the home or household.

A suggested spatial typology of home-linked urban personal movements is shown in table 5.2. While the typology inevitably involves subjective classifications and evaluations, in part because of the lack of comprehensive information, one merit of this approach to urban movement is that all purposes and distances of movements can be considered. Despite the simplification and generalization involved in its construction, the typology retains sufficient detail to highlight the great variety of movement characteristics, not least in space, and to demonstrate the effects of the six organizing principles on the spatial patterns of urban movements. As several recent research papers have demonstrated, systematic knowledge of the structure of activities and movements in space and time is still at a formative stage, and is being advanced at the present time as much by deductive and suggestive methods, as by empirical surveys (Taylor and Parkes, 1975; Carlstein et al. (eds), 1978). The typology evaluates journeys for different purposes and over varying distances, principally in terms of their frequency or number. No purpose or mode of a journey is regarded as unimportant, and even the shortest journeys making minimal demands for public facilities are included.

By putting forward four journey-distance categories, which have been termed neighbourhood, suburban, urban and metropolitan journeys, it is not implied that the statistical distribution of journey lengths exhibits a fourfold modal pattern, with identifiable low frequencies of journeys corresponding to the class divisions. It is suggested, however, that there are marked differences between the frequency, modal and purpose characteristics of the journeys which are most common within each distance range, defined by successive boundaries at one, five and fifteen kilometres. In other words, the characteristics which are most common among short-distance movements are unusual among medium- and long-distance journeys. An implicit feature of the typology is that the classes of journeys accumulate with increasing town size. For example, medium-sized towns will be the setting for both neighbourhood and suburban journeys and, following the same argument, the largest cities of over half a million people will exhibit all four categories of movement. Another implied feature of the larger towns is that only some of the journeys which in smaller places are over short distances are transformed into longer trips with altered modal characteristics. Inevitably, for example, in any part of a large town, some employed people will be within walking distance of their work, and only a proportion of workers will enter longer-distance categories when commuting. To take another example, the main difference in the distance and modal characteristics of trips to infants' schools between small towns and large cities may well be that, in the latter with their higher population densities, the average separation of homes from schools is less, and the proportion of journeys made on foot higher.

The typology attempts to express for each distance category and frequency group of journeys their purposes and modal characteristics in descending order of occurrence, and adds comments on the most apparent and consistent spatial feature of each group of movements. The typology, it is essential to emphasize, must be regarded as no more than a hypothetical scheme, for only elements of its structure can be tested against available empirical data.

Neighbourhood journeys

A large proportion of passenger movements in towns consist of short journeys in the vicinity of the home, an area which can conveniently be termed the neighbourhood and defined by a radius of one kilometre. Clearly in this category are most journeys to (and from) primary schools, many journeys to shops and to personal medical services, many to visit relatives and neighbours, and some journeys to work and to places of recreation and entertainment, e.g. public houses. There is, then, a considerable 'base' of short-distance movements which are spread throughout the residential areas of a town. Many of these trips are undertaken on foot, even by members of car-owning households and even by those with independent access to cars.

There is very little information about these journeys in British towns, and they have only recently been included with any regularity in the scope of travel surveys. A 1973 report of the Transport and Road Research Laboratory which sought to collate the available information could draw upon only seven sources, mainly being surveys of walking trips within central commercial areas, although several other similar studies do now exist (Mitchell, 1973: the most comprehensive sources of information available to Mitchell were data on all walking trips in Leamington Spa, and on work trips by this mode in Stevenage and from the 1962 London Traffic Survey. The latter two sources appeared to unrecord very short-distance – less than 300 metres – walking trips. There have been many local and partial surveys, and more than one bibliography is available: Lewis, 1974; Hassall, 1974; Copley and Matter, 1973). In the 1966 and 1971 population censuses, the mode of the journey to work was asked, and walking and cycling were separately distinguished, although the published tables do not include cross-tabulations with distance or destination. Walking and cycling journeys of over one mile have been included for some time in the National Travel Surveys, but shorter journeys were included for the first time only in the 1975/6 Survey. These show (figure 5.2) that nearly one-half of all trips covered less than two miles, that over 30 per cent were shorter than one mile, and that approximately 80 per cent of these were on foot. In other

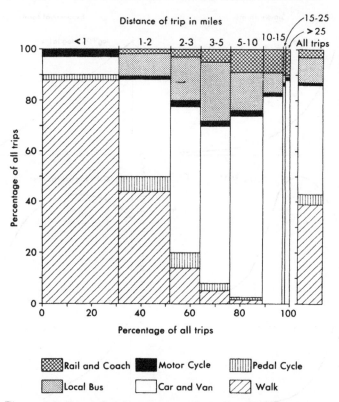

Figure 5.2 *Distance and mode of travel of urban personal journeys, 1975/6 (Source: Potter, 1978, fig. 2, p. 5)*

words, 35–40 per cent of all urban trips are non-motorized journeys of less than two miles (Potter, 1978, fig. 2, p. 5).

The actual frequency of these journeys, and the method by which they are undertaken is likely to vary not only between individuals, but also according to the details of the local physiography and human geography. On the one hand, attitudes towards walking, waiting for buses, or car use vary markedly among the population. Some make a point of being efficient shoppers and of minimizing their trips for this purpose: others, in great contrast, see a shopping trip as a diversion or as a positive use of time. Walking is esteemed by some as being healthy and economical, but detested by others to the extent that they will not travel beyond their street except by car. These personal variations are rarely considered in travel studies. Implicitly they are assumed to vary randomly throughout the population and therefore to have no consistent spatial variation within a city. On the other hand, it is likely that micro-topographic variations will produce ordered departures from average behaviour in different parts of a city. One physiographic element,

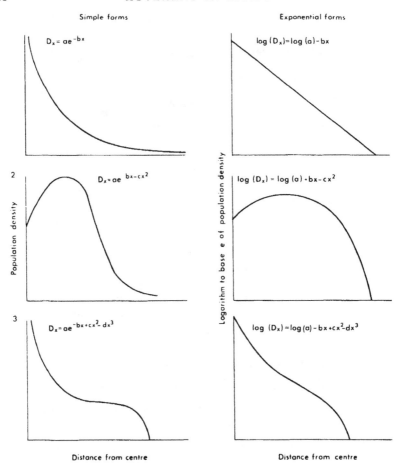

Figure 5.3 *The exponential density gradient of urban population densities*

relief, is particularly important, for adverse gradients have a deterrent effect upon walking. Most towns are also latticed by more or less severe topographic barriers to movement, such as watercourses, railway routes and land, and hill ridges. These barriers tend to be radially disposed, and therefore tend to contribute to sectorally biased movements within towns. (Sectoral bias has been most throughly researched in the case of intra-urban migration: see Donaldson, 1973; but has also been shown to characterize commuting and shopping tavel: see Taaffe *et al.*, 1963, and Davies, 1973a.)

Aside from these fluctuations, the distribution, or density, of this base of local circulation can be described by the gross residential density of population in different parts of towns. There will not be a perfect relationship between the two, for by now it will also be apparent that the age structure, marital status, and socio-economic group of the population are related to the frequency of its journeys. The earlier discussion of the relationship between the density or accessibility of opportunities and the frequency of trips, also suggests that in areas of extremely high or low population densities, the relationship between the size of the local population and the frequency of journeys will change. In areas with an unusually low density of population, for example, the operation of the minimization of travel effort principle is likely to curtail severely the generation of local journeys, as when the nearest shopping opportunities are beyond characteristics walking distance.

The general distribution of population within towns has been successfully described in mathematical terms by the negative exponential function. This states that the gross density of population at distance x from the centre of a town is related to an estimated (and maximum) density at the centre minus a quantity exponentially or geometrically related to x. As distance from the centre increases, so the population density decreases, but at a decreasing rate (figure 5.3). The mathematical expression, written as

$$y_x = a.e^{-bx}$$
or as $\log y_x = \log a - bx$,
where y_x is the gross density of population at x,
 x is the distance from the city centre, and
 a, b are constants,

was shown in 1951 by Clark to model densities in a large number of towns at a variety of dates from the mid-nineteenth century onwards (Clark, 1951, 1977). Since his work, a very large number of studies have corroborated the generalization, and some successful attempts have been made to specify more complex and accurate descriptions (for a recent review, see Haggett et al., 1977, pp.194–9). Perhaps the most convincing of these has been the suggestion of both Newling (1966) and Casetti (1967) that the low densities or the central areas of cities can be incorporated into the generalization by adopting a quadratic exponential function, of the form:

$$y_x = a.e^{bx - cx^2}$$

where c is another constant or coefficient.

This function (figure 5.3) represents gross population densities that at first increase at a decreasing rate when moving away from the centre. After reaching a maximum in the inner suburbs, density falls initially at an accelerating but latterly at a decreasing rate. This modified generalization has been substantiated by many studies, and is therefore employed here to represent the density of short-distance home-based, or neighbourhood, journeys.

The generalization describes the average of journey densities at a given

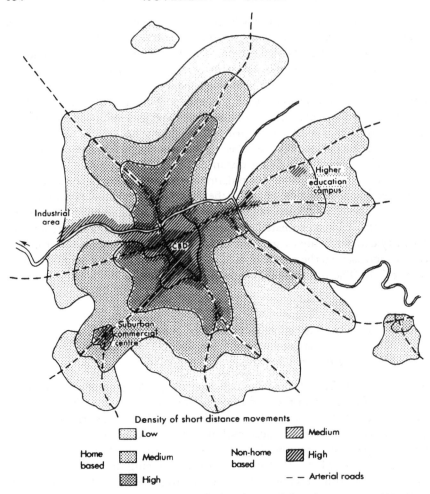

Figure 5.4 *Schematic representation of the distribution of short-distance personal journeys within a city*

distance from the centre, but in each and every direction from it. In a single radial sector of a city, the density of such journeys would not necessarily have a similar relationship with distance from the centre, for the locations of local shopping centres, industrial areas and large educational campuses would attract local concentrations of movement. The smooth quadratic exponential relationship results, in other words, from the aggregation of each sector's distinctive distribution of activity-opportunities. Although some attempts have been made by urban geographers to produce generalizations concerning the two-dimensional spatial distribution of service centres and of social characteristics within city areas, these descriptions are even less powerful than the one-dimensional gradient functions, and have much less empirical support. It is, for example, widely established that many socio-economic variables and mobility patterns tend towards a sectoral distribution in a city (Herbert and Johnston, 1976), and Johnston has argued that such sectoral variations can be associated with different shopping patterns and, therefore, local movement patterns. He suggests, with particular reference to Melbourne, that the shopping trips of residents in high-status sectors are distinguished by their low frequency, longer distance and greater number of purchases. These shopping patterns are associated in high-status sectors with a low density of isolated stores and small centres but a relatively larger number of medium-sized centres. The converse situation was found in low-status sectors and inner districts (Johnston, 1969). Although sectoral social and mobility patterns are commonly found, it is not possible to make generalizations concerning the number, angular or size distribution, and relative disposition of such sectors. At the moment, we can say no more than that where sectoral patterns exist, their precise disposition is related entirely to idiographic factors, among which physiographic influences in the orientation of valleys and hill-ridges, and historical growth factors are particularly influential.

By no means all short-distance journeys emanate from the home, for many shopping, personal-business and some social and recreational trips originate at places of employment and education. Even more complex trip patterns do, of course, occur. It is clearly insufficient to rely on generalizations concerning the residential or night-time distribution of the population, but one must take note of the adjustments to that distribution which occur through the day. Taking a broad view, it is clear that the day-time distribution of population in most cities is more centralized, with larger concentrations of people in the central areas and other commercial and industrial districts (Chapin and Stewart, 1959). The clearest example in England is the City of London, which was recorded in 1971 as having a resident employed population of 3,600 but an employed population of 340,000. Even more people will be present during the day-time of course, conducting personal business and carrying out business although employed elsewhere.

The most satisfactory nomothetic description of the distribution of short-distance journeys appears, therefore, to be one based on a one-dimensional generalization about the density of home-based local trips, but modified to represent the distribution of non-home-based trips and to indicate schematically the idiographic features of a city. The resulting two-dimensional distribution is presented in figure 5.4. The diagram reflects the influence of radial street patterns which would focus and channel local movements as well as provide many sites for trip attractors; as a result, the various choropleth grades take on a stellar rather than circular shape. It features the local shopping, employment and educational centres which produce scattered outliers of a higher density of movement. The overall picture suggested by the diagram is that the distribution of the shortest-distance personal journeys is one of intricate variation, and one that reflects both features of the population distribution and the distribution of activities within the city.

At the present time, there is a lack of empirical studies of local movement patterns, and of their response to environmental changes. Such studies offer a wide scope and would be inherently interesting to the human geographer. Local journeys raise, in a small way, most of the problems of conflicting interests and the best use of limited resources which characterize the wider field of urban transport. They are also of great interest because they illustrate at a scale which can be readily comprehended many of the characteristics of better-known urban travel patterns. For example, local pedestrian flows concentrate in time and space in like fashion to longer-distance journeys: many pavements of residential roads are at their busiest in the minutes before and after school hours, and it is often possible to identify platoons of pedestrian groups with distinctive composition and movement characteristics. A typical sequence is a peaking between 08.15 and 08.45 hours of secondary school children moving in groups to their schools; in the following half hour, younger children more often accompanied by adults are seen moving to junior and infant schools; and afterwards, smaller and less focused flows are found of mothers returning home or moving on to shops, friends, neighbours, doctors and clinics, alone or with pre-school children. In short, the number, purposes and the participants in local movements, as well as the flow map of local circulation, fluctuates through the day in an ordered manner and according to the characteristics of the local area and its population. Small-scale field observations of these movements can provide useful insights into the roles of distance, car ownership and personal characteristics in these movements. If such studies were already established, not only would there be a heightened awareness of the functions and importance of neighbourhood journeys, but this greater knowledge would itself contribute and foster more considerate attention to the requirements of local travel in general and pedestrian movement in particular when planning street alterations.

Longer-distance journeys

The remaining three categories of journey distances, termed suburban, urban and metropolitan journeys, identified in the spatial typology of urban movements (table 5.2) are those which, by involving over one kilometre of travel, are dominated by mechanically assisted transport. Cycles may be important for certain purposes such as trips to and from secondary schools in the suburban distance range, and rail journeys increase as a proportion of all commuting movements through the typology to the metropolitan class of journeys involving distances of 15 kilometres and more. By far the most important means of travel for the great variety of journeys in these three categories, however, are the private car and the service bus. The clearest distinctions between the categories are in the purpose and frequency profiles. Few school trips will extend beyond the suburban category of distance, and the only purposes stimulating daily trips in the urban and metropolitan distance bands of any magnitude will be work and, a long way behind,

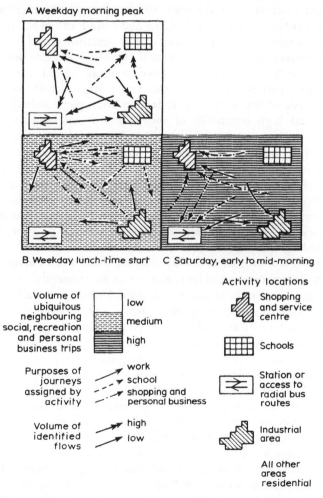

Figure 5.5 *Principal features of intra-suburban personal movement patterns*

higher education. Another distinction is that whereas many people will make weekly journeys for shopping, personal business, social or recreational purposes at the suburban distance level, the number making longer-distance journeys at this frequency is certainly very much smaller. Within the larger cities of the country, shopping, entertainment and social trips in excess of 15 kilometres will for the great majority of the population be considerably less frequent than once a week.

A description of the spatial patterns of these movements can effectively be gained by establishing the distributions of the activities which are attracting trips, and by relating these distributions to that of the residential population. In the first instance, it is desirable to disaggregate the total movement pattern, for otherwise the sheer volume and complexity of the trip patterns will be intractable. The scale of the tasks of description and interpretation can be illustrated through a hypothetical example, even one dealing with a simplified situation such as is illustrated in figure 5.5. This represents the likely journey-patterns within a suburb of a large British town at three different times of the week. Apart from a choropleth indication of low, medium, and high intensities of short-distance neighbouring journeys, journeys are represented as large or small flows to three activity agglomerations and to one point of access to the inter-suburban transport network. The activities are a suburban commercial and shopping centre, an educational precinct, and an industrial area. Only journeys to and from work, school, and shops including personal business establishments, are included on the diagrams, yet with all these simplifications the patterns of likely flows are seen to fluctuate considerably according to the time being represented.

Diagram A concerns the weekday morning peak, when movements to all the activity locations proceed from all parts of the residential area. There are also strong movements towards the arterial roads and railway stations linking the suburb with the rest of the town, and very often lesser reverse flows. Although many short-distance trips to infant and junior schools will be taking place at this time, the overall density of the ubiquitous base of movement is normally low because neighbouring journeys for other purposes are not prevalent until later in the day. Diagram B pictures the flow pattern at the beginning of a weekday lunch period, and shows movements away from the activity agglomerations into all parts of the residential area. In general the size of these flows is less than the complementary journeys during the morning peak, except that shoppers returning home from the commercial area will add to the numbers returning from work. Non-home based trips from schools and industrial areas to shopping centres reach peak levels at this time, as we saw in chapter 3 when we compared the diurnal patterns of trips in Norwich. On the other hand, relatively few movements will be evident between the suburb and the rest of the town. Finally,

Diagram C outlines the position early to midway through a Saturday morning. Flows at the neighbourhood level are expected to be unusually high with numerous local shopping, social and recreational trips taking place, and the dominant feature of journeys in the suburban distance band will be their flow away from residential areas. The principal destinations at this time will be the suburb's commercial and shopping centre, and the access points to the rest of the city, with the central city shopping facilities being a prime objective. There will be much less movement than on a weekday towards the industrial employment concentration, and a negligible number of journeys to the educational precinct. Further diagrams could be drawn to represent other distinctive flow patterns which occur at different times of the week.

The three diagrams show sufficiently that a map of aggregate movement flows over a week or a longer period would be the outcome of several contrasting patterns, and would be difficult to elucidate without breaking down the flows into more easily understood elements. Disaggregation according to time has been used in the example of Figure 5.5, but as the importance of this dimension has already been discussed in chapter 3, the method preferred in the following chapter is disaggregation by trip purpose. On the basis of the overwhelming evidence from the transportation studies that work, school and shopping journeys comprise around 85 per cent of all mechanically assisted journeys, the obvious next step is to examine the distributions of the locations of these activities, and of the journeys linking these locations to homes. As most journeys to school are short-distance and have been discussed in chapter 2 and in the earlier section on neighbourhood movements (additional papers to those already cited are Maxfield, 1972; Mensinck, 1973), in the next chapter attention is focused on commuting and shopping travel.

6
Spatial patterns of urban movement: empirical evidence

Spatial patterns of journey to work

Information about all aspects of commuting journeys is superior in quantity and quality to that available for any other category of journeys. This is not surprising given that they form such a large proportion of all mechanically-assisted and longer-distance movements and are so closely associated with peak demands upon facilities. Even the population census collects simplified information on the origin and destination of journeys to work, and has in fact done so in every census since 1921 except that of 1931 (for descriptions of these data, see Warnes, 1972). This source has been drawn upon by research studies of these movements ever since the 1920s (early studies include Hewitt, 1928; Liepmann, 1944). More recently, apart from the attention given to commuting patterns and trends in transportation studies, there have also been numerous special-purpose surveys and enquiries into these topic (see, e.g., Rowley and Tipple, 1974. Useful American summaries of such literature are Stutz, 1976; Wheeler, 1974).

A general picture of the geography of intra-urban commuting can be established by adopting a nomothetic approach, and inferring from the distributions of employment and population the pattern of dominant net flows. Just as the distribution of population within cities can be described by the negative exponential density gradient, so can the distribution of employment opportunities. Four general characteristics of urban employment distributions have been identified in recent publications (Warnes, 1975, 1977). First, employment densities have been found to decline with increasing distance from the city centre at an exponential rate in like manner, and in fact closely parallel, to the decline in population density gradients. Second, but in great contrast to the population distribution, employment densities reach unusually high levels in the city centre. This alone implies that the exponential rate of decline of employment density is steeper than that of population density. The first two characteristics are very clear and omnipresent features of British cities, although the same claim cannot be made for the next two generalizations. It is suggested additionally that the decline in the density of employment with increasing distance from the city

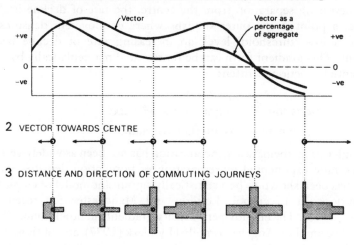
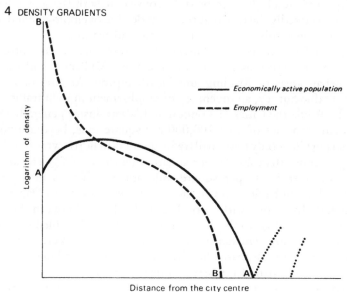

Figure 6.1 *Population and employment density gradients, and their commuting implications* (Source: *Warnes, 1975, fig. 2, p. 83*)

centre is at first very rapid, but thereafter at a decreasing exponential rate; and lastly that there is a tendency for employment densities to fall precipitately once they have declined to a low, threshold, level.

These four characteristics in combination point to a density gradient of the form illustrated in figure 6.1. Densities are extremely high at the centre of the town, but as one moves towards the periphery they fall very rapidly. With increasing separation from the centre, the rate of decline itself falls until, at a point in the outer suburbs where employment densities have reached a low 'threshold' level, an increasing rate of decline becomes apparent. This gradient form can be represented mathematically by a cubic exponential function, written:

$$y_x = a.e^{-bx+cx^2-dx^3}$$
or as log $y_x = \log a - bx + cx^2 - dx^3$
where d is another constant.

Although this mathematical generalization has not been as widely verified as the quadratic exponential representation of population densities within a city, it has been shown to be a statistically significant model of employment distributions in Liverpool, London and Manchester for recent years (Warnes, 1975). The employment density gradients in Stockholm, Chicago, Milwaukee and Los Angeles compiled by Clark (1977), and of those in Paris drawn by Bussière (1970), are broadly consistent with the generalization, and it is undoubtedly the case that the steep decline of employment densities through the inner suburbs is a very widespread feature.

The generalizations can be illustrated by reference to the intra-urban employment distributions in London, the West Midlands, and the much smaller urban areas of Reading and Northampton. As part of the 1962 London Traffic Survey, an inventory of employment in each traffic district was made. While the Cities of London and Westminster generally showed employment densities of over 100,000 per square mile, beyond two miles from Charing Cross only two small traffic districts in north-west London had densities of more than 20,000 per square mile (figure 6.2). The area with densities of over 5,000 per square mile approximated a cross, aligned north–south from Enfield to Croydon, and east–west from Dagenham to Southall and Hounslow, with an outlier at Heathrow. Between the arms of the cross, even as close to the centre as Camberwell, Wimbledon, Finchley and Leyton, employment densities were lower. There were several peripheral districts with less than 200 jobs per square mile, and in some of these areas, such as Northwood-Pinner and Buckhurst Hill, the gross densities of employed residents were in excess of 2,000, so that the employment deficiency exceeded a factor of ten.

In the West Midlands in 1964, there was a heavy concentration of employment around the centre of Birmingham (figure 6.3). One of every six jobs in the entire Transportation Study Area was in the most central four square miles (Sector 01); which also contained 40 per cent of all employment in the wholesale trade, 20 per cent of miscellaneous service employment and 12 per cent of retail employment. One-quarter of all employment in the Study Area was within two miles of the centre of Birmingham. In both

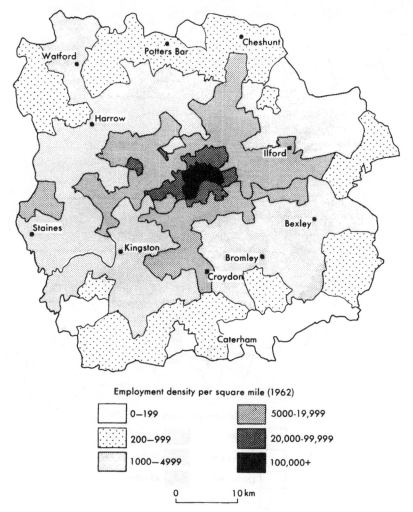

Figure 6.2 *The employment distribution of Greater London, 1962* (Source: *Greater London Council, 1969, fig. 2.4, p. 24*)

Reading and Northampton, 'the most striking feature ... is the concentration of employment in and near to the centre of each town. This reflects not only the concentration of non-residential land uses but also the greater density of employment per unit are of a given land use close to the town centres'. Perpective block diagrams of Reading from Taylor's study effectively show the contrasting features of the distributions of population and employment (Taylor, 1968, fig. 31, p. 82; Downes and Wroot, 1974).

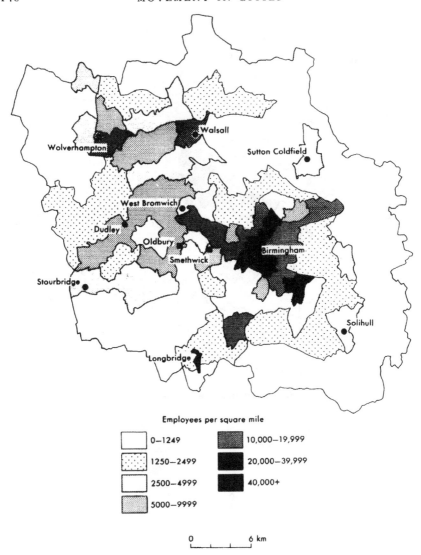

Figure 6.3 *The employment distribution of the West Midlands, 1964* (Source: Freeman Fox et al., 1968, vol. 1, fig. 3.4, p. 31)

Some British cities are well-known for their unusual concentrations of suburban employment, as in aircraft manufacture at Filton on the northern side of Bristol, or in motor manufacture at Speke on the south-eastern fringe of Liverpool, but these exceptions do not yet produce the densities of employment characteristic of central and inner areas except in highly localized areas. More general analysis of the distributions confirms the pattern of exponential decline from peak values at the centre, as has been shown in the case in Liverpool (Warnes, 1975; R. W. Thomas, 1977).

If these widely verified nomothetic generalizations of the uni-dimensional distributions of population and employment are considered together, the areas of a city with excesses or deficits of employment are identified, and from these, inferences about the broad patterns of intra-urban commuting may be drawn (figure 6.1). This is possible because in an isolated city region, the total of jobs must be the same as the number of employed people (except that some people have two or more jobs), and the net result of a morning's commuting is the alteration of the distribution of employed people from its night-time residential form to its daytime employment form (ignoring people temporarily off work). Near to the centre there is a substantial excess of jobs over the resident employed, but through the intermediate suburbs the two densities are very similar, with the employed population density being slightly greater. Further away, however, the difference widens as the rate of decline of employment densities increases.

These characteristics suggest that the dominant influence upon the intra-urban pattern of commuting is the central job surplus. These jobs must be taken up to some extent by residents from all parts of the metropolitan area and beyond, but, allowing for an accessibility effect, in large part by workers from the most adjacent residential areas. The net centripetal movement from inner areas, however, will mean that the jobs in these same suburbs will exceed the residual number of local workers, and so will be filled by a net movement from more peripheral suburbs. In other words, the net centripetal flow is a pattern which cascades from the innermost to the most peripheral suburbs. There will, of course, be centrifugal journeys, as well as trips in lateral directions (at right angles or orthogonal to the radials), but these are probably relatively least from the most inner and the most outer suburbs. Trips from the former areas will inevitably be strongly attracted by the abundance of potential destinations in the central area, while from the latter areas, the job opportunities in centripetal or lateral directions will be very low.

It is possible to test these and derivative generalizations because the actual spatial structure of intra-metropolitan commuting patterns in British cities is recorded by the census statistics on places of usual residence and workplace from the 1921, 1951, 1961, 1966 and 1971 enquiries. Apart from the *aggregate* figures of the numbers of employed residents and jobs in each local authority, to which reference has already been made, there are *specific* figures for the numbers travelling to other local authorities. Only those movements involving more than an unspecified number are given (approximately 25 in the census, and a sample figure of 4 in the 1966 10-per-cent sample census), resulting in a degree of inconsistency in coverage. All of the figures are available separately for males and females (for evaluations of the statistics, see R. E. Dickinson, 1964, p. 356; Warnes, 1969).

The *specific* data provide information on the size, direction and distance of commuting trips from each local authority. All three items of information define a *vector* of movement, and each flow from an authority can be represented by unique vectors (for a more detailed introduction to the nature of vectors, see Spain, 1965). These, in turn, can be summarized in a resultant vector, and conversely, any vector can be analysed in terms of its components of movement in different directions. It is therefore possible not only to describe the net flow or resultant vector of each areas's recorded commuting, but also to specify the centrifugal, centripetal or lateral components of the flows. For most areas of a city, the resultant flow will be centripetal, although it is possible either that there is no net radial movement or that the resultant is dominated by a minority of peripheral movements of exceptional distance.

From the alignment of employed population and employment density gradients, and from basic geometrical features of a uni-nuclear and circular city, it can be hypothesized that the strength of the centripetal component of commuting movements from residential areas will exhibit an ordered relationship with distance from the centre. Adjacent to the centre, although a large proportion of all movements will be strongly centripetal, the net component will not attain high values because the distances involved will be less than the metropolitan average. On the other hand, from the most peripheral parts of the city, the predominance of journeys with a strong centripetal component, and their above-average length, will produce a very strong resultant movement towards the centre. This is likely even if only a low percentage of the journeys extend to the city centre, because the high trip distance results from the low density of jobs in outer suburban areas. It is expected, therefore, that the centripetal component of commuting from the outer suburbs will be greater than from any other part of the urban area. The predicted relationship between the centripetal component and distance from the centre is illustrated in figure 6.1. The second part of this diagram illustrates a derived relationship, between distance from the centre and the ratio of the centripetal component to the arithmetic total of all commuting movements. The principal contrast between this relative relationship and that with the absolute value of the centripetal component is that it will produce high values in the inner suburbs as well as near the periphery (Warnes, 1972).

Many of these predicted features have been found in the larger British metropolitan areas, as in Greater London and Greater Manchester. Diagrams of the 1971 density of employed residents and of jobs, constructed by organizing the census data for local authorities by 2 km zones from the centre, show clearly the population excess in the outer suburbs and the greater relative employment excess in the central and inner areas (figure 6.4). The parallel declines in density of employed population and jobs through the intermediate suburbs are also clearly shown.

Further analysis of the data and corroboration of the above hypotheses is

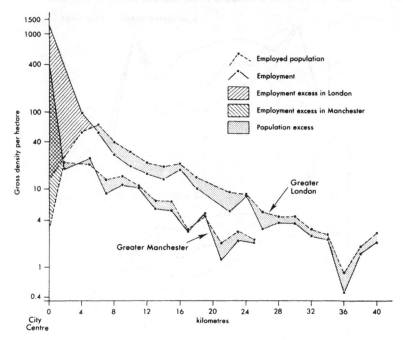

Figure 6.4 *Population and employment densities by 2 km zones in Greater London and Greater Manchester, 1966*

possible. For each 2km-wide anulus or zone, weighted mean centripetal commuting vectors were calculated from the *specific* commuting data. These are plotted as moving means of three contiguous zones, and do in fact reveal the unusual importance of centreward movements in two areas: the innermost and the outermost suburbs (figure 6.5). It is seen, therefore, that fundamental geometric characteristics of a metropolitan area in combination with the concentration of employment towards the centre impose a gross spatial order on the pattern of daily commuting movements.

Other more specific hypotheses can be drawn from a consideration of the distributions of employment and population within urban regions – for example, that commuting distances are positively correlated with the distance of an origin area of residence from the city centre. This argument is based on the difference in the local densities of employment opportunities in inner and outer areas. Other things being equal, the number of job vacancies within 5 km of an unemployed person's suburban home will be much less than for an unemployed person in the inner city: the former is therefore much more likely to travel further to find employment. When socioeconomic variations between the inner and outer areas are considered, this relationship is likely to be reinforced. Among the more important

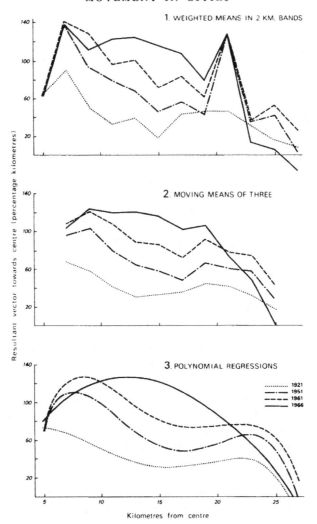

Figure 6.5 *The relationship between distance from the centre of Manchester and the strength (centripetal vector) of commuting movements towards the centre.* (Source: *Warnes, 1975, fig. 6, p. 93*)

reinforcing variables are the higher car-ownership rates of suburban as compared to inner districts (table 6.1), the higher level of educational attainment and professional qualifications of suburban residents which implies a more specialized occupation and a lower probability of any individual vacancy being attractive to the suburban resident, and, possibly, the larger average workforce at suburban as compared to inner-city places of

Table 6.1 Car ownership by household and area

	Country village	Small town	New town	Outer suburb	Inner London
	Percentage of sample households				
Carless	33	29	36	54	57
Owning 1 or more cars	67	71	64	46	43
Owning 2 or more cars	18	12	4	6	5

Source: M. Hillman *et al.* 1973, table 3.1 p. 53

employment. There is some evidence for this, at least with respect to the low mean size of employing units within the inner areas of a town. It would mean, of course, that the low density of employment in the suburbs was associated with an even lower density of places of employment which would increase average commuting distances. (Similar conclusions were reached by Greytak, 1974. One study adds the qualification that because journey speeds are low in the inner city, commuters from peripheral areas often have quicker journeys to work despite their longer distance: see Morgan, 1967.)

Some impression of the contrasting spatial patterns of commuting from inner and outer districts is shown by selected 1966 data for the recorded movements from suburbs of Greater Manchester (figure 6.6). The inner suburbs – Swinton and Pendlebury, Stretford and Droylsden – generated movements of relatively large percentages of their workforces towards the conurbation centre, but journeys in other directions were very weakly represented. The pattern from the outer suburbs, however, was one of relatively smaller numbers travelling longer distances to more diverse destinations. The centripetal movement, it is interesting to note, is much more prominent from the southern 'dormitory' and middle-class suburbs, than from the northern industrial settlements.

A rather different picture is arrived at if the commuting patterns *to* different parts of a city are examined. The most dispersed and comprehensive distribution of origins is that of the large number of journeys to the central area. All non-central destinations, however, tend to attract people dominantly from the side or sector of the urban region in which they are located. It was also found in a detailed study of Chicago that suburban places of employment were associated with large numbers of short-distance journeys to work, particularly for female employees (Taaffe *et al.*, 1963, table II–6, p. 17). Twelve separate distributions of origins were described, six of which were arranged on a west–east axis from the Loop or CBD to the 'West Suburban District', an area on the fringe of the continuously built-up

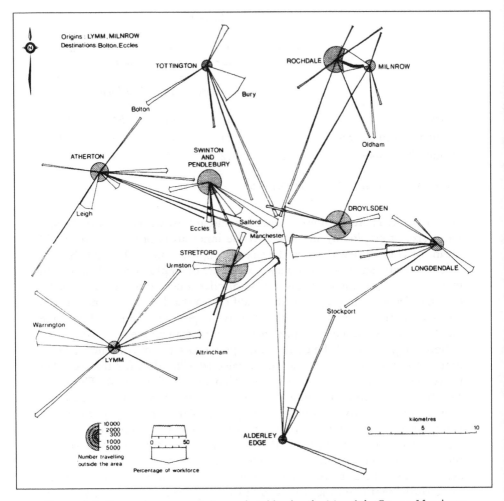

Figure 6.6 *Commuting movements from selected local authorities of the Greater Manchester Region 1966*

area in 1956 but less than halfway to the then current boundary of the Chicago Area Transportation Study (Taaffe *et al.*, fig. III, pp. 25–6; fig. IV, pp. 26–7). The remaining six were all more peripheral areas, arranged circumferentially from the north through the western suburbs to the south.

It was found that the distribution of commuter origins to the CBD covered most of Chicago's urbanized area (grid squares were used and a dot appeared if there was at least one commuter in the sample), and a strongly developed radial disposition of recent suburbs in Chicago was also clearly identified. On a second map, of commuters to the fringe of the CBD, there was a very similar pattern, but the third, representing the origins of commuters to a zone just to the west of the CBD, was markedly different. The transport advantages of the CBD have been lost in this area, and a clear sector pattern of origins results, with very few commuters coming from any but the western outer suburbs. Moving through the next two zones, essentially the same pattern prevails, of a 'general dispersion with something of a concentration on adjacent areas'. The chief difference by the 'West–Suburban' zone was the importance of a single radial, the Oak Park–West radial, and the fact that there was a sprinkling of origin-cells in the interstitial areas between the well-developed outer radials.

In summary, the sectoral traverse indicated an increased frictional effect of distance, as destination districts further away from the centre were considered, and this general conclusion was supported by the distributions to all six of the outer suburban employment zones.

Trends in the intra-metropolitan journey to work

Urban population and employment distributions undergo continuous change, and many writers argue that not only has the recent pace of change been faster than ever before, but that radical restructuring of the metropolitan area is now underway (Berry and Kasarda, 1977, pp. 248–67; Vining and Strauss, 1977). Ever since the mid-nineteenth century in London (with other large British cities not lagging long behind), population densities in the most central areas of British cities have been declining (Clark, 1951). At the same time, the areal extent of metropolitan areas has expanded, in many instances at an accelerating rate, partly to accommodate larger numbers of people and more rapidly growing numbers of households, but also as a reflection of the improving space and amenity standards of urban housing, itself enabled by improving standards of living. Lowering inner-area densities, areal expansion, and rising densities in the urban fringe of a generation ago, all imply that over time the gradient of density decline itself decreases. In terms of the mathematical generalization, both the estimated density at the centre (a) and the estimated rate of exponential decline (b) decline over time, as is shown very clearly for Liverpool, London and Manchester (table 6.2).

Similar decentralization trends have characterized urban employment, although, at least until recently, employment decentralization has lagged behind population shifts. The early date at which central area redevelopment displaced large numbers of jobs towards more suburban locations is frequently forgotten, but Kellett's study (1969) of the impact of railway extensions into British city centres in the mid-nineteenth century provides circumstantial evidence of the effect. Later in the century and into the twentieth, changes in the organization of manufacturing and retailing were

Table 6.2 The spread of Liverpool, London and Manchester as represented by population density gradients, 1921–71

City and date	Regression coefficients		r^2	$\dfrac{\ln(a)}{b}$	C(p)	F	d.f.
	a	b					
Manchester							
1921	57.40	−0.164	0.938	24.70	0.0689	120.65	1,8
1951	50.40	−0.156	0.918	25.13	0.0659	89.47	1,8
1966	45.61	−0.152	0.918	25.13	0.0634	89.00	1,8
1971	39.64	−0.145	0.905	25.38	0.0603	75.89	1,8
Liverpool							
1921	26.05	−0.166	0.633	19.64	0.0509	17.26	1,10
1951	23.05	−0.133	0.801	23.68	0.0491	40.21	1,10
1966	25.53	−0.141	0.876	22.98	0.0511	70.85	1,10
1971	26.58	−0.136	0.810	24.12	0.0600	42.71	1,10
London							
1921	96.16	−0.153	0.894	29.84	0.0821	151.06	1,18
1951	145.04	−0.141	0.938	35.30	0.0933	271.35	1,18
1961	134.16	−0.130	0.921	37.68	0.0911	208.77	1,18
1966	88.32	−0.112	0.925	40.01	0.0799	196.40	1,16
1971	80.72	−0.111	0.891	39.56	0.0775	131.09	1,16

Notes: $\ln(a)/b$ is the radius in kilometres, R, at which estimated density is 1 per hectare. $C(p)$ is the proportion of city employment estimated to lie within the central $p = 0.01$ of the city area. It is given by:

$$C(p) = \frac{e^{-b\sqrt{pR}}(b\sqrt{p}R + 1) - 1}{e^{-bR}(bR + 1) - 1}$$

Two-kilometre concentric zones of the conurbations were used for the analysis, and no separate figure for the city centres was included.

promoting considerable suburbanization of a wide range of manufacturing industries in London. Woodworking trades, for example, showed a considerable north-easterly displacement in late Victorian and Edwardian London (Hall, 1962; Martin, 1966).

These are local examples of very widespread trends, which have been described by Lawton in a series of papers (1959, 1963, 1968, 1977) which examine the ratio between the number of jobs and the number of employed residents in local authority areas. He amply demonstrates that the largest towns, and particularly the most central local authorities of the conurbations, have during this century steadily accumulated job surpluses. Complementarily, suburban local authorities have generally experienced falling job ratios, as their populations have increased faster than their number of jobs; and the hinterlands of the largest conurbations have expanded, as shown by the spreading extent of low job ratios. Other researchers have provided similar and more detailed evidence for particular urban regions of the country (W. K. D. Davies and Musson, 1978; Dewdney, 1960; Ellman, 1968; Fulleston and Bullock, 1968; Westergaard, 1957).

Since Lawton's early work, there has been a sequence of analyses of the changing distribution of population and employment within British urban regions, and of their journey-to-work effects. Successive studies have been more substantial and detailed than their predecessors: R. Thomas's analysis (1968) focused on the conurbation centres and conurbation remainders; the Political and Economic Planning study directed by Hall (1971,1973) adopted the United States census terminology and areal zoning of cores and rings in 'Standard Metropolitan Statistical Areas'; while the current Department of the Environment study (1976b) directed by Spence, Goddard and Drewett at the London School of Economics extends the standardized definitions to cores, inner rings and outer rings of Metropolitan Economic Labour Areas (MELA).

This latest research has defined in Great Britain 126 urban *cores* consisting of one or more administrative areas with 20,000 or more jobs or an employment density of at least five per acre. 119 of these cores were surrounded by *rings* sending 15 per cent or more of their labour force to the core; the combined area being termed the Standard Metropolitan Labour Area (SMLA). In 93 cases, an *outer ring* of the MELA was defined comprising those local authorities which sent more commuters to the MELA core than to any other (DoE 1976b, appendix, p. 58–9). Although the definitions are expedient rather than functionally meaningful some arbitrariness is inevitable when using standard definitions, and such divisions have had proven value in both the United States and Britain.

The changing absolute and relative distribution from 1951 to 1971 of population and employment by the three zones (which are constant and defined on the basis of 1961 figures) reflects the steady continuation of decentralization in the post-war period (table 6.3). Population dispersal has continued to proceed more rapidly than employment dispersal, resulting in rising core excesses and ring deficits of employment. As the authors state, 'the immediate inference that must be drawn . . . is that there is on aggregate an increasing separation of homes and workplaces in urban Britain with a consequent overall increase in the length of the journey to work'. A more detailed analysis shows the intra-MELA pattern of commuting and its change from 1961 to 1971 (table 6.4) (Gillespie, 1977). Within-zone

Table 6.3 Population and employment totals (millions) and ratios by MELA zones, Great Britain 1951-71

MELA zone	Population			Employment			Population proportion to employment proportion ratio		
	1951	1961	1971	1951	1961	1971	1951	1961	1971
Core	25.8	26.3	25.5	13.4	14.3	13.9	0.88	0.85	0.81
Ring	12.9	14.6	17.1	4.4	4.7	5.4	1.31	1.35	1.40
Outer ring	7.8	8.1	8.8	3.3	3.3	3.4	1.07	1.11	1.14
MELA	46.4	48.9	51.5	21.2	22.3	22.8	1.00	1.00	1.00

Source: Department of the Environment 1976b, tables 4, 5 and 6, pp. 10, 12.

movements accounted for over 67 per cent of all journeys in 1971, but those between zones showed a net movement towards more central zones, the difference between the share of journeys that were centripetal and the centrifugal score increasing from 7.4 to 9.4 per cent of the MELA total during the decade. In other words, the additional net flow of 356,000 centripetal journeys in 1971 represented a 23-per-cent increase on the 1,568,000 net centripetal flow in 1961. Further analysis shows that the dispersal of population and, more particularly, employment is taking place more rapidly in the MELAs with the largest total populations, and that in many of the smallest (with less than 125,000 people), both population and jobs continued to grow in the cores. Many of these were in the Home Counties, and although separately defined in the study, can be regarded as part of the periphery of the functional area of metropolitan London. This result of the study does demonstrate clearly the growing complexity of both commuting patterns and urban structure in the region. Although this can be partly attributed to the definitions used by the DoE study, with their emphasis on the direct links between homes and workplaces, the accumulated British research does raise very interesting further questions about the extent to which the topology of commuting patterns is changing as well as their scale. It cannot be taken for granted that the pattern is becoming more decentralized, more focused on multiple destinations, or more dominated by peripheral journeys. These elements of an intra-metropolitan commuting pattern have never been absent, but in the past were represented by smaller numbers travelling shorter distances commensurate with the more restricted extent of metropolitan areas. (Arguments that the changes are not only of scale have been advanced by, among others, Logan, 1968.)

Census workplace data not only enable the broad intra-urban analysis

Table 6.4 Intra-MELA percentage distribution of commuting journeys in 1971, and relative changes from 1961

Origins	Core	Ring	Outer ring	Destinations MELA	Centripetal	Centrifugal	+Centripetal −Centrifugal
A. 1971 percentage distribution							
Core	49.4	3.5	0.3	53.2	–	3.8	− 3.8
Ring	11.9	18.9	0.7	31.5	11.9	0.7	+11.2
Outer ring	1.2	0.8	13.5	15.5	2.0	–	+ 2.0
MELA	62.5	23.2	14.5	100.2*	13.9	4.5	+ 9.4
B. 1961–71 percentage change							
Core	− 10.0	+ 8.9	+18.2	− 8.8	–	+ 9.6	+ 9.6
Ring	+ 15.6	+11.8	+30.5	+13.5	+15.6	+30.5	+14.8
Outer ring	+ 51.6	+40.4	− 0.9	+ 3.4	+46.9	–	+46.9
MELA	− 5.3	+12.1	+ 0.6	− 0.9	+34.3	+12.3	+22.7

Notes: * Rounding error. Centripetal journeys are those to a more central zone, and centrifugal journeys are those to a more peripheral zone.
Source: Gillespie (1977), table 2, p. 3.

Table 6.5 Average journey-to-work distances (km) of the residents of selected urban areas of North West England, 1921–71

Area	1921	1951	1961	1971	1971:1951
Manchester CB	2.22	2.55	2.43	2.41	0.95
Droylsden UD	2.89	4.09	3.95	3.69	0.90
Stretford MB	2.42	2.57	2.43	2.31	0.90
Swinton and Pendlebury UD	2.67	2.98	3.06	3.08	1.03
Knutsford UD	4.29	4.78	5.61	6.96	1.46
Lymm UD	4.24	5.76	5.75	7.08	1.23
Milnrow UD	1.78	2.37	2.82	3.43	1.44
Liverpool CB	2.27	2.51	2.61	3.14	1.25
Lancashire CBs	2.14	2.56	2.55	2.64*	1.03*

Notes: *Figures for 1966 or 1966:1951.
The method of estimation is described in Warnes (1972).
Source: England and Wales Census, Workplace Statistics.

of the distributions of homes and workplaces, from which inferences about the gross patterns of commuting have been made, but also permit more direct inspection of the temporal trends in journey-to-work distance and orientations. The recorded information on the numbers travelling from a local authority to other stated authorities can be supplemented by estimates of the average distance of trips to these authorities and of trips within authorities, to produce serviceable estimates of overall mean commuting distances from each recorded local authority (table 6.5). These reveal that in North West England from 1921 to 1966, there was an average 1.0 per cent per year increase in trip length, although this figure conceals an acceleration in the rate of growth in successive inter-censal periods (for full details of the estimation procedure, see Warnes, 1972). Journey distances have consistently been lowest in the densest and largest urban areas, and over recent decades the differential has tended to increase. Indeed, in several intermediate suburbs of Greater Manchester, such as Droylsden and Stretford (for location see figure 6.6), the post-war trend has been for journey-to-work distances to decrease, mainly because a larger percentage of their declining workforces find work within the local authority. Stretford, for example, sent 54.4 per cent of its 31,800 workers in 1951 to other authorities, but by 1971 only 49.2 per cent of its 24,250 workers travelled beyond its boundaries. Recent trends can therefore be seen to be various: by no means all areas are recording increases in commuting distances (other evidence on journey-to-work lengths in Britain is provided by Beesley and Dalvi, 1973, 1974; see also Fullerton and Bullock, 1968). The very rapid increase in average distances during 1961–71 in the outermost suburbs are notable, however, and are in all probability associated with the increase in car ownership in these affluent peripheral suburbs, as well as the improvements in road communication through motorway construction. The latter have been extensively developed in North West England, to a greater extent since than before 1971, and will promote further changes in the region's traditional commuting patterns. Out-movements involving more than 1 per cent of the economically active population of Knutsford were far more diverse in 1971 than in 1961 (table 6.6), and included for the first time a flow over 45km to Liverpool.

Comparable changes in commuting flows and patterns have been occurring in industrial South Wales. Although the total employed population of 736,000 changed little during 1951–71, the number of women in employment increased by 44 per cent, and the number of males decreased by 13 per cent. In the same period, the overall volume of commuting increased

Table 6.6 The growing complexity of commuting from a peripheral urban area of the Greater Manchester conurbation: Knutsford, Cheshire, 1921–71

Destination	Distance (km)	Percentage of economically active population working at destination			
		1921	1951	1961	1971
Manchester CB	21.3	9.5	8.9	15.6	10.5
Bucklow RD	5.3	5.6	9.3	9.5	8.0
Altrincham MB	10.1	3.8	5.8	4.9	4.9
Hale UD	8.8	1.3	–	–	–
Sale MB	14.0	–	–	–	1.4
Stretford MB	18.8	1.3	1.4	2.2	2.4
Salford CB	21.9	–	–	1.7	2.2
Northwich UD	10.5	1.1	3.8	4.4	–
Northwich RD	13.2	2.0	1.5	1.7	1.2
Warrington CB	17.4	–	–	–	1.0
Warrington RD	18.2	–	–	–	2.0
Liverpool CB	44.9	–	–	–	1.2
Wilmslow UD	10.1	–	1.2	1.5	1.4
Macclesfield RD	16.4	–	–	–	1.4
Active population	–	2570	2920	4100	5910

Note: Only movements involving 1 per cent or more of the town's working population are shown.
Source: England and Wales Census, Workplace Statistics

substantially. The proportion of the total population crossing local authority boundaries in their journeys to work increased from 25.1 per cent in 1951 to 34.7 per cent in 1971 (males: 26.2 to 38.9 per cent; females: 21.6 to 26.6 per cent). W. K. D. Davies and Musson's study (1978, pp. 354–6, 365) of the changing commuting flows revealed a breakdown of 'self-containment' and the reorientation of many flows, mainly towards the larger employment centres (Swansea, Cardiff and Newport) of the south and east of the Region.

In the absence of other time series data on journey-to-work distances and distributions, the most valuable indications of temporal trends have been produced by a small number of studies of the situation before and after specific changes in residential or employment patterns, such as the closure of coal-mines in south Wales, Durham and the East Midlands, the planned decentralization of offices particularly within Greater London, or the planned decentralization of population to overspill areas and New Towns. (For commuting patterns in coalfields, see Dewdney, 1960; Holmes, 1968; Humphrys, 1962, 1965. The commuting implications of office decentralization are dealt with by Wabe, 1967; Daniels, 1973, 1975 pp. 208–15. Overspill and commuting is treated by Cullingworth, 1959–60, 1960; Rodgers, 1965.)

The special provisions made by the National Coal Board to transport miners to more distant continuing mines can account for the long and atypical journey patterns found in rationalized coalfields, but similar effects, which may more confidently be used as evidence of a general temporal trend, have been shown in industrial south Lancashire. The deep-seated changes in this area's occupational structure during the twentieth century, brought about by the decline in mining, textile and engineering employment, have been paralleled by a considerable alteration in the distribution of employment. The former relatively-dispersed distribution of employment, have been paralleled by a considerable alteration in the distribution of employment. The former relatively-dispersed distribution of employment (among a large number of mines, mills and small industrial settlements) has tended to be replaced by a more metropolitan and clustered distribution: industrial employment has concentrated in fewer, larger plants, and commercial employment has concentrated in the centres of the largest cities. As this region saw little redevelopment of its pre-existing housing stock during the first half of the twentieth century, the result, despite the massive extensions to the residential areas of the most thriving towns, has been a widening disparity between the distributions of housing and of population (T. W. Freeman *et al.*, 1966). One result has been that many people have become more remote from work, and that the average distance of journeys to work has increased. A lengthening of commuting time and travel, seen most clearly with organizational changes within the coal industry, is therefore also produced by more general structural changes in a modern industrial nation.

Daniels (1975) has reviewed British studies of the journey-to-work implications of redistributing office activities within and between cities, and himself investigated changes in the commuting patterns of 63 offices which moved from central London between 1963 and 1969. Such moves are essentially an intra-urban phenomenon: approximately one-third of the offices moved to other parts of Greater London, and another quarter to areas within the South East Region. The changed mode and duration of journeys to work were given particular attention. Prior to the relocation, 86 per cent of the respondents in the survey who travelled to offices in central London used public transport, mainly rail services, but to reach the decentralized locations less than 50 per cent used public transport, mainly buses, and almost 40 per cent drove their own cars. These averages concealed great variation, however, for 'an office at Mitcham, 3.3 km from the nearest suburban station and poorly served by bus routes, generated private transport trips by more than 60 per cent of its employees, compared with 16 per cent at a Harrow office with direct access to the underground and a good range of bus services'. Mean journey-to-work times were generally reduced as a consequence of relocation and for moves as far as 100km from the centre of London, a strong negative relationship was established with mean trip time, as well as a positive relationship between the distance of a move and the saving or difference in mean trip times. Former central London employees who moved to the suburban or peripheral locations had previously to spend twice the time on commuting as the mean for the new locations, but once they started travelling to the new destination they achieved trip times similar to locally recruited employees.

While the employees that continue to work for a decentralized office are likely to be a selected group, and will probably under-represent those living in sectors of the city opposite to the new location, this evidence does demonstrate that not all changes in the location of employment necessarily lead to a greater separation of homes from workplaces. Indeed, more recent evidence from a follow-up study conducted at the same office establishments in 1976 suggests that proximity of homes and workplaces in suburban areas as a result of employment redistribution improves as offices becomes fully integrated in local labour markets. There is a discernible trend towards more compact labour catchment areas, particularly in instances where staff turnover leads to replacement of former central London staff with local recruits. Average journey-to-work times have decreased since 1969 and more office employees can expect to take less than 30 minutes to reach their offices each morning. A lessening of the concentration of urban employment in the city centres is likely to result not only in improvements in travel time and comfort, but also in a potentially very large saving in fuel and in transport investment from reduced journey distances. As Daniels argues

(1975, 1977, 1978), these advantages will be enhanced by careful planning of the locations of suburban employment growth points in relation to the existing public transport networks.

There is no doubt, however, that during this century until the present time, economic growth and change have overall, and in many different ways, tended to produce longer and more complex commuting patterns. It has frequently been the case that the role of particular intermediate factors has been overstressed. Improvements in transport technology and speed have frequently been cited as *the* reason for suburbanization and the lengthening of journeys to work. If more stress is put on the widening availability of a superior transport form, rather than on its technical development and proving, then this factor is seen to be inextricably linked with social and economic trends. One demonstration of this argument is that the essential technology of the private car was proven by about 1910, and indeed widely adopted by the Californian urban population during the 1920s, but that it was not until the late 1950s that the British population acquired sufficient personal resources to adopt the form in large numbers. Increasing affluence appears, therefore, to be the necessary condition for the growing length and complexity of commuting patterns: it enables individuals to adopt more expensive travel habits, enables governments to invest more in the transport and road networks, and permits both the private and public sectors to implement improvements in housing, generally by constructing lower-density extensions to existing urban areas.

A consideration of the factors moulding in the long term the relationships between workplaces and residences points attention to deep-seated social and economic conditions of a country. A tentative identification of the most important factors is made in the form of a diagram (figure 6.7). This conceptualization is founded on the observation that in modern industrializing nations, economic growth and change is associated with increased real incomes, a reduction in the number of hours, days and weeks at work, and a number of social trends which increase the importance and incidence of spacious homes for small households. There is, in other words, a deconcentration of the residential distribution within the urban system. Economic growth and change is also associated with pervasive changes in the organization and scale of employing units. Domestic and small-scale units of employment in retailing, manufacturing, commerce, education and the professions all tend to decline and to be replaced by fewer, larger enterprises. This trend towards the concentration of employment by sites has been compounded in its locational effects by structural changes in the economy, for the growth of tertiary activity has been largely responsible for the concentration of employment into urban centres.

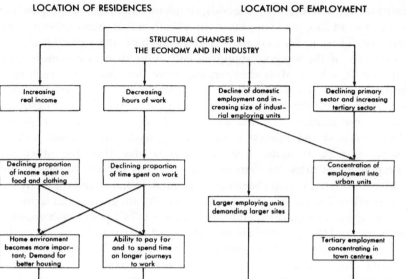

Figure 6.7 *Factors influencing the separation of homes from workplaces* (Source: *Warnes, 1972, fig. 1, p. 317*)

Spatial patterns of the journey to shop

Like journeys to work, shopping trips exhibit a wide range of characteristics. Many are short-distance, occur frequently and, being directed towards the ubiquitous newsagents and tobacconists, greengrocers, grocers and post-offices, are clearly neighbourhood trips. But others, including most of those trips made to purchase more specialized or durable goods, or food in bulk, extend beyond the residential neighbourhood and are, therefore, more commonly made by mechanically assisted means of transport. If all shopping journeys are considered, including those made on foot, their number in any urban area is probably roughly equivalent to the number of journeys to work. Those made by car, bus or train, however, constitute only about one-third to one-half of the number of mechanically assisted journeys to work (see chapter 2). This numeracy of shopping journeys, together with the considerable heterogeneity of goods, shops and shopping centres, makes the geography of shopping trips more complex than that of any other major trip purpose. There is, for example, certainly much greater variation according to the day of the week in shopping behaviour than in commuting or in journeys to school. Most shoppers use more than one centre each month, and in this period it would not be uncommon to use five centres of different size and quality; this is in great contrast to the constancy of the destination, mode and timing of work and school trips. Yet another point of distinctiveness and interest is that the aggregate intra-urban pattern of shopping trips is more dynamic through time than the other major components of urban movement. This is largely because of the dynamism and free-entrepreneurial character of the retail industry and the many small, independent units in its structure, but also relevant are demand factors such as the rapidly changing expenditure preferences of a population growing in affluence, and the spread of car ownership and use for all but neighbourhood shopping purposes. These alterations in consumer behaviour change the relative attractiveness and use of competing centres.

The greater complexity of shopping journeys is unfortunately not matched by data superior to that available on commuting. We have to rely, in fact, upon special-purpose sample surveys which are generally restricted in scope and areal coverage. Although retail and marketing geography has been a popular field of study among urban geographers (Davies, 1976a; Scott, 1970), the present authors have argued that the basic task of establishing the distribution of shops and shopping centres within towns has tended to be neglected (Warnes and Daniels, 1978, 1980). There has been much research into the more subtle variations in and influences upon shopping travel, and recent innovations and developments in urban commercial facilities attract great interest, but the everyday, humdrum web of local shopping activity excites few students. As a result it is not possible to state firmly and in any detail the nature and extent of the changes which this considerable base of urban shopping movement has been undergoing.

There are some cross-sectional surveys of consumer behaviour, including its spatial patterns, and two of the more comprehensive recent examples form the basis of this review. The most substantial was carried out in 1969 by Daws and Bruce (1971; see also Bruce, 1974) for the Building Research Station. 1,003 successful interviews were collected in ten sample districts within a four-mile radius of Watford town centre in Hertfordshire. Their questionnaire was effective in eliciting information about local, neighbourhood and town-centre shopping, and a useful range of socio-economic and demographic variables were also collected for analytical purposes. This was followed up by a travel-diary study of 1,672 people in the town (Davis and McCulloch, 1974). The second principal source is R. L. Davies' (1971) diary and home-interview study of a one per cent sample (487) of households in Coventry in 1969. This was carried out mainly for the City Planning Department to assist its formulation of a new shopping policy as part of the Structure Plan, but it was partly financed by J. Sainsbury Ltd and Marks and Spencer Ltd for their own commercial purposes. Supplementary information is drawn from the few other studies which, by focussing on the shopper rather than the use made of a single centre, produce a comprehensive or at least representative picture of intra-urban movement. These include C. J. Thomas's study in Swansea (1974) as well as surveys in Brighton, Leeds and the Outer Metropolitan Area of London (Ambrose, 1968; R. L. Davies, 1968, 1969; Hillman *et al.*, 1976).

Watford was regarded as a suitable area for a shopping study by the Building Research Station, not only because of its adjacency but also because the town and hinterland of 170,000 people shares many of the characteristics of the South East region as a whole. It has a higher than national proportion of owner-occupiers and of car-owning households, but a broadly representative demographic and occupational composition. Despite its proximity to London, in 1966 Watford Municipal Borough had relatively weak commuting ties with Greater London, for only 12.3 per cent of its workforce was employed there. A stronger functional tie existed with the hinterland, since 28.9 per cent of Watford Rural District's workforce was employed in Greater London. Its outer suburban position within the functional metropolitan area must be expected to affect and make unrepresentative shopping travel associated with high-order purchases (Daws and Bruce, 1971, p. 7). The survey was organized by selecting ten census enumeration districts, four in an outer zone, and three each in middle and inner zones. Within these enumeration districts, a random sample of electors was drawn, the sampling fraction being adjusted according to the numbers of households in the selected enumeration districts and to the zone populations that were being sampled (Daws and Bruce, p. 13). The questionnaire was addressed to the housewives of each household, or, if none was present, to the principal shopper.

A distribution map of shopping facilities and of the survey areas appears as figure 6.8. Unfortunately, the procedure used to grade the town's shopping centres was not specified. For the presentation of results, five areas were distinguished. A sample area in Oxhey, a post-war London County Council estate to the south of the town, and another in Rickmansworth, a high-status residential area to the west of the town, were separated from two other sample areas in the outer zone. The three sample areas in the middle zone were all between one and two miles of the town centre and, like the survey areas of the inner zone, were considered as one unit.

Both the household survey and the travel-diary studies collected information on a wide range of shopping characteristics, including the frequency, timing, and destinations of trips. Particular attention was given

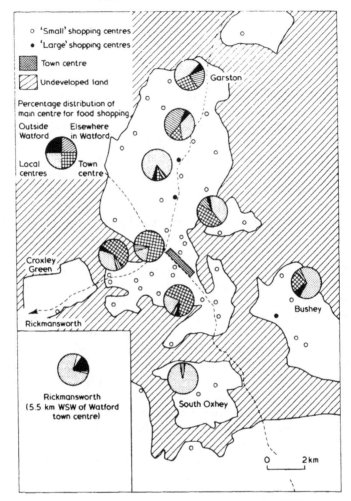

Figure 6.8 *Shopping centres and their use: Watford, 1969* (Source: Daws and Bruce, 1971, fig. 4, p. 26)

to the use of Watford town centre and to local shops, and attitudes towards shopping and shopping-centres as well as recent and anticipated changes in behaviour were also surveyed. A selection of the survey's results has been extracted and presented in a revised form in table 6.7, in which the seventeen sub-groups and five areas used for the presentation of results are also defined. In this table, the mean sample figure for fourteen characteristics is found in the first row, and each column of figures gives the per mille deviation of the sub-groups' means from the overall mean. For example, in the top left-hand corner of the table, it can be seen that respondents aged less than 35 years made 50 per mille more shopping journeys per week than the entire sample, or 4.2 journeys per week. Average per mille deviations are given for each shopping characteristic (columns) and for each sub-group (rows). The survey report also records the results which were significantly different from the expected frequencies, on the assumption that a sub-group behaved no differently to the entire sample, as established by a chi-squared test at the five-per-cent level of significance (the fullest description of these significance tests is in Daws and McCulloch, 1974, p. 27). The column and row totals of the number of significant results have also been given as another comparative measure of the variability of the characteristics and of the sub-groups.

Among the selected characteristics, the percentage visiting Watford town centre once a week or more yields the greatest number (10) of significant deviations, while the percentage of respondents shopping in Central London at least once a year yields the highest average (202) per mille deviation. More generally, it is noticeable that the characteristics with a spatial element show the greatest variation, and the variables which describe the timing and frequency of shopping trips least variation. There is, however, one exception to this generalization, because there is relatively modest variability in the use made of local centres.

Turning to the variability of the sub-groups, the row summary statistics show that none of the three age groups exhibited markedly different shopping behaviour from the total sample. The eldest group, aged 55 years or more, reported the greatest average deviation, with significant differences occurring in their avoidance of Friday or Saturday as their main shopping day, the high concentration of their shopping in the morning, the high percentage of households in which the housewife was the only shopper (which could be explained by the numeracy of single-person widow households), their above-average use of local centres for grocery shopping, and the infrequency of their visits to Central London.

As regards work status, the unemployed respondents and those in part-time employment produced very low average deviations from aggregate behaviour, although the latter sub-group yielded four significantly different frequencies. In contrast to the unemployed, they were less frequently shopping in the morning, and the part-time employed respondents also made a significantly high number of shopping trips, and significantly frequent visits to the town centre. On the other hand, full-time employed housewives (or principal shoppers) showed the fourth-highest deviation from the aggregate pattern, with significant differences in their very low frequency (1.8) of single-purpose shopping trips, the low percentage (32) of them making their most important shopping trip before noon, the low percentage buying a quarter or more of their groceries at the town centre, and the high percentage visiting Central London for shopping.

Table 6.7 Shopping behaviour in Watford, 1969

Sub-groups of sample	No. of shopping trips in a week	No. of single-purpose shopping trips	% regarding Fri. or Sat. most important for grocery shopping	% making most important shopping trip before noon	% households with housewife as the only shopper	% households where husband participated in shopping	% with main source of groceries within ½ mile
Unweighted mean	4.0	3.2	66	61	55	41	61
AGE							
under 35	+50	+31	+87*	−65	−36	+98	−131
35–54	0	−62	+7	−65	−18	−98	−82
Over 55	−50	+31	−20*	+148*	+218*	+98	+180*
WORK STATUS							
Not working	+25	+94	−13	+180*	+91	0	+16
Part-time	+75*	−31	+40	−115*	−36	−24	−16
Full-time	−25	−437*	0	−476	+54	+122	−14
SOCIAL CLASS							
I and II	−75*	−62	−13	−16	+91	−73	−180*
IIIa	+125*	−62	−7	−33	+182	−98	−16
IIIb	+50	+31	+53	−65	−109*	+73	−49
IV, V and retired	0	+31	+13	+115*	+164	+73	+230*
CARS OWNED							
2+	−75*	−62	+27	+82	+18	−220*	−230*
1	0	−31	+33	−65	−36	−49	−164
0	+25	+31	−13	+65	+200	−24	+279*
AGE OF YOUNGEST CHILD							
Less than 5	+100*	+156*	+20	+16	+36	−98	−49
5–10	0	+31	+33	+16	−164*	−49	−197*
11–15	+75	−62	+73*	−98*	−127*	−171	−197*
Over 15	−50	−62	0	−16	+146	+49	+82
Outer Watford	−75*	−125*	+7	−49	−73	+122	−394*
Middle Watford	+25	0	+20	−82	−54	+73	+82
Inner Watford	0	0	+40	+98	+127	+49	0
Oxhey	+75*	+94	−7	+98	+182	−244	+558*
Rickmansworth	−25	+94	−13	−65	+364*	−342*	−82
Significant results	7	3	3	6	5	2	9
Average 'deviation'	50	74	25	92	119	102	148
Source: Table, page	5, 24	5, 24	7, 30	8, 32	9, 34	9, 34	13, 40

Note: The mean sample figure for fourteen characteristics is found in the first row, and each column of figures gives the per mile deviation of the sub groups' means from the overall means. For example, in the top left hand corner of the Table, it can be seen that respondents aged less than 35 years made 50 per mile more shopping journeys per week than the entire sample, or 4.2 journeys per week. The survey report also recorded the results which were significantly different from the expected frequencies on the assumption that a sub-group behaved no differently to the entire

SPATIAL PATTERNS OF URBAN MOVEMENT II

Table 6.7 (Cont.)

Selected characteristics of shopping behaviour

% visiting Watford town centre once a week or more	% buying 25% of groceries at Watford town centre	% buying all or most of clothes at Watford town centre	% buying all or most of furniture at Watford town centre	% shopping in Central London at least once a year	% with main local centre within ½ mile	% shopping at main local centre more than twice a week	Significant results	Average 'deviation'
57	43	65	66	44	83	76		
+121*	+280*	−15	0	+159	−24	0	3	78
+18	−70	−46	−15	+250*	−60	−66	1	61
−71	−70	+46	+30	−364*	+60	−13	5	100
−54	0	−31	−30	−68	0	+26	1	45
+106	−47	+92*	+61	+68	−36	+26	4	55
+18	−210*	−15	+91	−250*	−24	−211*	4	142
−140*	−116	−169*	−197*	+431*	−180	−39	6	127
+121*	+116	+77	+182*	+137*	+48	−39	3	89
+121*	+116	+46	+137*	+68	+12	−52	3	70
−88	−47	+62	−15	−500*	+96*	−26	4	104
−106	0	−123*	−106	+409*	−156*	0	6	115
+18	+47	0	+15	+137	−60	−13	0	48
−18	−23	+31	+30	−227	+84*	0	2	75
+35	+70	+77	+45	+23	−36	0	2	62
−54	−139	−15	−45	+227	−72	+39	2	77
+175*	+23	+123*	+122*	+114	−132*	−13	8	108
−35	−186	−46	−15	−45	+24*	−52	0	77
−175*	−116	−46	+122*	+114	−253*	0	6	119
+18	−116	+77	+61	−23	+108*	+39	1	56
+649*	+861*	+108*	+152*	−91	+48	−145*	5	169
−386*	−651*	−169*	−91	−182	+193*	+105*	7	217
−810*	−814*	−492*	−547*	+545*	+253*	+171*	9	330
10	4	7	7 .	7	8	4		
152	190	87	96	202	94	51		
18, 52	22, 60	25, 68	26, 70	28, 74	29, 80	31, 84		

sample, as established by a chi-squared test at the five per cent level of significance. These results are indicated by asterisks. The column and row totals of the number of significant results have also been given as another comparative measure of the variability of the characteristics and of the sub-groups. The fullest description of the significance tests used in the reports is found in Daws and McCulloch, 1974, p. 27.

Source: Daws and Bruce (1971).

Among the four social-class sub-groups, as one expects the greatest departures from average behaviour were found at the two extremes, with the highest-status group consisting of the Registrar General's Social Classes I and II being the most deviant. They showed a significantly low frequency of shopping trips, significantly low use of their main local centre for groceries, and significantly low use of Watford town centre for each of the three identified types of purchase. They also had a significantly high number of shopping trips to Central London. Car ownership is often highly correlated with occupational status, but distinctive patterns of shopping behaviour were produced by the car-ownership sub-groups. There was a bigger difference in the housewives' behaviour between those from two or more car-owning households and those from one-car households, than between the latter and those in non-car-owning households. Two-plus car-owners had a low frequency of shopping, an unusually low rate of participation of husbands in shopping, and low reliance on both local centres and the town centre for grocery shopping, as well as a high frequency of visits to Central London.

The family-type sub-groups were probably too finely divided to bring out 'stage in the life cycle' differences in shopping behaviour, as table 6.7 shows that there was only modest deviation from average behaviour according to the age of the youngest child in the household. Those families with a young teenager as the youngest child were the most distinctive of the family-type groups, and indeed were the eighth most deviant sub-group with as many as eight significantly different characteristics. Friday or Saturday was regarded as unusually important for shopping by the housewives of these families, morning shopping had a low prevalence, there was low reliance on the housewife for shopping and on local shops for groceries, and a high rate of visiting Watford town centre for shopping, particularly for clothes and furniture.

Undoubtedly the most interesting findings from the Watford studies for the urban geographer are those relating to the areal differences in shopping behaviour. The deviations which have been described so far are for the most part modest in comparison to those associated with the area of residence. This is shown in the last five rows of table 6.7. The shopping characteristics returned by the respondents in Rickmansworth, Oxhey and Inner Watford were respectively the first, second, and third most deviant of all the sub-group patterns. Additionally, the shopping behaviour of Outer Watford's respondents was the sixth most deviant. Rickmansworth has a significantly high dependence on housewives and a low participation of husbands in shopping, an exceedingly low use of Watford town centre for all purposes, and a high frequency of shopping trips to Central London. Although in Rickmansworth the local main centre was more distant from the respondents than on average, a significantly higher-than-average usage was made of it.

On the other hand, to the south of the town in Oxhey, the respondents showed a significantly high frequency of shopping journeys, a high dependence on their local 'main' centre for groceries, a low use of Watford town centre except for furniture, and high accessibility and frequency of visiting their local centre. The significant differences of the shoppers' behaviour in Inner Watford were all associated with the exceptionally high use made of Watford town centre, particularly for groceries, and the correspondingly low frequency of visiting local 'main' centres. For the respondents in Outer Watford, the principal distinctive features are the low accessibility of local shops and the low frequency of shopping trips, particularly those which are single-purpose. An unexpected feature of this area's shopping is the significantly high reliance placed on the town centre for furniture shopping, although this may be related to a distinctive expenditure pattern in newly established suburban households.

These most interesting results indicate that even though in Watford, as elsewhere, there are correlations between area of residence and social class, car ownership, and stage in the life cycle, it does appear at this relatively simple level of analysis that the single most important conditioning influence on shopping behaviour is the area of residence. The importance of this variable is explained by the considerable differences in accessibility to shops and shopping centres of different levels. Proximity to large shopping centres such as that in Warford town centre stimulates their use. Similarly, the availability of a good local 'main' centre, or of other local shops, is a strong influence upon the way in which people organize their shopping and shopping travel. The evidence is sufficiently strong to allow the assertion that the local 'opportunity-set' of shopping facilities is a more important influence upon shopping behaviour than social class, age, car ownership or family type. It is rivalled only by employment status as a conditioning influence.

Clearly these conclusions call for a more elaborate multivariate analysis of the survey's data. This is not available at the present time, nor was any more detailed information on the spatial patterns of shopping trips in Watford reported. Despite the daunting volume of data involved in a simultaneous consideration of the locations and the characteristics of origins and destinations, there is manifestly a need for a more tenacious investigation of the spatial characteristics of shopping. A great deal has been revealed even by the limited tabulations of the Watford studies, and to conclude this review of the spatial patterns of shopping, some comparative evidence from other British towns is called upon.

Attention is focused again on the spatial characteristics of shopping travel – the trip distances and the use made of different centres – as the timing, frequency, mode and other aspatial characteristics were dealt with in chapter 3. All the published studies have established that shopping journeys

Table 6.8 Cumulative percentage distributions of shopping trip distances

Place and type of trip	Distance in miles						
	$-\frac{1}{2}$	-1	$-1\frac{1}{2}$	-2	-3	-4	n
1. Brighton, 1966							
Food	65.4	79.8	87.1	92.8	95.7	98.0	1395
Clothes	38.3	54.7	64.0	72.1	81.4	86.0	86
Household goods	45.8	65.4	76.6	85.0	88.8	95.3	107
Other goods	66.1	76.9	86.5	91.6	94.0	98.0	251
Services	61.8	78.7	88.8	93.3	94.4	97.8	89
2. Watford, 1969							
Meat	66	77	→	91			1003
Vegetables	66	78	→	91			1003
Bread	77	87	→	96			1003
Groceries	62	75	→	91			1003
Nearest 'main' local centre	81	90	→	→			1003
3. Coventry, 1969							
Monday	66	77	→	87			431
Friday	48	60	→	81			598
4. Broughty Ferry, Dundee: grocery supermarket, 1967							
Tuesday, 3–4 pm	→	55	→	76	→	85	
Saturday, 3–4 pm	→	37	→	55	→	66	
Friday, 7–8 pm	→	33	→	46	→	54	

Sources: Ambrose (1968), table 1, p. 328; Daws and Bruce (1971), table 13, p. 40, and table 29, p. 80; R. L. Davies (1973a), fig. 5, p. 25, and table 2, p. 19; Pocock (1968), pp. 113–15.

are characteristically very short. The fullest analysis of this aspect has been provided for the Brighton conurbation by Ambrose. The frequencies of journeys of different lengths for five shopping purposes have been recalculated as cumulative percentages in table 6.8. As he remarks, 'the fall off of journey frequency with distance is most marked in the services group and least marked in the clothes and household goods group. While the relationship between frequence and distance is an inverse one for all groups up to the $2-2\frac{1}{2}$ mile distance class, the relationship beyond that distance is less predictable . . . . There is little sign of a falling off in journey frequency between roughly $2\frac{1}{4}$ and $3\frac{3}{4}$ miles' (Ambrose, 1968, p. 329). Table 6.8 also presents comparative information for Coventry, Watford and for the customers drawn to a supermarket in the district centre of Broughty Ferry, Dundee. In Coventry, 53 per cent of all 'primary' (analogous to single-purpose) trips involved a straight-line distance of less than half a mile to the first shopping centre visited. The average trip length for convenience goods was 0.84 miles, and for durable goods 2.44 miles. There was a smaller peak of longer trips of between one and two miles, 'which reflects on motorised travel to the central area' (R. L. Davies, 1973a, p. 26). In Watford, a remarkably similar cumulative-distance distribution to that for food shopping in Brighton was found for four convenience-food commodities. The respondents purchased bread, of the four commodities, nearest to their homes. For comparative purposes, the distance distribution of the nearest 'main' local centre is also tabulated, demonstrating that even for purchasing basic foods, consumers travelled on average slightly farther than their nearest centre (Daws and Bruce, 1971, pp. 40–1, 80–1). The evidence from Dundee is not drawn from a comprehensive survey of shopping behaviour, but from Pocock's study of the utilization of stores in different parts of the urban area. He was able to relate the distance-distribution of journeys to location within the city: a branch of Woolworth's in an inner high-density area drew its customers from a much more restricted area than the branches in the city centre and in a peripheral high-status suburb. Another interesting feature of the Dundee study was the attention given to changes in the distance-distribution according to the time of the week. This was determined by sampling the customers of a grocery supermarket in Broughty Ferry during a Tuesday and a Saturday afternoon, and on a Friday evening. Table 6.8 indicates that during the first sample period, 55 per cent of the customers originated less than a mile away, but on the Friday evening a higher percentage of customers were drawn from longer distances (Pocock, 1968, pp. 113–15; for related studies of movement to particular centres, see Toyne, 1971; Downs, 1970. These differences were readily related to variations in the type of grocery shopping and the mode of shopping travel through the week.

Empirical evidence about the spatial distribution of these journeys is much more restricted, and indeed for intra-urban examples is largely confined to R. L. Davies' study of Coventry and C. J. Thomas's study of Swansea. The former took a one-per-cent sample of households throughout Coventry, and asked the principal shoppers of each household to make a diary of all movements during the course of a week in 1969. 487 usable records were collected, and these detailed 3,121 shopping trips (Davies, 1973a, pp. 16–19). Davies presents an extended series of maps of the distribution of journeys on each day of the week for convenience and

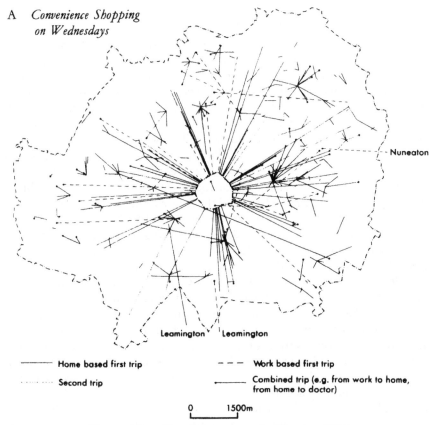

Figure 6.9a *Shopping movements in Coventry, 1972.*

durable goods. Two of these representing Wednesday movements for convenience goods, and Saturday movements for durable goods, are reproduced as figure 6.9. These illustrate the three types of shopping journey made by Coventry's residents: the short-distance movement to a local centre, the longer-distance radial movement to the town centre, and the much rarer movement to facilities beyond the town. The first type is particularly characteristic of weekday convenience-goods shopping, and generates in aggregate a pattern of locally focused movements with the expected asymmetry produced by centripetal movements having a longer mean distance than centrifugal movements. The second type, as well as dominating durable goods shopping on every day, also characterizes convenience-goods shopping towards the end of the week, notably on Friday. Weekend durables shopping is also distinctive in the number of cross-town and extra-urban journeys that are practised, and by the unusually low frequency of short journeys.

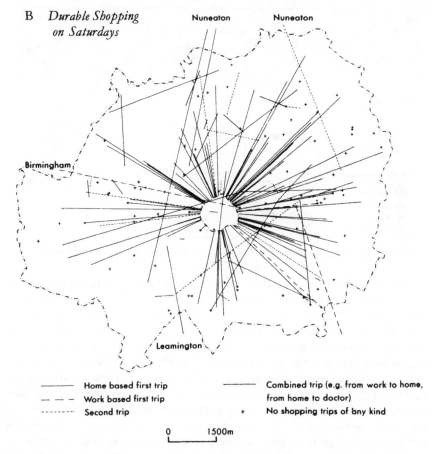

Figure 6.9b *Shopping movements in Coventry, 1972* (Source, a and b: R. L. Davies, 1973a, figs. 8a and 11b, pp. 32, 39).

The Swansea study interviewed between 83 and 100 households in twelve survey areas grouped in adjacent pairs in six contrasting locations within Greater Swansea (figure 6.10). Each pair comprised a local authority housing estate to represent low-status households, and a private residential area to represent high-status households, the intention being to control for the influence of location, or of accessibility to shopping opportunities, and to highlight the roles of social class and car ownership in generating the contrasts within each pair (C. J. Thomas, 1974, p. 102; for other studies emphasizing social-status variation in shopping behaviour, see Nader, 1969, Holly and Wheeler, 1972, and – though not entirely in an intra-urban setting – Thrope and Nader, 1967; a recent review is provided by Bowlby, 1977). Although there was some variation in the social characteristics

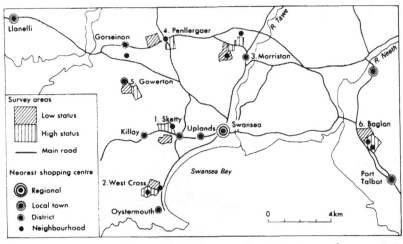

Figure 6.10 *Shopping centres and household survey district in Swansea* (Source: C. J. Thomas, 1974, p. 102)

among the six high-status and the six low-status areas, contrasting shopping behaviour within these two groups was closely related to locational factors. This is shown by the destinations of main grocery trips as presented in table 6.9. The differences between the six locations are considerable and in most cases are clearly related to the comparative distances to different types of shopping centre as shown in figure 6.10. Only in 10 of the 36 cases are the differences in the percentage points allegiance to the various centres of the two status groups greater than ten, and several of these contrasts must in part be attributed to location. For example, the greatest contrasts occur between the status groups of Penllergaer, but here the two status areas are not contiguous, and each stands in a different relationship (a) to the neighbourhood centre, this being adjacent to the high-status area; (b) to the 'local town' centre of Gorseinon, the low-status area being approximately one-third nearer than the high-status area; and (c) to the city centre, the high-status area being slightly nearer. Overall, the Swansea findings show that about one-quarter of all *main* grocery-shopping trips are to the city centre, and that the remainder are largely distributed between the nearest district centre, with a little over 45 per cent (unweighted mean) of all such trips, and the nearest neighbourhood centre, with about one-fifth of all trips. The remaining 10 per cent of main grocery trips are distributed among other district and neighbourhood centres, with the former clearly predominating (C. J. Thomas, 1974, pp. 107–12). This pattern is consistent with the pattern of convenience shopping on a Friday in Coventry and, as in that town, contrasts with a more dominantly centripetal spatial pattern of durable goods shopping. The distributions of the destinations of main clothing shopping trips are also shown in table 6.9. With the exception of shoppers from Baglan, who are most distant from Swansea's centre, and approximately a third of whom patronize Neath or Port Talbot, over 90 per cent of the respondents stated that they used the city centre for their main clothing trips. Over half of the remainder used the nearest district centre. A small amount of extra-urban movement for clothes and for furniture shopping was also recorded in Swansea.

Although Thomas's study did not consider all shopping trips, and tends to emphasize the longer-distance trips to the principal centres, it provides useful corroborative evidence for the Watford and Coventry findings. Although more detailed spatial investigations are urgently required, the evidence so far available on consumer movement in British towns points to the existence of a relatively simple hierarchy of intra-urban journeys. The Watford and Outer Metropolitan Area studies, among others, substantiate the existence of a very large number of extremely short-distance trips throughout the residential areas of the town. Above these are the less-frequent and longer-distance trips to 'main local' or neighbourhood centres, although it should be remembered that most of these are sufficiently short to be undertaken on foot. As in Coventry where these are the lowest order of trips considered by the survey, the spatial pattern generated by these trips is a scatter throughout the town of asymmetrical 'roses' of journeys focusing on the centres (R. L. Davies, 1973a, p. 15, explains that 'the questionnaire surveys were primarily concerned with female shoppers interacting with the nucleated centres'; earlier, 1969, figs 6 and 7, pp. 118–19, he showed the same spatial pattern to exist in two contrasting suburbs of Leeds: Middleton and Street Lane). A third order and spatial pattern of trips is also recognizable, and consists of trips made once a week or less frequently to the first-order shopping centre of the urban region, which produce not only a distinctive radial pattern but also the distinctive temporal and modal characteristics of more specialized shopping trips.

Without detailed survey evidence from the larger cities and conurbations of the country, it is only possible to speculate about the way in which these patterns are modified or extended in more complex shopping environments. R. A. Day's study (1973) of consumer behaviour in Crawley New Town includes some suggestive incidental comments about the complex distribution of destinations, even for food shopping, produced by the relatively low order of the town centre and the attractiveness of alternative places. It cannot be untypical of shoppers' behaviour in the outer suburban areas of the largest British cities to find a majority of consumers buying most of their clothes outside the town. From Crawley, 63 per cent of consumers went elsewhere, very frequently shopping on multipurpose journeys, and they divided their custom between the nearby centre at Horsham, the sub-regional level shopping centres of Croydon and Brighton, and Central London.

Table 6.9 Destinations of main grocery and main clothing shopping trips in Swansea, c. 1972

Respondents' area of residence		Percentage distribution						Sample size
		Swansea city centre	Local town centres	Nearest district centre	Other district centres	Nearest neighbour-hood centre	Other neighbour-hood centres	

1. Main grocery shopping trips

Sketty	low status	47	0	19	4	27	1	83
	high status	26	0	20	16	32	5	89
West Cross	l.s.	33	0	33	0	33	0	91
	h.s.	47	0	39	0	14	0	88
Morriston	l.s.	18	0	64	2	11	2	86
	h.s.	18	0	80	0	0	1	87
Penllergaer	l.s.	19	0	72	4	6	0	86
	h.s.	26	2	33	19	20	0	90
Gowerton	l.s.	12	0	74	11	1	1	100
	h.s.	31	3	58	6	1	1	100
Baglan	l.s.	5	38 (PT)	9 (N)	0	44	4	93
	h.s.	11	31	0	2	54	3	86

2. *Main clothes shopping trips*

					Other regional centre	Mail-order	
Sketty	l.s.	92	0	0	3	0	83
	h.s.	95	1	0	0	1	89
West Cross	l.s.	92	0	5	2	2	91
	h.s.	86	0	14	0	0	88
Morriston	l.s.	92	1	7	0	0	86
	h.s.	92	0	5	0	3	87
Penllergaer	l.s.	85	2	11	2	0	86
	h.s.	93	0	6	0	2	90
Gwerton	l.s.	93	0	1	5	0	100
	h.s.	94	0	1	1	3	100
Baglan	l.s.	50	15 (PT)	29 (N)	0	4	93
	h.s.	63	11	18	3	0	86

Source: C. J. Thomas (1974), table 1, p. 104; table 4, p. 108; table 6, p. 113.
Notes: PT – Port Talbot; N – Neath.

While, therefore, we are far from establishing a clear picture of the full complexity of consumer travel, especially in larger settlements, it is believed that the principal features of these movements can be discerned. They provide evidence of an ordered spatial structure which reflects, not only the very strong influence of the friction of distance on shopping travel, but also the effect of the other principles of personal movement that were described in the previous chapter, such as the temporal ordering of travel and the existence of relationships between journey distance and the mode and frequency of trips. A hierarchy of trip characteristics has been recognized, although there is no evidence to suggest the existence of the special case of a rigid, stepped, hierarchy. We find, in fact, an intra-urban distribution of shopping journey distances and characteristics that does not conflict in any way, and which provides broad support for, the features of consumer travel predicted by Davies, and others' theoretical developments.

Although it cannot be denied that our descriptions and generalizations concerning the spatial patterns of urban personal movement have not progressed beyond a relatively simple level of analysis, this is in part inevitable given the inadequate or inappropriate nature of much of the evidence. Their further development is essential, and this will demand travel surveys with a more elaborate and finely specified spatial content, of kinds that have been suggested at various points in the chapter. More explicit evidence on the spatial patterns of urban personal movements will be awaited with great interest.

7
Analysis and prediction of travel patterns

The discerning reader will have noticed that much of what has been discussed so far in this book has been presented within the framework of a static system. This has been necessary in order to bring out as clearly as possible the salient features of the patterns of urban movement. But it is important not to lose sight of the dynamic character of urban movement; patterns are in a constent state of flux in response to changes in the urban transport system or because of changes in the socio-economic characteristics of the population. The need to understand and predict the causes of change in urban travel should have been apparent in chapter 1, where it emerged that the net effect of the new transport technology described has been to loosen the fabric of cities, i.e. to lower residential densities, to create more dispersed patterns of urban industry, and to generate less centralization of retailing and other services. At the same time there have been great improvements in the speed of rail-based transport and the amount of motorway standard roads available for movement within the larger cities. This has emphasized even further the accessibility advantages of particular locations within cities, especially the CBD and certain suburban nuclei with good transport services. Therefore, there has been an increase in the length of trips within urban areas, particularly for the journey to work and social/recreational trips, while, conversely, some shopping and school trips have become shorter (see chapter 3). But common to these divergent trends has been the way in which public transport, which was so important in the nineteenth century and the early decades of this century, has been rapidly supplanted by private transport, both in terms of the structure which it has imparted on trip making in urban areas and its effect on the patterns of ownership of private vehicles, especially the car. It is therefore useful at this point to consider briefly some of these changes.

Growth of vehicle and passenger movements/volume since 1900

At the outset, it is difficult to obtain time series data over a sufficiently long period for the changes in the number of alternative forms of transport in cities. Most sources such as land use/transport studies or various government reports are difficult to compare because of the differences in the objectives and terms of reference of individual reports; differences in the classifications of data; or differences in the methods used to collect travel information, and the varying length of time series. One of the more useful sources which can be used to demonstrate the major changes in the structure of movement in Britain as a whole – and it seems reasonable to infer that this applies to urban areas in particular – is the information published annually by the British Road Federation (1973, etc.). But these data are incomplete in that they exclude rail transport with the exception of trams, which travelled along highways although guided by rails. It is therefore necessary to supplement the data with statistics on passenger transport published regularly by the Department of the Environment (1973, etc.).

In 1904 there were just 18,000 vehicles on Britain's roads (figure 7.1). By 1975 this had increased to 17.5 million vehicles, following particularly rapid growth in numbers during the period since the last war when the number of vehicles increased to more than five times the level which existed in 1946. This growth has not been evenly distributed between each transport mode; the only group of vehicles which have consistently increased in numbers at the same rate as the overall total are private cars and vans. They have increased their numbers by almost six times since 1946; from 2.3 million to 13.7 million in 1975. Public transport vehicles, such as the bus, initially started to increase at much the same rate as the private car but, as figure 7.1 shows, subsequent growth has been very slow and there have even been decreases in the number of motorcycles and trams. More recently, however, as the effects of higher energy costs are felt by more of the population, there has been an increase in the number of motorcycles, scooters and mopeds, some of which can travel more than 75 miles on one gallon of petrol. They also offer some advantages over other traffic in congested city streets. In 1975 their numbers increased by 120,000 to approximately 1.2 million. This is the highest annual total since 1968 but it is still 0.5 million below the levels reached in the early 1960s. The number of buses and coaches has changed little since 1950, but there is some sign of a downward trend which is symptomatic of the stranglehold of private transport, especially the car, on movement in Britain and its cities at the present time. The only other group of vehicles to continue increasing in number steadily since 1904 are goods vehicles, although the rate of growth has been much slower than for cars. Even so, their numbers have increased by three times since 1946 along with the capacity of individual vehicles. The pattern of urban growth described (chapter 1) has ensured a continuing and expanding role for goods vehicles, which can compete very effectively with, and provide access to areas not served by, the freight-transport facilities provided by the railways. Indeed

ANALYSIS AND PREDICTION OF TRAVEL PATTERNS

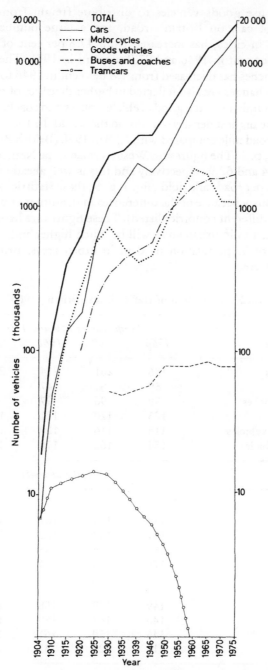

Figure 7.1 *Growth in the number of road vehicles: Great Britain 1904–75*

the railways use goods vehicles to distribute freight from stations. The proportion of cars on Britain's road, and a large number of them are concentrated in cities, has increased from 47.8 per cent of all vehicles in 1904 to 57 per cent in 1946 and 77.9 per cent in 1971. The proportion of buses and coaches has decreased from 1.8 per cent in 1946 to 0.5 per cent in 1971. These changes are also reflected in higher densities of traffic so that, in terms of the number of cars, goods vehicles, buses and coaches per road mile, Britain has the highest density of traffic in the world. In 1965 there were 52.8 vehicles per road mile compared with 63.8 in 1970 (British Road Federation, 1973, table 5, p. 6. The figures for West Germany, the Netherlands and Italy are 60.2, 60.4 and 62.9 respectively and this is well ahead of the USA with 29.1 vehicles per road mile (ibid., fig. 1, p. 9; these statistics should be interpreted cautiously because of the differences in definition of what constitutes a road in the different countries listed). These figures are based on total road mileage; densities in urban areas will be much higher and, with more than 80 per cent of its population resident in urban areas, Britain is amongst the worst affected.

Table 7.1 Growth of traffic: Great Britain, 1960-70

Types of vehicle	Traffic indices (1960 = 100) all roads				
	1966	1967	1968	1969	1970
Cars and taxis	186	201	214	222	247
Motorcycles	59	51	47	41	39
Buses and coaches	99	96	97	97	92
Light vans	123	120	121	123	129
Other goods vehicles	118	116	120	119	121
All motor vehicles	154	162	170	174	184
Class of road	All motor vehicles				
Urban					
Trunk	158	160	161	169	175
Class I	151	157	162	166	175
Class II	156	164	165	165	176
Class III	176	176	174	183	194
Rural					
Trunk	149	150	158	161	167
Class I	140	147	158	160	167
Class II	146	150	161	171	184
Class III	116	125	133	136	140

Source: British Road Federation (1973), table 7, p. 8.

ANALYSIS AND PREDICTION OF TRAVEL PATTERNS 179

The volume of traffic has clearly increased in parallel with the rising number of cars (table 7.1). Taking 1960 as 100, the volume of cars and taxis on all roads has more than doubled, compared with a decrease for motorcycles, buses and coaches. On urban trunk and classified roads the volume of all types of motor vehicle movement has almost doubled since 1960, and this is slightly ahead of the equivalent changes on rural roads. Other indices of the rapid rates of change in road transport since the war are the number of licence-holders, which have increased from 5.9 million in 1951 to 16.2 million in 1971, while private car mileage (including taxis) has increased from 20.7 million miles to 108.1 million miles, between 1953 and 1971 (Department of Environment, 1973, table 47, p. 33). This represented 77 per cent of all vehicles miles in 1971, compared with 40 per cent in 1953. Goods transport increased by 57 per cent from 1953 to 1971 (in million ton miles) and almost all of this was accounted for by road goods transport, which increased by 260 per cent.

Probably the most significant feature of urban travel since the war has been the steady decline in the number of passengers using public transport (table 7.2). The number of passenger journeys by rail refers to British Rail and London Transport in the London Transport area; this roughly corresponds with the built-up area. The statistics clearly demonstrate a more or less continuous decline in the number of journeys between 1956 and

Table 7.2 Trends in public transport use in main urban areas: Great Britain, 1956–71

	Passenger journeys (millions)					
	Rail			Bus		
	B.R.	London Transport Area L.T. Railways	Total	Stage services Main urban areas	Other areas	London Transport
1956	491	678	1,169	9,180	5,948	3,220
1961	500	675	1,175	7,764	5,213	2,626
1966	403	667	1,070	6,332	4,234	1,978
1971	364	654	1,018	4,992	3,121	1,480
% change 1956–71	−25.8	−3.5	−12.9	−45.6	−47.5	−54.0

Sources: 1956: Ministry of Transport (1968), Passenger Transport in Great Britain, 1966. 1961–71: Department of Environment (1973), Passenger Transport in Great Britain, 1971, table 27, p. 17; tables 38, 39, p. 26.

1971, and this is particularly true of British Rail which was carrying 25.8 per cent fewer passengers in 1971 than in 1956, although these figures do exclude journeys which cross the London Transport Area boundary. This probably leads to an overstatement of the rate of decrease in passenger journeys with destinations in the central area, particularly commuter trips from the Outer Metropolitan Area and long-distance business and other trips on inter-city services. Even so, the figures clearly suggest that for intra-urban journeys British Rail now has a less important role than the private rail companies had in the last quarter of the nineteenth century, but for certain trips, such as the journey to work, it remains pre-eminent in London, if not in the other conurbations. Passenger trips on the underground (London Transport Railways) have only decreased by 3.5 per cent since 1956 and this is probably the result of better network coverage of the urban area, particularly the inner areas, and the larger number of stations at more frequent intervals than those provided by British Rail. Overall, therefore, passenger journeys in the London Transport Area have decreased by 12.9 per cent between 1956 and 1971.

Competition from the private car seems, however, to affect road-based public transport more severely than the railways (see table 7.2). Patronage of bus stage services in our main urban areas has declined by 45.6 per cent since 1956, a trend which is only slightly better than in other (mainly rural) areas. London Transport buses have suffered particularly badly, with passenger journeys in 1971 at less than half the level in 1956. These trends have not been paralleled by a reduction in vehicle miles, so that stage services in the main urban areas still covered 665 million miles in 1971 compared to 908 million miles in 1956, a decrease of 26.8 per cent (DoE, 1973; the statistics are based on traffic counts: the category 'cars and trains' includes three-wheeled vehicles but excludes all vans, whether taxed for private or commercial use). The equivalent figure for London transport bus routes is 49.1 per cent. Therefore, the changes in the volume of movement on buses has been accompanied by a reduction in the number of passenger journeys per vehicle mile, and this means that services are also becoming less economic and steadily more difficult to continue. In the main urban areas passenger journeys per vehicle mile numbered 10.11 in 1956 but this had declined to 7.51 in 1971. The decline has been less drastic in London, which was already operating at 8.28 in 1956 and by 1971 had reached a very similar position, 7.47 passenger journeys per vehicle mile, to that which existed in other urban areas.

Typical of one of the main urban areas outside London is Coventry where the number of passengers conveyed by Coventry Corporation buses fell by 4.2 per cent in 1967/8 compared with the previous year (City of Coventry, 1972, table 10.01, p. 11.6). In 1971/2 the decline was 8.3 per cent and there is no sign of a halt in this downward trend. Indeed, since 1950 the use of buses has declined in real terms by 50 per cent despite an increase in the population of the urban area from 265,000 to 335,000 in 1971/2. Despite these changes the route mileage operated has only decreased by 2.7 per cent between 1966 and 1971 even though the number of passengers carried fell by 10 per cent. A substantial reduction in route mileage of 8.3 per cent was not made until June 1971. Unfortunately, bus operators seem to reduce services well after the fall-off in passengers has in fact taken place, so that uneconomic services must be cross-subsidized by more profitable routes or by subsequent fare increases designed to recoup earlier losses. Consequently, some fare increases may be higher than necessary and will cause further decreases in the number of passenger journeys, and this, in turn, may cause additional routes to encounter operational difficulties. The government has recently been pressing every major city to charge higher fares for its bus services, and as a result the number of passengers is continuing to fall. A 35-per-cent increase in fares in London in March 1972 resulted in a 6.3-per-cent decrease in the number of passengers; a 15-per-cent increase in Greater Manchester created a 13-per-cent decrease; in Greater Glasgow a 58-per-cent increase led to a 10-per-cent decrease; in Tyne and Wear the figures are 30 per cent and 5 per cent respectively (City of Coventry, 1972, p. 11.11).

It could be, however, that we are exaggerating the significance of the changing emphases in intra-urban movement which accompany the continuing development of modern transport technology and increasing affluence (Thomson, 1974, pp. 25-9). The growth of population is probably the main generator of more person and vehicle journeys in cities, but this is effectively counterbalanced by the reduction in frequency of shopping trips, for example because there are better storage facilities available in the home to keep goods fresh, etc. Hence between 1952 and 1966 Londoners showed a decrease in the number of vehicle trips from 13,799,000 in 1952 to 12,490,000 in 1966. The population of the LCC area also decreased by approximately the same amount, 10 per cent, so that this change may not itself be remarkable; more significant is that there was an extensive transfer from public to private transport (Thomson, 1974, table 1, p. 26). This transfer is significant, not only because of its consequences for public transport operation and economics but also because it has allowed substantial increases in the distances travelled by urban residents on individual trips.

Analysis and prediction of travel patterns

In order for transport planning to perform a useful role within the overall process of urban planning it must be possible to monitor and evaluate the changes created by the vehicle-ownership or public transport patronage outlined above. In a situation of continual change there would be very limited merit in planning for urban transport without knowing the characteristics of existing travel patterns and how they would be modified by the insertion of new suburban railways into a city's public-transport network, for example. It would also be difficult to justify major transport investments if their effects on travel and the community were not assessed through surveys and analyses which might indicate whether other similar, or alternative, schemes would be appropriate elsewhere in the city. Investment can then be carefully planned so that the changes which will inevitably occur can be pointed in the direction which most closely represents the policies adopted by individual cities. Transportation studies have led the way in fulfilling these objectives, perhaps because they deal with the most easily quantifiable of the complex relationships which come together to form the urban system. The very high levels of investment in urban and inter-urban highways in the United States during the 1940s and 1950s created an obvious need to evaluate their effects on urban and other traffic. Thus the initiative for transportation studies was taken in the early 1940s by the US Bureau of Public Roads, which became interested in understanding traffic flows and developed a method of simulating daily traffic patterns on the basis of sample data provided by households (Catanese, 1972, p. 2). Similar surveys were made of goods and taxi establishments and the foundations were laid for origin–destination surveys of the kind which have formed the framework for most of the subsequent transportation studies in British cities. This work indicated that there were relationships between traffic flow and the social economic characteristics of households and establishments which could be measured and expressed mathematically. These were first expressed concisely in the *Detroit Metropolitan Area Traffic Study* (1956) and developed more extensively, and with improved methodology, in the *Chicago Area Transportation Study* (1962). These two studies had a far-ranging influence on subsequent transportation studies in the United States and, later, during the 1960s in Britain. There have been a number of changes in the approach and techniques used but the basic methodology has remained remarkably consistent right up to the present time. Transportation studies have come to play a crucial part in the overall task of transport planning and the preparation of policies for urban growth as well as for transport. This should become apparent during the remainder of this chapter, which is concluded with some general comments on the value of transportation-studies for urban transport planning.

The transportation planning process

Transportation studies are made up of a series of stages, all of which revolve around a basic sequence of trip generation, trip distribution, modal split and trip assignment (figure 7.2). These could be described as the 'core' of the

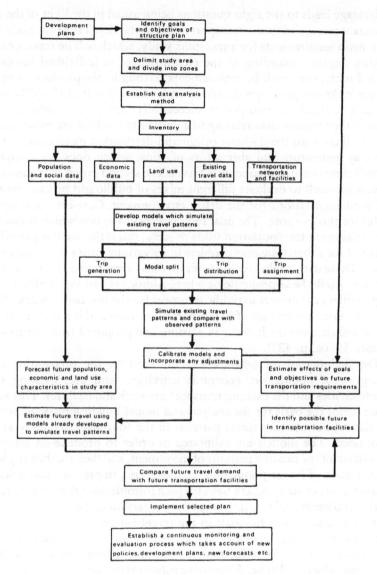

Figure 7.2 *Main stages in the transportation study process*

urban transportation planning process, but before such work can begin it is necessary to identify the goals and objectives of the study within the context of the transport system envisaged for the city in twenty or thirty years time. This will affect the design and planning of the study and, because transportation studies can take several years to complete, it is crucial that this initial stage leads to the right questions being asked in the light of the most important issues and future transport needs of the city. There are likely to be three main requirements for a transport study, which will be represented to varying degrees according to the circumstances in individual towns and cities. Firstly, there will be requirements relating to the production of data for use in the design of specific transport facilities (Lamb, 1970a). There are the most difficult to incorporate successfully into a transportation study because they require data relating to a particular mode of travel or land use that conflicts with the 24-hour origin-and-destination matrix, using traffic zones as generators and abstractors of trips, which does not adequately identify directional components in travel. The second group of requirements reflects the need to evaluate different mixes of public and private transport and their implications for the urban environment. Cost-benefit analysis is useful for this purpose. The final requirement is the one which is easiest to satisfy using the transportation study process; this is the need to quantify the extent of the urban transport problem in its various guises at the time of the study. These data can be used to indicate those areas in the city where traffic restraint would be appropriate or where additional parking facilities would help to direct car drivers to public transport for the journey to work. This is the most open-ended part of a transportation study and is a useful indicator of the imbalances which occur in existing and proposed land-use structure (Lamb, 1970a, p. 423).

The delimitation of objectives is followed by the preparation of an inventory of population, economic activities, land use, existing travel characteristics and the existing transport network and facilities. This survey phase forms the basis for the analysis and model-building phase which uses the data about existing travel patterns in the four-stage sequence already mentioned. The models are calibrated in order to produce an acceptable simulation of the existing patterns of movement, and they are then applied to the forecasts of future population, land use, etc., to produce predictions of future travel demand and as a basis for plan formulation which can be related to the availability of future transportation facilities. All this assumes that there is an underlying rationale in the travel behaviour of urban residents and workers which provides an adequate basis for forecasting future behaviour and the introduction of solutions which can deal with the problems which will arise. Alternative solutions can be tested and evaluated with respect to cost and other implications before a final transportation plan is selected for implementation. Ideally, as the proposals in the transportation plan are put into practice they should be the subject of continuous monitoring and evaluation that takes into account the effects of new policies, changes in population forecasts or new transport technology.

Transportation survey data are usually collected for a set of origin-and-destination zones delimited by a cordon line around the study area which contains all the important movements under consideration (for a good discussion of the limitations of these zones for travel surveys see Oi and Shuldiner, 1962, pp. 23–71). The latter does not usually coincide with administrative boundaries and is used as the framework for describing trips as internal (i.e. origin and destination inside the cordon); external–internal and internal–external (both of these trips are sometimes sub-classified into those by residents and non-residents); through trips, non-stopping, and through trips, stopping; and intra-zonal and inter-zonal trips. Information about travel within and through the urban area is obtained either by eliciting responses from travellers at their home addresses or by intercepting them during the course of a trip. Interviews or postal questionnaires are used to obtain the data required (for a more detailed discussion, see Bureau of Public Roads, 1954; Taylor, 1965). The kind of information collected usually includes reference to where each trip begins (origin) and ends (destination); how the trip is undertaken, or travel mode (e.g. car, bus); who makes the trip (e.g. by sex, status in household, socio-economic group or occupation); the purpose of the trip, or why it is made (e.g. to get to school or work); the time at which the trip is made; the cost of the trip; and what facilities are used *en route* (e.g. car parks, bus or railway stations, route choices). Some surveys will go on to establish reasons for choosing a particular mode, route or facilities, and try to identify the alternative trips which urban travelled would make if changes were made to the transport system, the cost of using it, or certain vehicles were banned for all or part of the time from selected areas of the city. This behavioural approach to transportation planning has become more important in recent years. The data produced can be aggregated or disaggregated, according to the particular objectives of the transportation study, to provide answers to a wide range of alternative proposals. These include: assessment of the travel and economic effects of alternative land-use configurations and transport proposals; estimation of the effects of different levels of public transport provision on the amount of travel by private and public transport modes; determination of the number of trips generated by different types of land use; economic assessment of the benefits to be expected from alternative transport projects; an estimate of the use of transport interchanges, or the best location for an urban motorway or ring road.

It is not intended to discuss the transportation planning process in detail in this chapter; this is more than adequately discussed in other places (see, e.g., Bruton, 1970; Creighton, 1970; Hutchinson, 1974; Lane *et al.*, 1971). It seems useful, however, to consider some of the main features of the four 'core' stages in the process which make a substantial contribution to an understanding and prediction of the spatial pattern of movement in cities. All transportation models and the stages within them are concerned with explaining and predicting the changes which will take place in transportation demand, particularly if the circumstances of city structure, population or socio-economic structure are changing through time. The accuracy of such calculations depends to a large degree on the accuracy of the four 'core' procedures and models, most of which make assumptions about the behaviour of transport users which have only recently begun to be analysed in detail. The models used have therefore been modified as our understanding of urban travel behaviour has improved. Trip generation, which is concerned with the relationship between the number of trips into and out of a zone and various socio-economic or land-use parameters, employs multiple linear regression analysis and, more recently, category analysis. Generated trips may be analysed by purpose and/or mode, using households or zones for the application of the models. Once trip generation rates have been calculated it is assumed that these trips will require destinations, and trip distribution is used to allocate trips between zones. A sequence through growth factor, gravity model and intervening-opportunities model represents the statistical evolution of this part of the transport sub-model. Various methods are also available for assigning the distributed trips to the appropriate network, ranging from the all-or-nothing method where trips between zones are allocated to the shortest route, to capacity-restraint models in which the effects of congestion are taken into account. The fourth sub-model deals with modal split (the division of trips between alternative modes) and this can be undertaken at any of the three other stages in the transport model. Ideally, it should be carried out at the generation and assignment stages, thus eliminating the disadvantages of confining it to one or the other. Once these models have been calibrated and are considered to simulate adequately the performance of the existing transport system, the transport model is ready for use in the forecasting phase. Some of the characteristics of the transport sub-models will now be discussed in more detail.

Trip generation

The movements generated by urban activities and land uses are the root cause of many urban transportation problems. Most of the activities which attract or produce trips cannot be spatially correlated, and hence people and goods must move from one point in the city to another; from home to work, from work to business appointment, from home to shop, from factory to warehouse or from warehouse to retailer or manufacturer. The range of possibilities is endless and the resulting web of travel is complex because it varies both in space and time (as demonstrated in Chapters 5 and 6). Some general relationships between changes in urban form and transport technology have been outlined in Chapter 1 and their net effect has been to increase the number of trips generated by urban land uses. During the trip-generation stage of a transportation study an effort is made to rationalize the apparently chaotic patterns of travel by developing mathematical models which reproduce the existing volumes of travel demand between locations revealed by the field surveys, and to forecast these patterns for the end of the study period, perhaps 1991 or 2001. Other inputs to the generation process are land-use, economic and demographic data which are used to explain the variations in numbers of trips generated by various land uses at different times and places in the city. At this point it is worth being clear about the terminology being used (Lamb, 1970b; Douglas and Lewis, 1970-1). A 'production end' is the home end of any trip that has one end at the home, or the origin of a trip which neither ends or begins at the home. An 'attraction end' is the non-home end of any trip which has one end at the home; it can also be the destination of a trip with neither end home-based, e.g. a trip from an office to a meeting at another office. The 'trip' is any movement between a production end and an attraction end made up of a single or any combination of modes for a specific purpose. The differentiation between generations and attractions reflects the emphasis in transportation studies on the household as the main determinant of traffic generation, so that the terms 'origin' and 'destination' (the start and end of trip respectively) are not really appropriate. About 80 per cent of all travel in an urban area either starts or ends at home, and household characteristics can be measured and forecasted more easily than non-household characteristics.

Information is required from very different sources for generations as opposed to attractions. For trip generation, data is required about household characteristics such as income, household size, age distribution, and the residential density of the area in which the household is situated. The trip-making characteristics of the household are also determined, including journey purposes, travel modes used, and origin and destination for a specified time period, usually a particular day of the week or the day immediately preceding the interview or questionnaire. Information about the attraction end of journeys is also required and, because it is not usually possible to question employees at individual establishments about their trip-making patterns, measures such as employment density or proportion of floor space in various uses are brought in. Except for places or work, these measures are really far from being satisfactory indicators of the trips arriving at and leaving land uses such as entertainment, shopping centres or recreational facilities.

Three main groups of variables influence trip production and trip attraction. The pattern of land use and development will be reflected in the mix of journey purposes and rates of trip generation which, in turn, will be influenced by land-use densities and spatial relationships between developments. Accessibility is the second variable, and this is directly dependent on transport systems' characteristics and the spatial separation of households from the opportunities they wish to consume. The third group of variables affecting trip generation and attraction are socio-economic parameters such as household size or car ownership, which influence the potential and the desire of households to make trips.

Therefore, the purpose of trip-generation analysis is to identify variables which have a significant and separate effect on trip generation. The most common technique used for this purpose is multiple linear regression, which is usually applied to aggregated household data for each of the zones in the study area in an attempt to explain variations in trip rates or number of trips using different modes. The general equation is as follows:

$$Y = k + b_1 x_1 + b_2 x_2 + \ldots \ldots \ldots b_n x_p$$

Where Y = number of trips (vehicular or person trips)
$x_1 - x_p$ = independent variables such as number of cars in each household, household size, etc.
$b_1 - b_n$ = coefficients of the respective independent variables
k = constant included to represent the proportion of the value of Y which is not explained by the independent variables.

Such an equation can be used to calculate trip generation in both the origin and destination zones of trips. For the results to be useful the sample required is large, while the forecasting ability of these models is very poor (Douglas, 1973). This may be a product of the basic assumptions which are used in any regression analysis but which are difficult to satisfy in practice. Firstly, a linear relationship is assumed to exist between the dependent-variable trip-generation rate and the independent variables, such as family size or housing density, used to explain the variation in the dependent variable. Secondly, it is assumed that all the variables are independent but in practice, frequently used variables such as income and car ownership are closely related while residential density is also likely to be negatively correlated with car ownership. Ideally, highly correlated independent variables should be excluded from the equation. Thirdly, there is an assumption that all the data for each variable are normally distributed; skewed distributions for variables such as income or household size are usual, however, and it is important to transform them before adding them to the analysis. Fourthly, the data for each variable should be continuous and it is undesirable to introduce car ownership in the form 0, 1, 2 or 3 + cars or household size as 2, 3 or 4 + persons. All these assumptions are rarely, if ever, reconciled satisfactorily and this limits the forecasting ability of the regression model. But such models are easily used and accessible in the form of package computer programmes and, provided that the limitations are understood and allowed for during analysis, it is likely that this model will be used for some time ahead.

There is an alternative technique for trip generation known as users' classification or category analysis (Winston, 1967). This assumes that the generation rates associated with particular groups or categories of the population remain constant. If three variables – car ownership, income and household structure for example – are considered important in trip generation, they are reduced to categories representing ranges, and trip-generation rates are estimated for each. The number of people in each category in some future year is then used to calculate the number of trips generated. Matrices are produced of the average number of trips per household per day classified against a set of variables such as income or vehicle ownership. These are calculated from a total set of survey data obtained for the complete transportation study area. The Greater London Council (1968) calculated average trip rates for 243 separate categories, or 81 categories by three journey purposes. The 81 categories comprised three levels of employed residents per household, three residential density groups, three ownership groups and three income groups.

Wootton and Pick (1967) classify the factors affecting trip generation by households into two groups: internal influences, which are disposable income, car ownership, family structure (particularly number in employment), and family size; and external influences, which are rail and bus accessibility, particularly the latter because the former is primarily a factor affecting trip generation for the journey to work (see also Wilson, 1974; South East Lancashire, 1971). The trips generated by the various combinations of internal factors are further divided into modes and purposes of travel, i.e. modes are drivers of cars or motorcycles, public transport passengers, other passengers (mostly car passengers); purposes are work, business, education, shopping, and social non-home-based. The internal and external influences on trip generation are grouped into 108 combinations of household types and are associated with 18 mode-and-purpose combinations.

Zonal characteristics can, then, be taken from the total data set, depending on the particular variables relevant to the zone or area concerned. The important difference between regression and category analysis is that the household is the basic unit for the latter, rather than the zone (Douglas, 1973). The basic premise is that each household, which has a specific set of characteristics, makes an average number of trips per day. If the household moves into a different category, by increasing its income for example, it then takes on the travel habits of households in that new category. This assumption is likely to operate well at one point in time but is less reliable over longer periods of time. Despite this, the forecasting ability of category analysis is better than for regression analysis, although its major weaknesses are that it is not possible to test the significance of individual variables, that it is very difficult to add new variables, and that defining categories can also be troublesome. It also requires a large set of sample data, particularly if a very fine set of categories is to be used and if observations are to appear in all of them. For these reasons, multiple regression analysis of trip generation at household level is more common.

The way in which the data from category analysis can be used to calculate trip generation for a specific zone can be illustrated by using some data for Dublin (An Foras Forbartha, 1973, pp. 13.11–13.13):

Base year generation:
Let us assume that vehicle-ownership levels by socio-economic level and medium residential density (20–39 persons/acre) are as follows:

	Vehicle-ownership level	Socio-economic level (%)						
		1	2	3	4	5	6	7
Medium	0	92	58	40	31	29	27	25
residential	1	8	41	57	60	56	49	40
density	2+	—	1	3	9	15	24	35

and the total person trip rate per household per day, by socio-economic level and vehicle ownership (excluding walking and cycling) is:

Socio-economic level	Vehicle ownership		
	0	1	2+
1	2.00	4.40	—
2	2.87	5.50	8.50
3	3.78	6.60	9.70
4	4.85	7.70	11.00
.	.	.	.
.	.	.	.
n	.	.	.

These trips can be reduced to purpose and mode according to the level of household vehicle ownership. Only home-to-work and home-to-shop examples are given here (PT = public transport):

Purpose	Level of vehicle ownership of households					
	0		1		2+	
	% trips	% PT	% trips	% PT	% trips	% PT
Home to work	40.90	80.99	30.10	30.43	23.22	22.50
Home to shop	10.56	88.46	8.44	42.07	10.37	18.50

ANALYSIS AND PREDICTION OF TRAVEL PATTERNS 191

And, finally, vehicle-occupancy rates for home-to-work trips are 1.19 for one-vehicle households, 1.14 for 2+-vehicle households. For home-to-shop trips the equivalent figures are 1.67 and 1.47.

To estimate the total number of trips generated by a zone with 200 households with socio-economic level 4 and medium residential density, the following calculations are necessary:
(a) Household vehicle ownership – the table above shows that there will be 31 per cent with '0' vehicles = 62 households; 60 per cent with '1' = 120 households; and 9 per cent with '2+' vehicles = 18 households.
(b) Total trips generated:
from '0'-vehicle households   = 62 × 4.85 = 301
from '1'-vehicle households   = 120 × 7.70 = 924
from '2+'-vehicle households = 18 × 11.00 = 192.
Hence total trips per day generated by the zone = 1,417.

These trips can then be divided according to purpose and mode, e.g. for home-to-work trips:
(a) from '0'-vehicle households – 40.90 per cent of 301 = 123 trips, of which 80.90 per cent will be by public transport. Hence, there will be 100 public-transport and 23 private-transport trips from households not owning a vehicle.
(b) from '1'-vehicle households – 30.10 per cent of 924 = 281 trips, of which 30.43 per cent will be by public transport, which means that there will be 85 public-transport and 196 private-transport trips.

The total number of public-transport trips from these two categories of households will be 185, and the total for private transport will be 219.

This process can be repeated for other socio-economic levels and residential densities, as well as for other purposes.

Once the base-year model has been calibrated and tested it is possible to produce the design- or forecast-year trip-generation model. The forecast model is similar to that used for the base year except that it uses estimated activity levels in each zone instead of the observed levels from the sample survey. The output from the design-year model will then show where, for example, the pressure on parking facilities is likely to be focused; the changes in the direction and volume of movement as a result of changes in employment levels; or the likely modal split and its consequences for the effectiveness and efficiency of public transport.

Trip generation by non-residential land use

While a great deal is known about trip-generating characteristics of households, until recently relatively little has been known about the relationship between non-residential land use and trip generation. Most transportation studies are generally weak on commercial-vehicle trip-generation data, for example, and they assume that the land-use/trip-generation relationship is linear for all activities and that urban land uses have similar trip-generating attributes (see Wootton and Pick, 1967; Williams and Latchford, 1966; Thomas *et al.*, 1966; Leake and Gan, 1973; Smyth, 1974). In the same way as household size and composition influence the number of trips generated in a given time period, so the size of a factory or office measured by the number of employees or floor space may affect the volume of trips generated, particularly non-work trips involved with the collection and distribution of goods and services. It is also possible to introduce location into the equation and to examine the way in which this factor influences the structure of trips according to transport mode, distribution of origins, and trip length. Most studies consider both work and non-work trips, so that total trip generation of an activity can be accurately assessed over a 24-hour period. The value of these studies is that they focus attention on the intensity, distribution and type of road traffic generated by non-residential uses in cities, in a way which will provide more objective guidance than household trip-generation data alone on the likely impact of new development or restructuring of land-use patterns on the local transport system and the environment generally. The ability of the private and public transport networks to handle additional or new demands can also be assessed more accurately, and the data is also useful for an evaluation of location policies for individual land uses. Hence, siting an office block in the CBD too far from public transport services may generate more car trips for work and non-work purposes than adjacent parking facilities can accommodate, while also adding to congestion. These undesirable conditions might be better controlled if the building were located in a way which encouraged as many public-transport trips as possible.

With reference to work trips generated by industrial plants, Maltby (1970) found that the number of cars used for the journey to work decreased as plant size increased. Car-occupancy rates to a group of small steel-manufacturing plants averaged 1.16 persons, compared with 1.36 persons for a group of larger steel-manufacturing plants in Sheffield. The occupancy rates for journeys from work to home were also higher than for home-to-work trips, so that the demand for public transport is likely to be lower at the end of a shift than at the start. It will also vary according to whether normal day-working or shift-working is involved. For the former, Maltby found that for travel to work, bus trips represented some 60 per cent of the total for 15 steel-manufacturing plants, and for travel from work the figure decreased to 55 per cent. For shift-workers, particularly the 06.00–14.00-hours shift, public transport was used less because of the greater inconvenience of using services at that time of the day. The proportion of public-transport trips increases with the size of the plant, however, mainly because large numbers of employees using bus services will make the special provision of frequent and convenient services a viable proposition for municipal or even company transport undertakings.

The size of an undertaking as measured by number of employees is clearly the best measure for estimating work trip generation by non-residential land uses. It has also been found that this is a better explanatory variable for non-work trips than floor space of an office, manufcaturing plant or warehouse. In his study of commercial-vehicle trip generation by manufacturing activity in the Medway towns, Starkie (1976b, pp. 32–3) found that the number of trips per plant per day was best explained by a logarithmic function as follows:

$$Y = a + b \log X$$

where: X = number of employees per plant or floor-space area per plant
Y = number of commercial vehicle trips.

This non-linear regression equation is based on the observation that the trip-generation/floor-space relationship is basically curvilinear rather than linear as has been assumed in many other similar studies (see, e.g. Maltby, 1970; Latchord and Williams, 1965–6; Latchford and Dobson, 1964; Townsley, 1974). Hence, Williams analysed total traffic generation by industrial premises in North East England using the assumption that there is a linear relationship with plant size or employment.

A study by Redding (1972) of the traffic generated by industrial premises in London also utilizes a linear relationship. But it appears reasonable to assume, and empirical evidence produced by Starkie (1967b) supports this, that as industrial plants become larger the rate of increase in commercial vehicle trips will decrease, i.e. there are economies of scale in commercial-vehicle generation at larger industrial plants as well as internalization of linked processes which in smaller plants would generate trips. For a plant employing 25 persons, the ratio of employees to trips per day would be 2:1; for a plant employing 400 it would be 8:1; and for 10,000 it would be 40:1.

In Starkie's study the above equation explained 54 per cent of the variance in commercial-vehicle trips in relation to employment and 60 per cent in relation to floor-space area. It is dangerous, however, to use aggregate data in order to estimate trip generation for all manufacturing industry, particularly if an untransformed linear equation is used because this will greatly overestimate trip generation of the smaller manufacturing plants. Analysis of data for 37 engineering, metal-working and allied trades plants out of the total sample of 77 units in the Medway towns suggests that the log-transformed relationship between total employment and number of trips generated is not simply the result of a combination of linear sub-sets with varying regression coefficients, because 87 per cent of the variation in trip generation at these plants was explained by employment in the non-linear regression. This would give a large measure of accuracy for estimating the overall number of trips generated by all plants in that group, and Starkie uses the results of this analysis as a 'standard' against which to compare the trip-generating characteristics and potential of other types of manufacturing so that some could be described as high and others as low trip-generators when compared with the standard trend (figure 7.3). Plants with trip volumes exceeding two standard errors on either side of the standard line are high or low trip-generators, while those between are described as medium trip-generators. Food, drink, tobacco and building manufacture are high trip-generators, with the ready-mix concrete plants especially important. Precision engineering, clothing industries and certain types of light manufacturing are low generators, while industries in the medium group include printing, publishing, chemicals, timber products and paper manufacturing, and most of the engineering and metal plants amongst the sample establishments. Redding found, however, that clothing premises generated three times as much non-work traffic (trips per 1,000 sq. ft per day) as plants in electrical engineering, with a mean of 10.52 and 3.65 trips respectively. This is a rather different conclusion to that arrived at by Starkie and it underlines the difficulty of translating traffic-generation studies of individual plants or groups of similar plants into aggregate relationships which can be applied universally in land-use/traffic planning and forecasting. This can be easily demonstrated by comparing data for two industries in London and Manchester (table 7.3). Although the differences between the traffic-generating capacities of the two groups are roughly the same in each city, the actual number of trips involved in Manchester is almost double the figure for London.

Some of the relationships demonstrated for vehicle traffic generated by manufacturing plants have also been shown to exist for pedestrian traffic generated by retail stores in central London (Hassall, 1974; see also Saunders, 1972). Nine large and medium-sized stores in high- and low-tourist concentrations, i.e. the Oxford Street and Holborn/Strand/Victoria

Table 7.3 Commercial vehicle and car (non-work) trips generated by clothing and electrical engineering plants in London and Manchester

	London		Manchester	
Trips per 1,000 sq. ft	Elec. Eng.	Clothing	Elec. Eng.	Clothing
By commercial vehicles	2.16	4.90	3.30	6.85
By car	1.16	2.82	1.90	5.00

Sources: Redding (1972), pp. 170–3; Mellor, (1970), pp. 384–8.

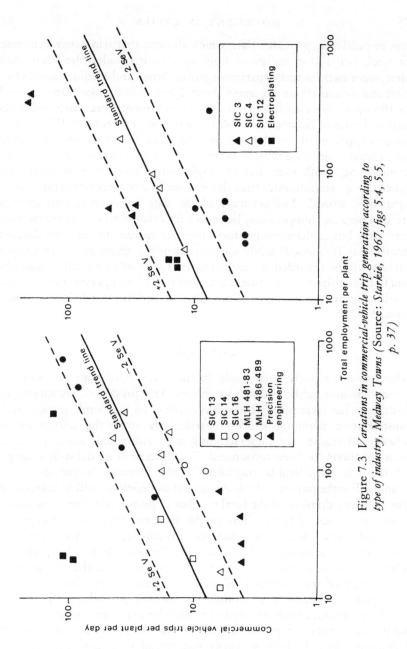

Figure 7.3 *Variations in commercial-vehicle trip generation according to type of industry, Medway Towns* (Source: Starkie, 1967, figs 5.4, 5.5, p. 37)

areas respectively, provided data which showed that trip-generation rates per week per 100 sq. m. gross floor area varied widely between chain stores/supermarkets and department stores. Small and medium-sized chain stores and supermarkets generated about 2,500–3,000 pedestrians per week per 100 sq. m. floor area, while the four department stores included in the study had rates of between 330–600 pedestrians per week per 100 sq. m. It was also apparent that within the two classes of store type trip-generation rates tended to decrease with increase in store size. The data produced must be interpreted with care, but the high tourist-concentration stores had higher trip-generation rates than those in the low concentration (at least for department stores). The accuracy of the data collected is also open to question because shoppers may be expected to visit several stores in the same retail area, but in department stores internal movement between different departments is 'missed' while all movement for chain stores and supermarkets will be recorded at the entrances. Large office buildings may also tend to internalize trips because it becomes feasible to operate staff canteens, printing or photocopying facilities, so that pedestrian trip generation is reduced.

Modal split

Reference has already been made to the division of trips generated by households into public and private modes. This process of identifying the modal split has increasingly been considered as an independent stage in the transportation planning process, particularly since the importance of behavioural studies for understanding why individuals select particular modes of travel has been recognized. The concept of modal split is simple, but there is considerable confusion in the literature about its precise meaning. For the purpose of this discussion its definition will be taken as the proportionate division of the total number of person trips between various means or modes of travel. The modal split can therefore be expressed numerically as a fraction, or a number of fractions, of the total trips; as a ratio; or as a percentage division between, for example, car, bus, tram and other modes. One of the other variable elements in the modal split literature is the range of travel categories used. The range is not infinite but the combinations in terms of groupings and number of modes are numerous. Usually, however, a basic distinction is made between 'individual' or private modes and 'mass' or public modes of travel. The importance of this distinction has only been seriously recognized in recent years as it has become clear that dependence on individual transport can neither be absolute or desirable in either contemporary or future cities. The actual division of trips into their component modes can be undertaken in one of two ways: an origin-split in which the percentage of trips by each mode entering or leaving a given area is calculated; or a trip-split in which the total number of trips made between two given areas is split into the proportion for each modal category. In a transportation study area with 100 zones there would therefore be 100 origin-splits or 9,900 trip-splits.

The term 'modal split' is also used to describe the process of arriving at a decision about which mode to use in any particular set of circumstances. This approach is usually described as modal choice analysis and it is increasingly used during the preparation of transportation plans, in which three approaches can be identified. Firstly, there are studies which seek to demonstrate that a relationship exists between city size, age, and population density, for example, and the use of public transport. This is of rather limited value for short-term prediction of modal split. These studies also exclude any *a priori* causal hypotheses about people's travel behaviour. Secondly, models of mode choice have been developed to forecast the volume of public and private transport trips between any pair of zones in a transportation study area. These models are derived from observations of existing patterns of mode choice and its relationship with other variables such as land-use patterns or density of bus routes. Finally, there are now available models which explain and predict individual choice of mode by reference to variables relating to household characteristics and travel behaviour of the kind used in trip generation analysis. Some households, mainly those without cars or other forms of private transport, cannot exercise choice of travel mode. These are the 'captive' users of public transport. In the light of the trends already described, captive users are becoming fewer in number, but eventually their proportion will stabilize as the old, infirm, young and other groups consciously choosing not to use private transport emerge as 'residuals'. The needs of the residual groups must be allowed for, or even increased, in future urban transport planning, but this should not be at the expense of private transport users. As Hillman *et al.* (1973) have contended, most transport studies ignore this basic fact and captive users of public transport have, almost traditionally, suffered at the expense of travellers with a more wide-ranging choice of travel facilities. These are usually thought of as the car-owning households whose travel mode decisions are crucial to the balance between public and private transport movements in cities.

At a higher level, planners and politicians must also make decisions about the most appropriate modal split to aim for in alternative transport strategies for cities or as a guide in the selection of the best strategy. The term 'modal split' has also been used in the context of the growing number of attempts to influence the modal split by using various kinds of traffic restraint. Some of these measures and their effects are discussed in the next chapter.

A model is therefore required which is sensitive to changes in the variables considered to affect modal choice, in that any changes in key variables should

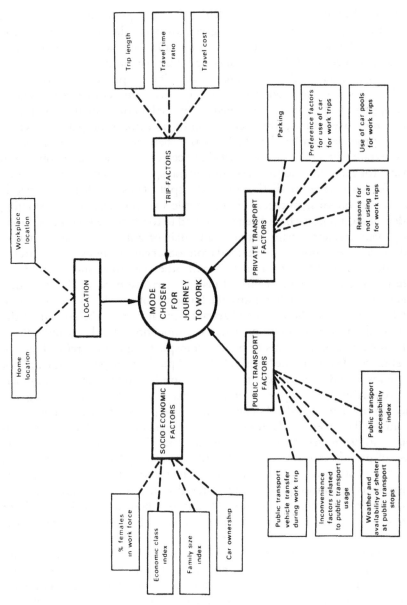

Figure 7.4 *Some factors affecting choice of mode for the journey to work*

produce realistic and accurate changes in modal split prediction. The outcome of modal split analysis is closely related to the way in which modal choice factors are incorporated in the models used to allocate trips to various modes (figure 7.4). As an example, some of the variables which may influence choice of travel mode for the journey to work can be grouped into five major factors. The first relates to the location of the individual making the decision in relation to his place of work; if he lives 400 yards from his office there is every possibility that he may choose to walk or cycle, but if he lives 25 miles away from the office without suitable access to a bus service he will have to choose between travelling by car or by train. Location is relevant to the values placed on the second set of factors, which are the characteristics of the trip which must be made – its cost, length and time. If the trip length is taken as constant (which it will not always be because the commuter may not use the most direct route to work every day), then the time and cost will vary according to the variables described as private and public transport factors. If the commuter decides to travel to work by car he may do so simply because he prefers the privacy and comfort of using his own transport rather than the uncertainty and discomfort of waiting for public transport. His employer may provide parking facilities at no extra charge, and this may encourage the commuter to use his car even though the public transport journey takes only marginally more time and costs 20 per cent less than the car journey. Permutations of this kind are almost endless and reference has not yet been made to socio-economic factors which control the extent to which individuals can actually choose between private and public transport, whether they can only choose between public transport alternatives, or whether they cannot afford public transport and must rely on their own energy and therefore reside as near to their places of work as possible. Choice of travel mode for shopping trips or recreation trips is also influenced by the five major trip factors shown in figure 7.4.

Apart from journey and traveller characteristics, transportation system attributes are also included in modal split models. A number of indices have been developed for this purpose and two examples are given here. Relative travel times such as the travel time ratio (door-to-door travel time by public transport/door-to-door travel time by car) indicate the possible advantage of using either mode to get from one zone to another. Relative travel cost, represented by the travel cost ratio – public transport fare/(cost of petrol + oil + $\frac{1}{2}$ cost of parking ÷ car occupancy) – is a second example, although only the costs of operating the car have generally been known to have a significant effect on modal split.

The way in which many of the variables affecting mode choice actually operate depends on the commuter's perception of the attributes of the private and public transport factors (table 7.4). Although based on data for Philadelphia the results could be applied to a commuter's evaluation of

Table 7.4 Modal choice: average scores, grouped by transport factors, for systems attributes as perceived by users in Philadelphia

Factor	System attribute	Perceived satisfactions	
		Public transport	Private transport
Reliability	Arrive without accident	6.0	6.1
	Avoid stopping for repairs	5.2	5.5
Travel time	Arrive in shortest time possible	2.5	5.8
	Travel in light traffic	3.9	4.8
	Arrive at intended time	4.4	5.9
	Arrive by shortest distance	4.3	6.4
	Avoid changing vehicle	4.3	6.2
	Ride in safest possible vehicle	5.6	5.9
	Travel as fast as possible	4.3	5.6
Weather	Protection from weather while waiting	3.0	5.2
	Vehicle unaffected by weather	4.6	6.0
Cost	Total trip cost	4.5	5.7
	One-way cost of $0.25 instead of per mile cost of $0.03	4.9	5.5
	One-way cost of $0.25 instead of	4.6	5.7
		4.9	5.6
Vehicle condition	Clean vehicle	4.8	5.4
	New modern vehicle	5.1	5.8
Personal safety	Avoid unfamiliar area	5.0	5.7
Self-esteem	Ride in uncrowded vehicle	4.0	6.0
	Feeling of independence	4.0	6.2
	Avoid waiting more than 5 minutes	4.0	6.3
	Comfortable ride	4.8	5.9
	Pride in vehicle	4.9	4.8
	Avoid riding with strangers	5.0	5.7

Diversions	Listen to radio	5.0	5.5
	Take along family and friends	4.0	6.1
	Ride with people who chat	4.7	5.6
	Look at scenery	4.5	5.6
	Ride with friendly people	4.8	6.4
	Ride with people you like	4.6	6.4
Convenience	Avoid walking more than a block	3.9	6.1
Packaging	Package and baggage space	4.4	6.2
Fare payment	Need not pay fare daily	5.3	6.1

Notes: Data are for work/school 'car available' trips for family household with income of $4,000–$5,999. Weighted averages are on a 7 point importance scale that increases with increased importance.

Source: Paine *et al.* (1967).

transport systems in most cities. The table clearly shows that the car provides greater perceived satisfaction in relation to almost every factor and system attribute than public transport. The study uses a seven-point 'importance scale', and the difference between public and private transport is equal to two points on the scale. Although the ratings do not indicate the relative importance of the choice factors listed, it is perhaps surprising how much weight is given by users to prestigious or personal-comfort attributes such as the ability when using private transport to avoid travelling with strangers, to have a feeling of independence when travelling, or to be able to listen to the radio. Unfortunately, if these are really crucial to the mode choice decision, then there are various problems of measurement to be overcome before they can usefully be incorporated in modal split models.

Most people do not undertake a comprehensive search of the best mode for their particular trip purpose, especially the journey to work, because it costs too much in terms of information retrieval and time. The amount of information required to resolve a problem increases approximately logarithmically with the complexity of the problem. Hence most people make do with some 'acceptable' level as represented by the perceptual space in which habit is very significant and within which decisions are formulated. The costs involved in the search process, in terms of effort required, are often so large that a 'satisfactory' selection is really all that many consumers aspire to.

This 'perception gap' can be related to the existence of two periods, the 'habit period' and the 'decision period'. During the habit period the alternative means of transport is not seen as directly relevant, in the sense that the individual is aware of it but does not consider it. It does become relevant during the decision period when the individual is considering a change. Because individuals are not constantly changing their commuting patterns, the majority are in the habit period at any given time. But once the conditions of his present mode change to an unacceptable level or he changes his workplace, the commuter enters the decision period. He begins by looking at the alternative(s) and in the light of his discoveries he may revise his view as to the desirability of changing mode at that time. Alternatively, the commuter changes mode, simply assuming in advance that the alternative is suitable for him. Hensher (1975) therefore doubts whether the trade-off theory of mode choice used in many behavioural studies is really appropriate. He suggests that during the habit period the traveller does not really take account of the time attributes of the alternative method(s) of travel for the journey to work, even though he may or may not be aware of the true attributes of the alternative mode(s).

A good example of an intangible variable is unreliability, which has been recognized as a significant component of the generalized cost of trip-making and modal choice, but few attempts have been made to quantify it. This neglect may be explained by the difficulty of observing a suitable trade-off situation in which transport users can trade money directly or indirectly for improved reliability of their transport modes. The characteristics of a possible trade-off which might be made by commuters have been investigated by Knight (1974). This is an allowance of extra time for travelling in order to avoid unpredictable lateness at their destination.

The degree of system unreliability varies considerably from mode to mode. It is most severe in the case of bus travel where congestion makes an important contribution, exacerbating both the problems of bus control and those arising from staff shortages. This has been demonstrated in various studies (Opinion Research Centre, 1971; Social and Community Planning Research, 1973). Car users are not immune from unreliability as a result of breakdowns or servicing, but flexible route choice and the absence of the various rigidities implicit in a scheduled public transport service easily compensate for this. Fixed-track systems also have lower levels of unreliability than buses; such problems mostly arise at peak periods when headways are at a minimum. Problems of unreliability are also likely to occur during multi-mode trips, and late arrival resulting from one stage of a trip may have a greater than proportional effect on total journey time due to, for example, the interdependence of timetables such that a missed connection may add 15–30 minutes to a trip. A wide range of transport improvement projects may, therefore, be instrumental in reducing transport system unreliability – improved signalling, bus-priority schemes, improvements to interchange facilities, or road improvements in heavily congested sections of the network. Apart from reducing mean trip times, these may also reduce mean trip-time variance.

One of the fundamental problems facing a commuter with a fixed time of arrival at work is how to select the departure time from home which minimizes the gap between arrival and starting time at work, but which also allows for the possibility of time lost en route. A study is underway at the Department of the Environment which is attempting to examine possible allowances made by commuters on different transport modes in London for delays in their journey to work. The relationship to be tested is:

$$T = u + f(x)$$

where T = time on any one trip between departure time and starting time at work
u = mean trip time
x = some measure of variation of the trip-time probability density function.

An attempt will be made to identify independent variables such as the degree of overcrowding or other activities undertaken before arrival at the workplace.

A final example of an attempt to quantify an intangible variable is given by efforts to develop a set of quantitative measures for describing the comfort variable which is usually incorporated qualitatively in behavioural and probabilistic mode-choice models (Nikolaidis, 1975). The measures, one for each mode and individual, are developed using modern psychometric techniques, initially as an aid to the explanation of travel mode-choices for urban-area work trips. The data used were provided by a questionnaire survey in Ithaca, New York, of the faculty and staff of Cornell University, employees in downtown Ithaca and housewives. A total of 100 questionnaires were used in the analysis. The questionnaire dealt with the measurement of attitudes towards the comfort variable, the collection of information on travel time and costs, and the collection of data on the socio-economic characteristics of the travellers. Attitudes of respondents towards the comfort variable were measured and converted into similarities data. These similarities were then analysed by the use of disaggregate multidimensional scaling techniques, and comfort indices derived for incorporation as a variable in mode-choice models. Nine characteristics were assumed to describe the comfort variable and were used as semantic scales ranging from one to seven (Nikolaidis, 1975, p. 56). The comfort characteristics were:

(a) means of travel well-protected from weather conditions;
(b) possibility of adjusting the temperature;
(c) plenty of storage space for parcels, shopping bags, etc.;
(d) few stops due to pick-up stops, traffic lights, etc.;
(e) immediate environment clean;
(f) good visibility of the surroundings;
(g) no fatigue felt when using a particular means of travel (constant attention, glare, uncertainty, etc.);
(h) feeling of privacy;
(i) ease of entrance and exit from means of travel.

It would be inappropriate to discuss the quantitative background to Nikolaidis' analysis here but his results support the argument that comfort, as measured in his work, is related to mode choice and that its effect can be measured. If this is the case, then more research is needed to quantify other qualitative variables such as convenience, reliability, safety or accessibility. The data used were for a very small sample from a small urban centre and it remains to be seen whether the techniques of measurement used by Nikolaidis would be suitable for assessing the value of comfort as a mode-choice variable in the more complex transport systems of large metropolitan areas.

Behavioural models of mode choice

Most behavioural models of mode choice attempt to explain the decision on the basis of the more tangible factors and variables included in figure 7.3. In studies of actual modal choice, time emerges as the most important variable and its measurement has exercised transport economists extensively during the last twenty years. Time is important because to some degree it is a product of one of the unique features of passenger transport, which is that it involves the user (or consumer) in a quantity of time which he is usually more interested in minimizing for most trips than the money cost (Thomson, 1974, p. 56). For car travellers there is the added consideration that few of them can adequately estimate the true money costs of, for example, their journey-to-work trips, though they can arrive at a time cost rather more easily. Modal choice by individuals therefore involves some consideration of the value of time, and this has led to questions about the values which people actually place on their time and what values they should place on time savings produced by transport improvements or time losses for the journey to work caused by congestion (see Mansfield, 1971).

It is important to appreciate that time has no 'value' in the conventional sense; it simply controls an individual's ability to be in the right place at the right time to take advantage of the opportunities he wishes to utilize. Equally, if choice of car as opposed to bus for the journey to work creates a time 'saving' for the individual, all it does is to allow him to spend more time in the way that he prefers. Therefore the value of time is the 'value, or net utility, of spending time in one way rather than another' (Thomson, 1974, p. 56). Provided that the alternatives are known, the individual is in a better position to decide their consequences for time consumed and to translate these into a decision on the appropriate mode of travel. The value placed on travel time will depend not only on differences between individuals in the way in which they value time but also according to trip purpose. The consequences of a delay of 15 minutes caused by traffic congestion during a recreational journey will be assessed rather differently to a similar delay likely to take place during the work journey (the measurement of the value of time is disucssed by Thomson, 1974, pp. 57–60). Waiting for public transport vehicles is often considered a major disutility in the use of time and is given twice the weight of walking, which is less of a disutility than waiting. Most people prefer to creep along in a car than wait for a bus or a train, even though the last has a high probability of arriving on schedule so that it is less frustrating than waiting for a more unpredictable bus. The value of time to an individual may even vary according to the length of the journey to work (Wabe, 1966).

Money cost, although its role should not be overstated, also plays its part in modal choice decisions. It may well be that by choosing to travel to work by bus, an individual will save money, but only at the cost of a longer journey time. This trade-off problem has been examined by Beesley (1965) for 387 office workers at the Ministry of Transport who were able to choose between a car or a bus for their journey to work. Respondents provided information on their usual travel mode for the journey to work, and the method of travel which represented the next best alternative for the same trip. Beesley aimed to find the value of time which successfully explained or rationalized the maximum number of choices for all the office workers. Both their normal and alternative journeys were costed, along with the time differences between them. From this material it was found that the office workers could be divided into four groups according to the combination of travel time and cost chosen:

Group A – journey faster but more costly,
Group B – journey faster but less costly,
Group C – journey slower but less costly,
Group D – journey slower but more costly.

A and C are the only choices in which a trade-off is possible, in that A implies a lower limit to an individual's value of time and C an upper limit. After dividing his respondents into three income groups, Beesley found that by using a different value of time for each group it was possible to produce the best explanation of modal choice by using a value of time equivalent to one-half their average hourly earnings. Such conclusions should be interpreted with care, however, because the sample is small and the method cannot be relied upon to give good results. Consequently, the value placed on time tends to vary between studies, so that Quarmby (1967), for example, arrives at a figure of one-third of hourly earnings for time spent in vehicles. Time spent in buses was valued twice as highly as in cars and almost three times as high as in-vehicle time for walking or waiting time. In an American study (Lisco, 1968, p. 68) it has been found that commuters who drive to work in the CBD are prepared to pay 30 cents a block or 12 cents a minute just to avoid extra walking. Hence, an office more than three blocks from an underground or railway station is likely to attract more car-driver trips than buildings which are nearer.

In most behavioural models the choice of mode by an individual or by a group of individuals with similar characteristics of location, economic status, or car ownership is viewed as a problem of binary choice (Warner, 1962; Wilson, 1967; Lisco, 1968; Stopher, 1969; Quarmby, 1967). Warner, for example, first examines the typical or average choice behaviour of individuals and then tries to estimate the probability that an individual will choose one mode of travel rather than another. The concept of binary or hierarchical choice of mode for any kind of urban travel is useful because it greatly simplifies the analysis problems which confront model builders and, secondly, it is possible to rationalize that most individuals' choice of mode does in fact reduce to a binary situation (figure 7.5). Let us assume that an individual can choose between bus, car, bicycle or walking modes for a particular trip which he wishes to make. He is likely to divide them into motorized and non-motorized forms as shown in figure 7.5. (A), and he will

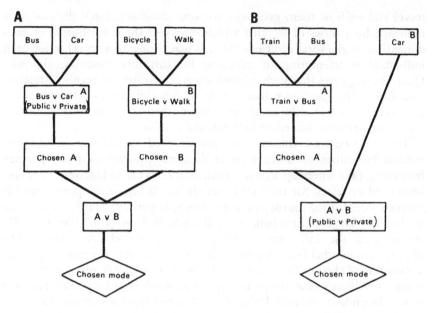

Figure 7.5 *Schematic model of binary choice of travel mode for urban travel*

then choose between the two choices in each of those groups, selecting the preferred mode on the basis of the initial choice (Watson, 1974). Similarly, an individual with a threefold choice between car, bus and train may well divide them initially into public and private transport and proceed to make his final choice along the lines shown in figure 7.5 (B). It is likely that most people do not possess knowledge about more than two modes, once they have decided what their basic choice is. They will obtain information on what they consider to be the best alternatives and not bother to elicit costs or times of third- or fourth-choice alternatives. The type of trip to be undertaken may also affect mode choice and Warner, for example, distinguishes between non-work and work trips before assigning probabilities that mode x will be chosen in preference to mode y.

Most behavioural models are concerned with mode choice for the journey to work. Stopher (1968) has studied a group of London commuters and using field evidence attempts to identify the variables which transport users considered important in their mode-choice decision. Four factors – time, cost, convenience and comfort – emerged as most important, but only the two which could be quantified, time and cost, were actually used in the journey-to-work model. In a more complex behavioural model, which rests more heavily on theoretical constructs, Quarmby (1967) introduces the concept of disutility in mode-choice decisions. He assumes that walking time, travel cost, travel time, or inconvenience comprise the dimensions of travel and each of them gives rise to some disutility. Each dimension is indicated by a series of weights which represent the contribution of each dimension to the total disutility (D_{ij}). When choosing a travel mode each individual is attempting to minimize his disutility function, although Quarmby suggests that such minimization is marginal in terms of relative disutilities. The absolute values of time or cost differences between modes are less important in influencing mode choice than a relative measure such as a ratio or per-cent difference between these values.

Time and cost are the most commonly used variables in behavioural models but other variables such as distance, journey purpose, journey frequency, car ownership, age/sex, availability of a car and income, have also been used or tested for their effect on choice. When developing a modal-choice model for the journey to work which, hopefully, could be applied to all but the largest conurbations in Britain, Wilson (1967, pp. 94–170) investigated the influence of 17 possible modal-choice factors. The objective, as in other behavioural studies, was to identify the most important parameters of choice. These can then be used in a multiple regression model using, in Wilson's case, the percentage of all work trips by public transport as the dependent variable. Using data obtained from surveys in Coventry and London, Wilson found, for example, that parking availability, or the cost of it, did not influence the choice of mode for travel to work. One of the main reasons for the limited influence of this variable is that most firms provide parking facilities for their employees, usually free of charge, so that it is relegated to a minor position in the decision-making process. Other factors such as the effect of vehicle transfers during journeys were also found not to be sufficiently significant on their own to be considered as separate parameters for inclusion in the final model, even though they introduce cost or time advantages/penalties into work trips. Wilson resolves this by building an allowance for them into the equations for travel time and travel cost which were included in the final model.

Various measures of travel time have been used in modal-choice models. Some use excess travel time, i.e. the total time excluding time spent travelling in a vehicle, and this includes parking, walking or waiting times (see, e.g., Quarmby, 1967, pp. 280–1). Alternatively, only the time spent in transit in the vehicle, including delays caused by traffic, can be used, or thirdly, total travel time including in-vehicle and extra-vehicle time (Stopher, 1968; Wilson, 1967). Wilson used total trip-time observations to calculate travel-time ratios between private car and bust trips; these representing the most common form of binary choice in all but the largest urban areas in Britain (the method for calculating the travel time ratio is discussed in detail by Wilson, 1967, p. 160; ratios have also been used by Warner, while Quarmby and Stopher use travel time differences between

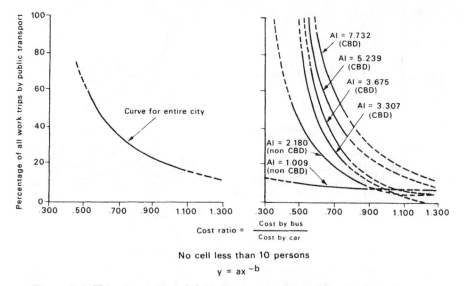

Figure 7.6 *Trip-time ratios and their effect on travel-mode choice, Coventry* (Source: Wilson, 1967, figs 8.27, 8.28, pp. 166–7)

modes). The influence of trip-time ratios on modal choice can be demonstrated by using the percentage of work trips by public transport as the dependent variable and producing groups of diversion curves (figure 7.6). These enable the proportion of trips made by public transport for a particular purpose between any pair of zones to be predicted, given the ratio of travel times by public transport to private car; the service ratio (a measure of walking and waiting times involved); the cost ratio; and the economic status of travellers. The higher the value of the travel-time ratio the more superior will be car trip times, and there is clearly a negative correlation between door-to-door travel-time ratios and percentage of all journeys to work by public transport. This is the case for both CBD and non-CBD employment centres. The accessibility index (AI) for employment centres, which measures the ease with which an activity within an urban area can be reached from a particular zone on a specified transportation system, does lead to a variation in public transport use, while detailed location characteristics of workplaces can also lead to high or low levels of public transport choice. ($AI =$ [service level index for zone in seat miles] [% of employment centre's workers originating in that zone]/[labour force in immediate destination area], where the service level index = [number of miles of bus routes in zone] [number of buses in peak hour] [number of seats per bus]/[labour force in zone].) Cost has similar effects on mode choice

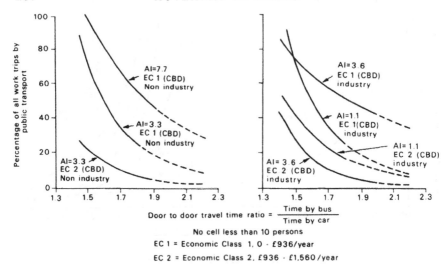

Figure 7.7 *Relationship between travel-mode choice and cost ratio between bus and car, Coventry* (Source: *Wilson, 1967, figs 8.25, 8.26, pp. 163–5*)

(figure 7.7). The diversion curves for use of public transport for the journey to work are certainly sensitive to the costs of competitive modes. Figure 7.7 also shows that the proportion of work trips by public transport is generally smaller for lower values of the accessibility index at any given cost ratio. CBD locations also reveal more sensitivity to the effects of the trip-cost ratio than non-CBD workplaces, particularly if the *AI* value is low.

Costinett (1973) has suggested that modal choice by the individual traveller can be generalized into a behavioural cost (*BC*), which comprises the actual cost of travel (petrol, fares, tolls), the time taken while undertaking a journey, and the income of each traveller. The behavioural cost can be calculated from:

$$BC = a_3 XT + a_2 XE + a_1 XF$$

where: BC = behavioural cost in equivalent minutes,
T = riding time in vehicle (minutes),
E = excess time waiting and walking (minutes),
F = out-of-pocket cost, e.g. fare (pence),
$a_1 a_2 a_3$ = constants where $a_1 = 1$, $a_2 = 2$ and $a_3 = \dfrac{1}{\text{wage rate}}$ (mins/pence).

Using this equation, the 'cost' to each individual of using alternative modes can be measured in compatible units (equivalent minutes), although there is no indication as to how the constants critical to these calculations are derived. The behavioural-cost measure is likely to be useful in relation to the

Table 7.5 Calculation of the behavioural cost of a 5 km journey to work

BC component	Mode	
	Bus	Car
Riding time	15 (at 20km/h)	10 (at 30km/h)
Excess time: walk 10		7
wait 5		0
15		7
Perceived value	30	14
Out-of pocket expenses at 60 p/h in equivalent minutes	10	12
Total (equivalent minutes)	55	36

Source: Lesley, (1975), table 1, p. 127.

journey to work where a commuter has the choice between car and bus (a standard binary choice problem). Calculation of the BC for each mode should indicate the choice which the individual will actually make. The procedure is summarized in table 7.5 for a commuter journey of 5 km length in the provinces. It is assumed that the commuter earns 60 pence per hour, and this can be converted into equivalent minutes in the way shown. The behavioural cost of the bus is some 35 per cent higher than that of the car, and even if a parking charge of 10 pence per day is included for the car its behavioural cost of 46 pence still remains lower than that of the bus. Therefore for commuters with cars there is a high probability that they will use them for the journey to work. The difference between bus and car behavioural costs is sustained for journeys in excess of 5km in provincial cities, and for journeys over 10km the difference tends to increase even more in favour of the car. It will be shown later that the fundamental problem facing cities which want to improve the level of use of public transport is how to reduce the behavioural cost differential between public and private travel modes.

Regression techniques and modal split analysis

Once the variables to be used in a modal split model have been selected and translated into a suitable numerical form, they are usually examined by least-squares multiple regression in the form:

$$Py = b_0 + b_1 \log X_1 + b_2 \log X_2 \ldots \ldots \ldots b_n \log X_n$$

where Py = proportion of trips by public transport or by private means,
 $b_0 - b_n$ = coefficients,
 $X_1 - X_n$ = independent variables.

Hence, Wilson chose the travel-time ratio, cost ratio, car ownership in origin zone, family-size index for origin zone, trip length, and percentage of females in the workforce in origin zone, as his independent variables. The major difficulty with this approach is that it evaluates the effects of the independent variables on the use of one particular mode of travel available in an urban area (see, e.g., Watson, 1974, pp. 102–18). Wilson uses bus trips as a proxy for public transport and therefore must assume that the independent variables will operate in a similar way on the choice of other public transport modes such as trains or the underground. Regression and other models are also inflexible, in that it is difficult to estimate the effects of varying the values of the independent variable(s) on the values of the dependent variable. Because of the interaction between many of the independent variables, e.g. between car ownership and income class, any change in the value of one variable will be reflected in the values of the others in the equation. Hence, although Quarmby, Stopher and others seek to produce models which are universally applicable, in practice it is usually difficult to transfer a model for mode choice and the journey to work from one city to another. Data must be available for each city or town so that the appropriate regression coefficients for the area can be derived.

It must also be apparent that although modal-choice models attempt to explain individual travel behaviour, they are applied, in the final analysis, at some aggregate level using zones. The finer the network of zones, the more useful the regression equations are likely to be and, conversely, high levels of aggregation will lead to models with a poor predictive/analytical capacity. Behavioural models of modal choice have been heavily focused on the journey to work, mainly because in proportional terms it is so important in urban travel and, in terms of the transport needs of communities, it has a significant effect on the evolution of proposals and policies. It is also the most predictable form of movement in cities, and this is an aid to model confidence limits. The binary choice problem is also likely to be more clearcut for the journey to work than it is for social, recreational or shopping trips where modal choice, particularly in relation to the use of public or private transport, is perhaps less carefully considered by individuals and more difficult to rationalize. Households without cars will use public transport, those with cars will use them for most non-work trips. Perhaps the problem of binary choice is mainly a problem of isolating which form of public transport non-car owning households will utilize in a given set of circumstances and, for car owners, which form of private transport they will utilize.

Modal split may be incorporated at the trip-generation stage or after the trip-distribution stage (the latter is discussed below). If it is carried out during the trip-generation stage, a multiple regression equation is used in which the dependent variable is the percentage of all work trips by public transport, and the independent variables are those which are considered to influence modal choice. The disadvantage of this is that transportation system characteristics cannot specifically be taken into account. On the other hand, modal split at the distribution stage of the transport model does not allow the influence of traveller characteristics on modal choice to be fully incorporated.

Trip distribution

Trip distribution deals with the problem of locating the estimations of the trips undertaken between all pairs of zones in the transportation study area. The distribution is obtained from estimates of the number of trips generated and attracted by each zone, and these are usually derived from the trip-generation phase. In the early models this was the second stage in the process of simulating travel patterns, but some more recent studies have placed modal split analysis second and then considered distribution, following the argument that modal choice has consequences for where individuals choose to travel to in urban areas. Trip-distribution procedures provide the transportation planner with a number of systematic ways of estimating the pattern and number of trips between zones in a city, in relation, if necessary, to alternative land-use and transport strategies. This information about the destinations and directions of person and vehicle travel in the future is vital initial information necessary for adequate planning and provision of a wide range of future transport services and facilities in cities.

Three methods of trip distribution have been used in British transportation studies; growth factor methods, gravity models, and intervening opportunity models. As with modal-choice procedures there have been improvements through time in the methods used for estimating trip distribution so that growth factor methods, for example, are used most frequently in the earlier transportation studies, such as those for Leicester and Belfast (City of Leicester, 1964; Travers Morgan and Partners, 1968). Gravity and intervening opportunities models appear rather later in the 1960s in, for example, the West Midlands, Merseyside, and Harlow transportation studies (Freeman Fox et al., 1968a; Traffic Research Corporation, 1969). Some of the reasons for this change in the techniques used for distribution will become apparent during the following discussion of each of the main procedures.

Growth factor methods

In the growth factor method, trips are distributed by an expansion of the existing distribution pattern, as derived from the data-collection phase of the transportation study. Hence, existing trip origins and destinations are expanded to the forecast year by utilizing a growth factor which reflects the anticipated land-use changes and development in each of the study-area zones. This method is therefore 'a rather mechanical approach divorced from notions about how persons actually behave' (Starkie, 1973b, p. 374). The key element in the growth factor equation is the value assigned to the expansion term, which may be described as mechanical or analytical (Catanese, 1972, pp. 73–4). The mechanical method is the least satisfactory because it involves extrapolating the composite elements of past trends – assuming a direct relationship between past experience and future distribution of trips and trip ends. The analytical method, on the other hand, is more comprehensive in that it incorporates an allowance for the relative contributions of the different factors which go to make up previous trend patterns, as well as taking into account any new factors which might become operational or more influential in the future.

The best-known growth factor method is the Fratar (1954) technique. This relies on the basic premise that the distribution of trips from any zone i is proportional to the present level of movement out of zone i modified by the growth factor of the zone j to which the trips are attracted. The volume of trips is determined by the expansion factor of zone i. In order to be able to use this method a complete set of origin-destination data is required, along with an adequate data base upon which to forecast the growth of residential or industrial land use, for example, in each traffic zone. The basic procedures are outlined in figure 7.8, which illustrates the basic data required to calculate the trip distribution between three zones, A, B and C. The following equation is then used to calculate the future distribution of trip destinations:

$$T_{AB} = \frac{1}{2}\left[A_b \cdot G_A \left(\frac{AB_f \cdot G_C}{AB_f \cdot G_B + AC_F \cdot G_C}\right) + B_G \cdot G_B \left(\frac{BA_f \cdot GA}{BA_f \cdot G_A + BC_f \cdot GC}\right)\right]$$

Where T_{AB} = trips from A to B at $t+1$,
$A_b - C_b$ = base-year trip ends in zones $A-C$,
$G_A - G_C$ = growth factors for trip attractions to zones $A-C$,
AB_f = present trips from zone A to B.

The forecasted trip distribution using the above equation is shown in figure 7.8, in addition to the forecasted design-year trip ends which result. It

ANALYSIS AND PREDICTION OF TRAVEL PATTERNS 215

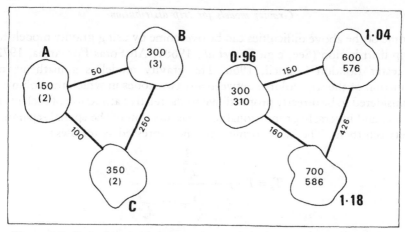

Figure 7.8 *Basic data for distributing trips between zones using Fratar Method*

is apparent that in zone A the forecasted trip ends exceed the design-year figure, while in zones B and C there is a deficit. A more balanced distribution of trips can be achieved by substituting a correction factor for the original values of G_A–G_C in the above equation and applying them to the forecasted flows between zones instead of to the original flows.

The Fratar method allows the growth factor for each traffic zone to be varied so that trips to and from a traffic zone which is entirely residential can be estimated on the basis of the forecasted population in the zone; but for a traffic zone which is entirely industrial the future trips to and from it must be estimated in relation to future employment levels or anticipated levels of car use for the journey to work. Zones with mixed land-use structures are more difficult to handle but estimates can, according to Fratar, usually be made for a representative growth factor. Growth factor methods simply forecast changes in the scale of movement between traffic zones without taking adequate account of directional changes triggered by new land-use distributions or changes in trip-making behaviour, such as those caused by the introduction of a new urban motorway which may encourage longer journeys to work than before. In addition, no account is taken of zones where change cannot take place, i.e. there is nothing to expand or there may be trip reductions. Underestimation of long-term changes caused by rapid growth in those zones at some later date may then result. There is also an inability to take into account changing land-use distributions in zones which may cause readjustments in trip-generating characteristics. But the growth factor methods do have some advantages: there is no calibration required and, secondly, equilibrium can be reached quite rapidly with only a few iterations so that a good, rapid approximation of trip distribution is provided.

Gravity models for trip distribution

Some of the above difficulties can be overcome by using gravity models for trip distribution (See, e.g., Fox *et al.*, 1968a; An Foras Forbartha, 1972; Greater London Council, 1966. The gravity model is a mathematical function of the trip volume between any two zones in which the volume is considered to be directly proportional to the relative attraction of each of the zones and inversely proportional to some function of the spatial separation between them. The basic formula can be expressed as follows:

$$T_{ij} = P_i \frac{\dfrac{A_j}{d_{ij}*}}{\dfrac{A_1}{d_{ij}*} + \dfrac{A_2}{d_{ij}*} + \cdots \dfrac{A_n}{d_{in}*}}$$

where T_{ij} = trips produced in zone i and attracted to zone j,
P_i = total trips produced by zone i,
A_j = total trips attracted to zone j,
d_{ij} = spatial separation between zones i and j,
$*$ = a measure of the frictional effect of distance such that if $* = 2$ the volume of interaction is conversely related to the square of the distance between i and j. The factor is empirically derived.

The application of this equation to a simple example involving distribution of trips from a residential zone to three centres of office employment can be

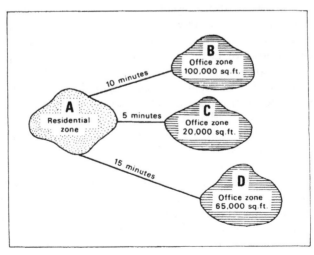

Figure 7.9 *Basic data for trip distribution using a gravity model*

illustrated using the basic data given in figure 7.9. The trips between any pair of zones can be calculated by using the following equation:

$$T_{AB} = \frac{k \cdot P_A L_B}{t^*}$$

where T_{AB} = number of trips between A and B,
  P_A = total trips produced by zone A,
  L_B = total trips attracted to zone B,
  t^* = a measure of travel time where the value of * will vary according to trip purpose.

The distribution of trips (per cent) from the residential zone A to all three office centres can then be calculated from the equation:

$$T_A = T_{AB} + T_{AC} + T_{AD} = 100$$

$$= k \left[\frac{P_A L_B}{t^*_{AB}} + \frac{P_A L_C}{t^*_{AC}} + \frac{P_A L_D}{t^*_{AD}} \right]$$

where $* = 3$

If office floorspace is taken as the measure of attraction (L_B, etc.) and trip times between zones are as specified in figure 7.9, then it can be calculated that $T_{AB} = 36$ per cent, $T_{AC} = 57$ per cent and $T_{AD} = 7$ per cent.

This is clearly a very simplified example of how the basic form of the gravity model distributes trips produced by a particular zone in relation to competing centres of attraction. In a full transportation study many measures of production and attraction, as well as a variety of trip purposes, will be incorporated in the gravity models used. It is usually necessary, however, to introduce certain constraints which ensure consistency within the model between the estimated distribution of trips and externally derived trip totals that are used to derive these estimates (for a good description of gravity models, including worked examples, see Masser, 1972, pp. 94–109). One of the major problems is how to determine the exponential value of spatial separation which tends to vary both within and between cities, and according to trip purpose. Much of the problem derives from the various ways in which distance can be expressed: airline distance, route distance, or travel time. Most transportation studies use travel time; examples include the London Transportation Study, SELNEC, and the West Midlands Transportation Study. The way in which these alternative measures operate may also vary according to the time of day, and it should be apparent, too, that a travel-time exponent on its own does not completely explain the propensity to travel between two points. Travel patterns can be affected by a variety of economic and/or social linkages which are not properly quantified (sometimes referred to as the K factors). Even a very basic distinction between white- and blue-collar workers, for example, may well improve the reliability of gravity model trip distribution. More recent gravity models also try to take into account the effects of physical barriers to movement such as rivers and the number of bridges across them. Usually, time penalties are introduced for links across physical barriers, such as bridges, ferryboats or tunnels.

Gravity models have been widely used in transportation studies and can give excellent results when properly used. Provided that we can accept that a law of physics can be adequately applied to human behaviour and suitable K factors can be derived, the advantages of gravity models for trip distribution probably outweight the disadvantages.

Intervening opportunities models of trip distribution

A third way of simulating the distribution characteristics of trips in urban areas is to use the intervening opportunities model developed by Stouffer (1948) in the United States. This is a probabilistic model, originally used by Stouffer to analyse migration patterns, which utilizes similar concepts to the gravity model except that it uses an unique independent variable – the effects of intervening opportunities. Hence, rather than using spatial separation measured in terms of travel time as an absolute value in the equation, it is used to consider the possible destination zones in order of increasing travel time from the origin. Stouffer (p. 846) expresses the concept as 'the number of people going a given distance is directly proportional to the number of opportunities at that distance and inversely proportional to the number of intervening opportunities'. The formula for the procedure is as follows:

$$T_{ij} = P_i(e - LiA_i - L_i(A + A_j))$$

where T_{ij} = trips produced in zone i and attracted to zone j,
P_i = total trip productions from zone i,
A_j = total trip attractions to zone j,
A = cumulative trip attractions considered prior to zone j,
L_i = basic probability that random destination will satisfy the needs of a trip produced from zone i,
e = base of natural logarithm (2.71818).

The model is based on the hypothesis that total travel time from a point is minimized, subject to the condition that every destination point has a stated probability of being accepted if it is considered. Secondly, the probability of a destination being accepted, if considered, is a constant, independent of the order in which destinations are considered. The destination closest to the origin zone, in time, is therefore considered first and has the highest probability of being accepted. The same basic probability of acceptance exists for the second-choice destination, but the actual probability of acceptance is decreased by the possibility that the tripmaker has accepted the first opportunity as his destination. This procedure is applied to each successive destination from the origin with the probability of acceptance decreasing as the number of intervening opportunities increases.

Calibration is an important stage in the development of both gravity and intervening opportunities models. This involves finding 'the value(s) of the parameter(s) which provide the best fit between the model and the observed situation' (Lee, 1973, p. 23), so that they can be used with maximum confidence in a predictive situation. In the case of the gravity model this means adjusting the values of the exponent or friction factor x (or *) until it produces a trip distribution curve which is relatively close to the observed data from the origin–destination survey (figure 7.10). Most calibration procedures attempt to ensure that the difference between the durations of the average trip lengths does not exceed ± 3 per cent. For the intervening

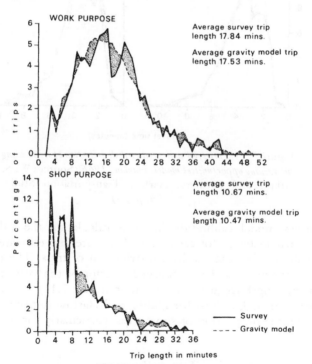

Figure 7.10 *Observed and predicted patterns of trip distribution for work and shop purposes using a gravity model, Dublin, 1970* (Source: *An Foras Forbartha, 1972, figs. 15.4, 15.5, pp. 15.5–15.10)*

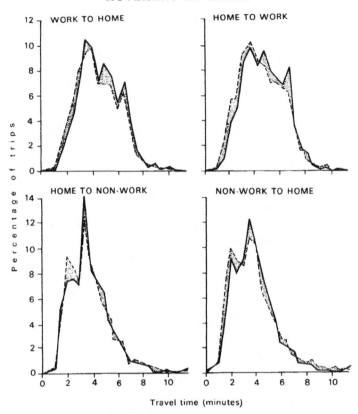

Figure 7.11 *Observed and predicted distribution of trips using intervening opportunities model, Harlow, 1966* (Source: W. S. Atkins and Partners, Harlow Transportation Study, *fig. 7.6, p. 25*)

opportunities model calibration involves calculating the values of the probability parameter L_i for each zone. These are first estimated and used in the model to simulate the base distribution of trips. The distribution produced is compared with the observed situation on the basis of total travel time and trip-length frequency distribution (figure 7.11). It is notable that both types of model, at least for Dublin and Harlow respectively, tend to overestimate the proportion of short trips, particularly for the journey to work.

Trip assignment

The final stage in the travel model is the procedure referred to as trip assignment, which is the allocation of the estimated zone-to-zone trips (derived during the trip-distribution stage) to specific links in the highway network or to the public transport network. The result is an estimate of the volumes of traffic on all sections of the road or public transport network for the design year of the study. This permits an assessment of deficiencies in the existing transport system; an evaluation of any improvements and extensions to the existing system; testing of alternative system proposals and the development of construction priorities. Techniques used for assignment have not changed greatly as they rest on a number of fundamental assumptions about the behaviour of transport users, whether they are private individuals or companies planning freight or service delivery routes (for a discussion of techniques, see Chu 1971). A major assumption is that the minimum time or cost route through the city's transport network will always be chosen. Implicit, therefore, is the assumption that all users are fully informed about the alternatives open to them on a transportation network. They will, it is assumed, always know the shortest route to take for any particular journey which they make. It is also assumed that all types of vehicles show an equal preference for the same route, irrespective of difference in vehicle size, the length of route involved, or the traffic conditions which may be encountered. Multi-path assignments, where the shortest three or four routes are identified and trips assigned to them probabilistically, permit some of these problems to be overcome.

These assumptions may not be as unrealistic as they appear at first sight—given that a large proportion of intra-urban trips are short and to destinations very familiar to the trip maker. The most obvious examples are the journey to work and the journey to shop or school. Comparatively few trips are made at random to locations throughout or across an urban area, so that the problem of system knowledge may be exaggerated. But assignment procedures do not take account of different preferences expressed by various socio-economic groups, for example. Families in low-income groups may value time less highly than high-income families who may well be prepared to use more costly transport modes or make more roundabout trips in order to be able to save time, via a motorway bypass, for example. The assignment process simply selects routes on an all-or-nothing basis irrespective of population characteristics, the significance of which has already been discussed in relation to trip generation. In view of the changes in socio-economic structure of urban populations or new transport technology, it is likely that future transport networks will have completely different demands placed upon them from those experienced by existing systems. Assignment procedures are based on their ability to produce existing travel patterns with reasonable accuracy.

In recent years the dependence upon travel time or cost for assignment has been reduced to some degree by including other costs such as parking charges (SELNEC), fares paid for public transport, or the out-of-pocket costs of running vehicles. There have also been attempts to assign trips to the highway and public transport network separately (An Foras Forbartha,

1972, pp. 16.1–17.11). This may include reference to timing and frequency of services and the effects of waiting times at interchanges. It is interesting to note, however, that there remains a basic paradox in the assignment process compared to other parts of transport models; the latter recognize the possibility of sub-optimal behaviour, e.g. in mode choice, but the former only allows for optimal behaviour (Starkie, 1973a, p. 377).

The highway network, consisting of interconnecting links to which the centroids of traffic-generating nodes are connected, comprises the framework for trip assignment. The centroid of each zone is taken as the origin and destination of traffic into and out of it. Between any pair of centroids there are clearly many routes through the network, but one is the optimum if the assumptions discussed above are used. These optimum routes are computed for all pairs of zones as time or cost paths which are referred to as 'trees'. Once these have been calculated, all the zone-to-zone trips are assigned to each appropriate tree and the traffic volume on each link in the tree recorded. Assignment may be made on a directional or non-directional basis or it may also be made on all-or-nothing basis.

The all-or-nothing approach can produce unrealistically high assigned loads, and increasing efforts are now made to relate assignment to the ability of the network along any link to carry the load. Hence, capacity restraint, or the ability of routes to handle the traffic, is introduced and this allows travel speeds, volumes, distances and capacity to be taken into account after the first all-or-nothing assignment (Steel, 1965; Irwin *et al.*, 1961). Additional data relating to the capacity and speed/flow characteristics of each link will be required but the extra effort is perhaps rewarded by more realistic trip assignment. Minimum time path calculations can be adjusted if the practical capacity of individual links is exceeded and assignment on an all-or-nothing basis again undertaken. This can be done iteratively until reasonably balanced link loads are obtained.

How valuable are transportation studies?

The work involved in producing results based on the 'core area' analysis in transportation studies is formidable and much more complex than the brief review given here can adequately convey. Although the coverage has been deliberately kept at a general level, it should be apparent that modern transportation studies, however sophisticated the mathematical techniques and models used, cannot resolve at the first attempt all the transportation problems which confront cities. Many aspects of modal choice remain imperfectly understood, for example, and least progress in modelling urban transport movements seems to be made in the most important areas, such as modal split analysis. It would seem that 'the ultimate objective is to produce a model which gives a plan as an output rather than requiring one as input. It may seem unlikely that this will be achieved in the foreseeable future' (Greater London Council, 1968). Transportation studies have been a standard part of the analysis and management of urban transport problems in Britain for more than 15 years but, to date, they only allow the testing of alternative proposals, not the production of plans. The methods used still need to be refined even further and a large measure of judgement is still required, so that two of the main ingredients of transport planning are still 'the old fashioned qualities of intuition and imagination' (Thomson, 1969, p. 49; see also Holmes, 1973).

Policy-makers in local and national government play a big part in shaping the transport and land-use structures of cities, using the evidence of transportation studies. There can be no doubt that, to date, transportation studies have been largely concerned with improving urban mobility and reducing congestion through investment in new roads and road improvements. This has certainly improved accessibility in cities, but only for those sections of the population who have cars. Hillman *et al.* (1973, pp. 75–80) have attempted to draw attention to the fallacy that most people will soon have a car for their personal use. The viewpoint is explicitly articulated in the Ministry of Transport's *Traffic in Towns* (1963b, para. 9) and *Public Transport and Traffic* (1967e, p. 113). Yet in 1971, 48 per cent of the households in Britain did not have regular use of a car and, moreover, this includes nearly 73 per cent of households in the lowest-third income group and 90 per cent of the elderly persons' households (Hillman *et al.*, 1973, p. 76). It must be obvious that all households will not own a car in future because age, income and general ability influences individual desire to own one. Although forecasts vary, something of the order of 20 per cent of households will be non-car-owning by 2001 (DoE, *Highway Statistics 1971*; Central Statistical Office, *Annual Abstract of Statistics 1972*; Tulpule, 1973). Hillman also suggests that household car ownership does not confer equal mobility upon all members of that household for all purposes, so that members of a household who do not possess a licence will often depend almost as much on public transport as members of non-car-owning households. Transport studies generally manage to overlook this hidden, 'captive', market for public transport. They have therefore underestimated the opportunities for improving public transport in the interests of the substantial body of apparent and 'hidden' captive users, as well as the potential for converting consistent non-users of public transport, following a programme of carefully planned improvements. Hence, there has been a trend towards partial use of the transportation study process in relation to particular transport problems such as the balance between public and private transport use in cities.

A further limitation of transportation studies is that they have been almost exclusively concerned with mechanically assisted trips. Trips on foot are usually treated as insignificant, which they certainly are in terms of distance and their effects on land use or the environment. But walking trips are numerous, even amongst car-owning households, and are easily substituted by mechanically assisted modes, especially cars (Mitchell, 1973). Any tendency to do so could be enhanced by the combined decline of small, dispersed retail units which are easily accessible on foot; by the trend towards larger schools more remote from the homes of pupils; or by a reduction in the number of small-scale manufacturing or service employers. As soon as people must travel further than acceptable walking distances they are likely to use mechanically assisted modes. Existing transportation study data are not especially useful for the purpose of mitigating these effects.

Transporation studies and the models they use have been too introspective, dealing with only part of the problem and 'benefiting' only some urban dwellers. The time has come for the models to include the possibility of solutions other than those which are primarily road-orientated. One example is the need to consider more accurately the effect of actually promoting the use of public transport in a much more positive way instead of standing idly by while it struggles to remain viable. Secondly, transportation studies must take account of those solutions which involve restraint of some kind on the use of, and movement by, cars – parking controls, pedestrianization, road pricing, or supplementary licence fees of the type proposed by the Greater London Council for central London (see Ministry of Transport, 1964, 1966, 1967d). Thirdly, it is important to calculate the economic case for proposed urban transport investments along the lines of the cost-benefit study for the Victoria Line underground railway in London (Foster and Beesley, 1963). Some of these issues will be looked at in more detail in the next chapter.

There is also some doubt about the ability of transport models to reproduce with acceptable accuracy the effects of any changes in the transport system. It must be borne in mind that almost all transport models are based on a cross-section of data taken at a point in time; yet they project with confidence over a 20- or 30-year period in most cases. It is difficult enough to produce a model which accurately depicts the state of the transport system at the time of an origin–destination study, but even more difficult to create a model which reproduces the behaviour of the system through time. This has led Starkie (1973, p. 332) to suggest that less emphasis should be placed on calibration, because there are very few time-related variables actually included in transport models; therefore what is the point of assessing their accuracy for forecasting purposes? Instead, models should be assessed for their ability to reproduce past trends in travel behaviour. As time passes and it becomes necessary to repeat transport studies in the same cities, the right kind of data will become available. There might even be a case for reducing to some degree our dependence on transport models and utilizing 'iconic simulation', whereby traffic restraint

ANALYSIS AND PREDICTION OF TRAVEL PATTERNS   225

procedures are actually tested on the ground in a selected city or cities (Starkie, 1973, pp. 332–3). This demands both the will and the finance on the part of central and local government to adopt such tests in a situation where the outcome could well negative. But in situations where the outcome is positive the cost will probably be outweighed by the benefits extended to movement in cities throughout the country. Experimental bus lanes and roundabouts are now being tested in a number of cities and, as the next chapter shows, the number of such exercises seems to be increasing in scope and investment.

Forecasting future car ownership

One of the most crucial forecasts to be taken into account by transportation studies is that of car ownership. Decisions relating to the provision of additional road capacity in cities rest heavily on ownership forecasts which have now become the object of closer scrutiny as the environmental effects of roads and traffic have become more fully appreciated during the last decade. It is therefore important that car ownership forecasts are accurate, and they usually rely on past trends and relationships (figure 7.12) in an effort to arrive at the saturation level for the population as a whole. This is expressed as the number of cars per head, and the intercept suggests that the level is 0.54 on the basis of the percentage growth in the number of cars per head for the counties of Great Britain during 1965–70. The counties with

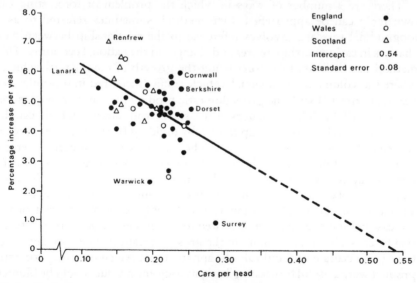

Figure 7.12 *Growth rates of cars per head, Great Britain 1965–70* (*Source*: A. H. Tulpule, 1973, figure 1)

the highest growth rates are those with the lowest levels of car ownership per head, so that the calculation of the regression intercept is heavily dependent on a wide distribution of growth rates and cars per head. A very closely bunched group of data would make this method of calculating the intercept rather suspect. It is also necessary to make assumptions about the effects of social/behavioural changes which may affect the way in which disposable income is used for car ownership in the future. Many of these may in fact reflect differences in income, for example differences in availability of leisure time, or differences in social behaviour and family contacts between various socio-economic groups. It may not be reasonable to expect people with equivalent income in the future to those who own a car today, to generate similar patterns of travel behaviour, attitudes to car ownership and use, or to participate in activities which create similar demands for mobility.

Despite these difficulties, it can safely be assumed that the growth of car ownership will eventually slow down as the number of individuals eligible to own and/or drive a vehicle reaches saturation point. The number of cars per head of population has been increasing at some 8–9 per cent per annum in recent years, but it would be inappropriate to project this rate as a means of arriving at future levels of car ownership. The problem is more complex because of the influence of underlying factors, such as residential density, household size or household income, which have been discussed earlier. In addition, car ownership may already be near saturation in some households, such as those with a single wage-earner for example.

There are a number of ways in which the problem of forecasting car ownership can be approached. One method, sometimes referred to as a longitudinal forecast, involves reference to the relationship between past changes in car-ownership figures and changes in the national economy. This method has been used for forecasting the growth of commercial vehicles where the volume of commercial-vehicle traffic (measured in ton-miles per year) is correlated with the gross domestic product for the year in question (Tulpule, 1969). This produces forecasts of ton-mileage which can be translated to vehicle-ownership figures by dividing by the average carrying capacity of each commercial vehicle. The link between commercial-vehicle ownership and the economy is like to be much stronger than the equivalent relationship for car ownership, in that 'non-economic' criteria such as the status value of a car, the costs of motoring or changes in social attitudes have an important part to play (Tanner, 1965). Only some 50 per cent of the increase in car ownership can be related to changes in national prosperity, so that forecasts based on changes in the gross domestic product are not likely to be very accurate. It should also be apparent that even if the gross domestic product were a useful forecasting base, its long-term value would be blunted as the saturation level for car ownership was reached. Other forecasters have turned to individual households in order to make their estimates. Wootton and Pick (1967) have used household income and assume that the functions relating the probability of a household owning n cars to the household income will remain stable through time. If it is then assumed that car prices will rise more slowly relative to household income, it is possible to derive a growth curve for car ownership. The main difficulty is that the method leads to a saturation level of two cars per household (0.6 cars per person) and this is unlikely to match reality for a long time ahead. Mogridge (1967) assumes that a household at a given level of income will devote a certain proportion of that income to car purchase, but inflation and the effects of changes in the relationship of car prices to income make this method of forecasting somewhat hazardous.

A second forecasting method involves the projection of trends by fitting a smooth curve to past rates of car ownership. The logistic curve is often used and this assumes that the relative growth of car ownership is proportional to the amount by which car ownership falls short of the saturation level (figure 7.13) (Tanner, 1962; Tulpule, 1969, 1973). Saturation level is based on the evidence for approaching saturation from statistics for various areas in the United States and Britain in which weight is given to car ownership as a function of population density and income. The figures shown in figure 7.10 suggest a likely saturation level of 0.45 cars per head but, in reality, the figure will vary between different parts of the country or within

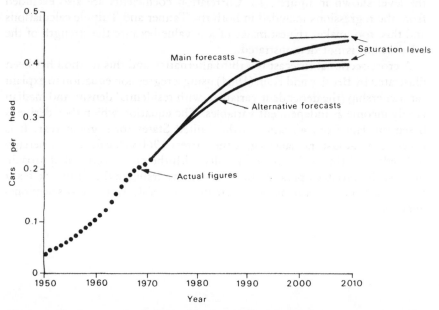

Figure 7.13 *Cars per head, Great Britain, 1950–71 and forecasts for 1972–2010* (*Source:* A. H. Tulpule, 1913, figure 4)

different areas of a city. Tanner (1965) has suggested that an appropriate scale could range from 0.50 cars per head for a rural county, to 0.40–0.45 cars per head for a medium-sized town, 0.35–0.40 for the major conurbations and 0.30–0.35 cars per head in Inner London or the equivalent areas in major provincial cities. Such differences reflect variations in population density, age, structure, income and congestion for the sub-areas listed. This scale has been used by Hermann (1968) to estimate car ownership saturation levels for counties and county boroughs in Britain.

The methods used by the Road Research Laboratory for forecasting car ownership are not foolproof (Adams, 1974). The saturation level is estimated by using the zero-intercept method which 'depends on the bold assumption that a relationship derived from the spatial variation in growth and ownership can be interpreted as a temporal trend which will remain constant over the next forty years' (Adams, 1974, p. 551). The basis for such an assumption is indeed weak but even more important, Adams (p. 553) draws attention to the way in which the TRRL estimates reverse the normal way of using a regression equation, i.e. car ownership is used as the independent variable even though it is the value which is being predicted. The predictor variable, the rate of growth, is used as the dependent variable. A regression conducted in the conventional way produces a saturation level for car ownership of 0.28 cars per head, which is over 50 per cent lower than the level shown in figure 7.13. Correlation coefficients are also excluded from the regressions included in both the Tanner and Tulpule calculations and this, too, makes the estimates of less value because the strength of the relationship is not demonstrated.

A cross-sectional forecast is a third alternative and this method has been illustrated by Beesley and Kain (1964) using a regression equation to explain car ownership (the dependent variable), with residential density and median family income as independent variables. The equation which they derive is based on data from 45 cities in the United States for a given year. It is necessary to adjust the data to give the correct 1960 value for car ownership in Leeds, and this is the main difficulty with the cross-sectional approach, because the variable operating to affect car ownership in the United States at time n will not appear as an explanatory variable in the cross-sectional analysis.

8

The development of legislation concerning urban transport policies and planning

Brief reference was made in chapter 1 to the spate of proposals for railway construction in central London in the 1840s and 1860s, and to the government's unavoidable intervention and large-scale prohibition of such schemes. It was argued that although from this time governments have been increasingly involved in regulating and guiding urban transport investment and services, until the Second World War this involvement was intermittent, partial and, it is probably fair to say, reluctant. Normally it was a reaction to a particularly pressing problem which had become a volatile and topical political issue. In other words, while there has been at least a century of government regulation in specific fields of urban transport, only during the last two decades have more positive management and planning policies developed. Gradually, but with increasing pace since the early 1960s, policies have extended from dealing separately with the different forms of urban transport, or even with one element of a mode, such as the construction of roads, towards a more coordinated policy for the different forms. Very recently, with the increased recognition of the needs of non-vehicular movements, of the social benefits and disbenefits of transport by various means, and of the environmental and social implications of transport policies, it can be argued that, at least in intention, we are closer than ever before to a system of comprehensive planning, in which transport is treated as an interdependent part. Furthermore, the 1974 Local Government Act gave heightened transport planning responsibilities to the new metropolitan and 'shire' county councils which, by retaining their role as the strategic physical planning authorities and through reformed and more flexible exchequer financing, gave them the means to practise more comprehensive and integrated environmental planning. As will be discussed towards the end of this chapter, what remains to be seen is whether these wide-ranging institutional reforms can be effective. This will depend on the very considerable questions of whether sufficient resources will be available for investment and to support unprofitable elements of the transport system, whether both planners and 'the planned' can reach sufficient consensus to enable the necessarily long-term plans to reach fruition, and whether professional, administrative and executive skills can be raised to enable the effective operation of the greatly expanded planning agencies.

The long-term extension of government urban transport policy has been encouraged, if not made essential, by many facets of the nation's changing economy and society, but the fundamental stimulus has been the increasing mobility of people and goods. Growing mobility has been both the result of socio-economic changes – such as increasing income and leisure time, lower housing densities and falling real costs of transport – and the cause of changes. Similarly it has been both a stimulus for, and an effect of, new and improved means of transport. The complex web of interrelationships that is adumbrated by a reference to increasing mobility has been examined in earlier chapters, but the principal implication for successive governments of this trend deserves emphasis. Just as individuals have spent larger proportions of their incomes on transport and travel, so the nation has devoted an increasing share of its income to these items. The largest element of the increasing national expenditure on transport has for decades been the capital cost of new roads, but in recent years government grants to cover the capital and working costs of railway and bus operation have become increasingly involved in the financing of transport, so inevitably has their role in planning and policy-making expanded. In this chapter, the principal benchmarks in this process will be described and the present-day legislative and financial framework for urban transport management and development reviewed.

The coordination and control of London's public transport

Partly because of its size, relative prosperity, and attractiveness for experimenting and innovation, urban transport changes and problems have tended to arise and to achieve public recognition in London first, and in other British cities later. This was clearly the case during the later nineteenth century when unprecedented volumes and concentrations of movement grew up in the largest cities of the economically advanced countries. London may for centuries have been congested, but its circulation problems took on new dimensions and spread over a much wider area during Victoria's reign. There were many reasons for this. It has already been noted that the period was one of unusually fertile technological development in urban transport: tramway construction dates from the 1870s and electrification from the 1890s, and railway electrification, underground construction and major street improvements were also taking place. Transformations in the geography of London were also instrumental in changing the nature of the problem. Its accelerating areal growth and developing economy, and in particular the accentuating dichotomy between a commercial or business core and the spreading, dominantly residential suburbs, stimulated the demand for movement. Shorter working hours, more regular employment and a greater separation of homes from workplaces were combining to magnify the problem at more clearly distinct 'rush' hours. When, in the first decade of the present century, these problems acquired a new shuddering dimension with the growing number of motor-cars and buses, the need for, and desirability of, more embracing regulation and coordination of facilities became obvious and a live political issue.

The debate which began at that time and continues today, about the objectives which London's transport should have, and about its control, management and plans, raised issues of continuing relevance to all sizeable towns. It is therefore an appropriate topic with which to open an account of the evolution of urban transport policy. The timescale of the institutional and policy response to the public identification of a major transport problem is itself instructive. It was not until 1933 that a single body was formed to be responsible for urban public transport planning. At the time of writing, the subject remains topical, and focuses on the means of taking over the responsibility for running British Rail's suburban services within its area. The case of London clearly illustrates the fact that legislation, policies and institutional arrangements to deal with such large-scale subjects as urban transport evolve slowly and progressively in Western democracies. This is an inevitable consequence of not only consensus politics but also of the steadily changing nature of towns, of movement within them, and of the means available for transport.

Among the questions that were most energetically debated at the beginning of this century was that of the allocation of responsibility for regulating and coordinating urban public transport services and facilities between, on the one hand, the private and public sectors, and on the other, between elected and nominated local bodies and central government. Public attention was raised by the fierce competition between two rival financial groups to promote and construct deep-level underground or 'tube' railways in London around the turn of the century. The fact that both groups were backed mainly by American finance may have been the first cause for popular interest, but it was not jingoism that led many people to worry that the provision of tube railways in the Capital was being determined less by an objective analysis of traffic needs, than by the strength, influence and political skill of financiers. 'Many Londoners rallied to the support of those who were advocating some kind of London passenger transport authority to scrutinise the various transport schemes in the public interest' (Barker and Robbins, 1974a, p. 85). Some wished to see a body with powers to control as well as scrutinize, similar to the New York Rapid Transport Commission.

At this time the maturing London County Council was the largest metropolitan government unit in the country, and unlike many of the larger county boroughs elsewhere it had authority over a considerable proportion of the functional urban area based on its territory. In 1902 delegates from the LCC urged the President of the Board of Trade to view the tube proposals in terms of the requirements of London as a whole, and not just in terms of possible profit to the promoters (Barker and Robbins, 1974a, p. 86). Reaction by the government was dilatory and of a well-established kind. In February 1903 it decided in favour of a time-consuming Royal Commission on London Traffic. Its report of June 1905 recommended the setting up of a Traffic Board, of no more than five members, to produce for Parliament an annual report on all transport and traffic matters in Greater London. The government, however, did not act upon even this very limited recommendation, although in 1907 a London Traffic Branch was set up in the Board of Trade with similar functions but no independent authority (Barker and Robbins, p.89).

As has been clearly shown by Barker and Robbins in their admirable *History of London Transport*, it was not at all surprising that the LCC's requests to be given greater authority over public transport were not granted. There were clear and disappointing limits to its willingness to act in the widest possible interest. It was inconsistent in wishing to be both an independent traffic tribunal and the sole tramway authority within its area, and gave far from convincing replies when asked by the Royal Commission how it intended to secure coordination between railways and tramways. Another difficulty was that the LCC did not control the rapidly expanding private and municipal tramway networks in the peripheral county boroughs and urban districts, and these local authorities not surprisingly were against the LCC having any rights over transport facilities in their territories. Nor did the LCC have complete control over affairs in its own territory, for it was not its own street or highway authority. These responsibilities were lodged in the smaller Metropolitan Boroughs as successors to the parish vestry committees, and under the 1870 Tramways Act they possessed the power of veto over any tramway proposal. The late reform of London's government, and the perpetuation of 'vestry rule' until the creation of the LCC in 1888, may have exacerbated and certainly did nothing to alleviate the nineteenth-century transport problems of the city (for an outline of this prevarication, see Briggs, 1968, pp. 319–26). It was a power that was often used and which, by preventing the extension of tramways into the central area, perpetuated the separation of the LCC's tramways into northern and southern networks and, incidentally, therefore enabled the horse bus to flourish longer in London than in most cities. Finally, but not least important, the omnibus and railway companies were by and large hostile to the LCC being given wider powers; their attitude to the proposal was strongly influenced by the natural assumption that it was little more than a means to featherbed the LCC's tramways. The government's reluctance to establish a coordinating authority can therefore be partly attributed to the hostility of private companies and the lack of agreement and trust among public bodies. There was, in effect, insufficient consensus that the benefits of a planning body charged to maximize the common good would outweigh the disadvantages to sectional interests, and perhaps most tellingly, insufficient confidence that the LCC would put aside its own vested interest (Barker and Robbins, 1974a, p. 88).

Renewed calls for the more effective integration and control of London's public transport services were to come frequently but with little response from the government until the 1930s. Although in the year after the First World War the Ministry of Transport was founded, its first concerns were putting the ailing railway companies on a secure peacetime footing, and on treading an uneasy legislative and fiscal path between motor manufacturers and users on the one hand and highway authorities, ratepayers, pedestrians and landowners on the other (for an intricate review of these topics, see Plowden, 1971). It is interesting to note in connection with railways, however, that its first 'Outline of Proposals as to the Future Organization of Transport Undertakings in Great Britain and Their Relations to the State', published in 1920, suggested that railways purely local to London should be amalgamated into one of seven national groups. (It even proposed that the boards of the groups should be composed not only of shareholder's representatives but also workers' representatives.) When the Bill was introduced in May 1921, an advisory committee on London traffic was still due to report, and the government explained that the London railways were to be dealt with later by separate legislation, 'as part of the London traffic problem'. In practice, 'later' was to mean ten years afterwards (Bagwell, 1974, p. 249; Barker and Robbins, 1974a, p. 203).

Among the many committees and bodies to recommend a new body for overseeing London's transport were a 1919 Select Committee on Transport in the Metropolitan Area, a 1920 Advisory Committee on London Traffic, and a Royal Commission on London Government of 1923. Barker and Robbins (p. 204) cite the LCC, gloomily commenting: 'These different commissions and committees, while not always in accord on questions of method or machinery, were in complete agreement in the view they took as to the necessity of some new body being set up.' The only result, however, was the London and Home Counties Advisory Committee of 1924, formed under the London Traffic Act of the new Labour government. Although the committee was advisory rather than executive, its nineteen members, including local and central government and police nominees, produced a series of reports that strongly influenced later steps towards unification and coordination. Its geographical area of concern extended throughout the pre-1968 London Transport area, but it worked chiefly by studying specific local difficulties. These characteristically highlighted the wasteful competition upon certain routes, the illogicalities of different pricing systems by the various operators, and the need for new, often expensive, railway facilities. According to Barker and Robbins's account (pp. 211–13), by 1927 the committee was increasingly commenting upon the difficulties of securing major improvements without the unified management of the local passenger transport agencies. As a step towards this end, they proposed a common fund of earnings and a common management, while leaving ownership with the existing proprietors, except for some of the smaller bus operators. Although the proposal gained the support of the LCC and the Underground combine, the early steps in enacting an LCC-sponsored bill to this effect were thwarted by a change of government in the summer of 1929. The new Minister of Transport, Herbert Morrison, envisaged a more radical solution to the problem.

By December of that year Morrison had outlined his proposals, which featured four points: unification under public control; management by a non-political body; participation by the main-line railways, without transfer of ownerships; and a self-supporting, unsubsidized system to be run commercially. After preliminary consultation and financial investigation, a parliamentary bill was published in March 1931, and this began to pass through committee in the following month. Although the Labour government fell in August, to the great surprise of many and with surprisingly muted opposition, the Minister of Transport, P. J. Pybus, of the incoming National government soon announced that the bill would be carried over into the new Parliament from the stage at which it had been left. This is yet another sign of the widespread agreement on the need for integration of the various elements of London's transport under a single management and even single ownership. After negotiations with the less enthusiastic and occasionally hostile directly affected companies, notably the Metropolitan Railway and the independent bus operators, the remaining stages of the Bill were completed, and the London Passenger Transport Act came into force on 1 July 1933 (Barker and Robbins, 1974a, pp. 270, 272–82).

The London Passenger Transport Board which was established by the Act was responsible for all local public passenger transport until the end of 1947. Its principal achievements were the steady and usually undramatic coordination of the various facilities under its control, such as the restriction of the numbers of buses plying along tram routes, and the improvement and extension of underground railway services, facilitated by the revenue pooling agreement and other signs of greater cooperation from the main-line railway companies. (The widespread electrification of the Southern Railway's suburban services during this period was not, however, due to LPTB initiatives.) Other achievements were the integration of the previously independent tram companies' services, and their partial rationalization and conversion to trolley-bus operation, and the standardization and marked improvement in buses, in part due to an early decision to convert to types with diesel engines and to commission a small number of bespoke designs of high quality. The rapid progress of the 1930s was inevitably halted during the war years, not least in financial terms. At the end of the war, to the problems of dealing with the backlogs of maintenance, new equipment purchases and modernization, were added more structural political and planning changes. Of immediate consequence was the new Labour government's fulfilled intention to nationalize the major transport undertakings of the country under the British Transport Commission. Under the 1947 Transport Act, therefore, the functions of the LPTB were transferred to the London Transport Executive (LTE). This survived (and was the only BTC Executive to do so) the partial dismantling of the 1947 Act by the Conservative government's 1953 Transport Act, in the view of Barker and Robbins (1974a, p. 335), because 'there was too much political risk in taking away from London a distinctly responsible transport authority which it had had since 1933'. Instead, the new government set up in April 1953 a committee of inquiry into London Transport. When this reported in 1955, it was generally favourable about the operations of the LTE, although it identified the relationship with British Railway's London suburban operations as a structural weakness and difficulty.

By this time the strides towards developing a more effective system of managing a large city's transport services which had been taken in London during the 1930s appeared to be faltering. This was not entirely unwarranted, and several changes in political, planning and transport attitudes in the post-war period were in useful directions, and after gestation for a decade or more, were to lead to further important steps in the evolution of urban transport legislation, policy and machinery. For one thing, it will not have escaped notice that the account so far has focused entirely on London. While the transport problems in the other large conurbations were neither so severe nor so complicated – except to a very limited extent in Glasgow and Liverpool no other city had underground railways, the role of the surface railways in urban transport was generally much less, and generally the municipal or city corporations operated both bus and tram services – nonetheless post-war governments were more readily willing to examine transport problems on a national basis. It began to be seen that there were urban transport problems rather than solely a London transport problem. Secondly, with the growing interest in, and effectiveness of, town planning, and with the rapid rise in private car ownership that the 1930s had indicated and the 1950s fulfilled, the severe limitations of a purely public transport authority, such as the LPTB, were increasingly evident. More than ever, the principal transport problem was bound up with the role of the private motor vehicle, and the largest public cost of urban transport was the construction, improvement and maintenance of roads.

There have been a number of important changes in the organization, control and planning of London's transport since 1953, but despite the continuing unusual scale and complexity of London's transport problems, and despite the earlier timing of local government reorganization within the capital, a clear feature of recent legislation in the urban transport field has been to treat similarly all the large conurbations of the country. Further references to London will therefore be made when we come to deal with such topics as the creation of conurbation passenger transport authorities and local government reorganization. At this point, it will be sufficient to note that the 1947 LTE survived in title until only 1962, and in the effective range of its responsibilities until the 1970s. Under the 1962 Transport Act, which dismantled the British Transport Commission for railway management and finance reasons, the LTE was replaced by the London Transport Board. Then, with the creation of the Greater London Council (GLC) in 1965, the opportunity at last arose for combining transport planning and operation under a single body. A new LTE was set up, to be known simply as London Transport, with a separate statutory existence to the GLC, but with the same area of operation, which broadly was the previous central or 'red' bus area. The GLC was to have powers of direction over policy and fare levels, but not over the day-to-day management of the tube and bus services (Ministry of Transport, 1968a, p. 10). To date, the relationship between the two bodies has not been entirely cordial, particularly with regard to financial planning, or stable, in that negotiations with British Rail to take over responsibility for suburban services within the GLC area are still in progress.

Although, therefore, a great deal has been achieved over a long period in coordinating London's transport services, awkward divisions of executive and political responsibility remain. Very recently, the DoE has initiated arrangements for setting up an advisory committee to coordinate the two London public transport operators (BR and LTE), and their political masters (DoE and GLC). This follows the recommendations made by the London Rail Study Group, but the clinching argument for this development appears to be the 'constraints on public expenditure that have developed since the London Rail Study was completed' (DoE, 1976a, vol. 1, p. 45). The reader will by now be familiar with both the tactic and the justification.

If one long-standing theme of the evolution of urban transport policy has been the control and coordination of public transport investment, operators and services within the larger cities, and particularly London, the other major influence has been the evolution of the legislation and policy pertaining to roads. It is to this complex story that we now turn.

First steps towards the modernization of highway administration

A major strand of transport policy in all Western countries at the present time is that pertaining to highway development and maintenance. This is as

THE DEVELOPMENT OF LEGISLATION

true of urban policy, where the cost and impact of new road construction is unusually severe, as it is of policies for inter-urban movement. In the latter sphere, it can be argued that in the United Kingdom, the task of adapting our transport system to the change from rail to road and constructing a national network of modern motor roads is well underway. Not all the motorways that will be built are yet under construction or even in the planning stage, but the general nature of the task and of the country's solution of it has been clearly specified, and the framework of the end-of-century road network already built. This is not to say that final decisions have yet been reached on the scale of twentieth-century road construction or the respective end-of-century roles of rail and road, but the questions have been clearly been asked and carefully examined (Foster, 1963, pp. 11–28).

In towns and cities, however, the situation is quite different. A decade ago, the separate conurbation and large-city transportation studies were usually recommending large-scale urban-motorway and other road-construction schemes for their study areas. Partly because the total investment required to carry out all these schemes reached untenable proportions, but also because of a change in attitude among both the public and the decision-takers against untrammeled provision for the motor vehicle, central government began to urge conurbation and planning authorities to give greater priority to public transport as a means for alleviating urban transport problems. Very few urban motorway schemes have proceeded in the last few years, but it is quite clear that no consensus path for urban transport policy has yet emerged. The Greater London Council is keen to proceed with selective road improvements, but elsewhere these continues to be considerable opposition to large-scale urban road expenditure: the place of roads in urban transport policy is still an open question. In this section, we shall briefly trace the evolution over the last century of highway administration to its present form, and review the present financial, legislative and planning structures in which the current issues are being debated and, to a greater or lesser extent, will be resolved.

The ancient system of community or parish responsibility for roads survived until surprisingly late in the nineteenth century. For a time, from about the middle of the previous century, it was supplemented by the turnpike trusts which, until railway competition ruined their finances, became particularly important as a means of new construction and major improvement of inter-settlement roads. By 1837 their combined revenue had reached over £1.5 million and they covered over 22,000 miles of roads, although this was only a fifth of the mileage of parish roads, which covered most urban streets. Their national coverage was very uneven, some counties such as Cornwall, Cumberland and Norfolk at no time having more than 8 per cent of their roads turnpiked (Dyos and Aldcroft, 1971, p. 222). As they declined, responsibility for their routes passed, and in many cases returned, to the parish authorities, although in a few instances Parliament voted sums for road construction, as with the Holyhead Road, when the turnpikes mismanaged their affairs (Foster, 1963, p. 168).

The first major reforms in highway construction and administration came, not in 1835, when the General Highway Act confirmed the parish as the basis of hired labour and salaried officials, but with the creation of Urban and Rural Sanitary Authorities in the 1872 Public Health Acts. Although some amalgamation of parishes into larger highway authorities had been lethargically achieved under earlier legislation, this was the first major step in the progressive reduction in the number of highway authorities from the 15,000 local vestries in 1835 to the 5,000 remaining highway parishes in 1894 (Dyos and Aldcroft, 1971, p. 225; Savage, 1966, pp. 84–5). The 1872 Acts, and the succeeding Local Government legislation progressively accomplished the wholesale replacement of the parish system, by 1888 making the new County Councils into Road Authorities responsible for all main roads, comprising about 22,000 miles, by transferring Home Office jurisdiction over roads to the newly created Local Government Board, and by introducing grants-in-aid from the Exchequer to highway authorities. Many small changes culminated in the Local Government Act of 1894 which abolished Highway Districts and Highway Parishes and merged them into the Rural Sanitary Districts, thus erasing 'the last trace of the immemorial responsibility of the parish for its highways' (S. and B. Webb, 1920, p. 214).

While the late nineteenth-century changes considerably rationalized highway administration and led to a higher and more consistent standard of road maintenance, they maintained the principles of local control and, with minor exceptions, local finance, through the rating system. Reflecting the local government structure of the time, in 1902–3 highway administration was controlled by nearly 1,900 separate local authorities: 132 county or county borough councils, 279 non-county and metropolitan borough councils, 812 urban district councils and 661 rural district councils. This structure was not suited to the instigation of new construction, or to road planning at the regional or national level: only 4 per cent was added to the total road mileage over the period 1899 to 1936 (Dyos and Aldcroft, 1971, p. 370). Yet, almost as soon as the late-Victorian local government structure had been established, the nuisances and requirements created by the rapidly growing number of, first, cyclists, and later, motorists, began to reveal the deficiencies of the road administration system. Interest groups emerged remarkably quickly, and each began to lobby the government to adopt its own, often narrowly sectional policies. As Plowden (1971, pp. 60–83) has shown in a detailed study, the more perceptive observers already saw in this first, formative decade of the century that the problem of the motor-car was not simply a short-term difficulty of speed limits, law enforcement, or even of dealing with the ubiquitous dust by road surfacing, but a question of

THE DEVELOPMENT OF LEGISLATION 239

structural reform to encourage and allow greater investment and more rapid construction of roads. In 1901 the Royal Commission on Local Taxation had recommended that main roads should be considered 'a subject of national importance which need not be entirely financed by the ratepayers, and that they should be in part supported by grants out of the central exchequer' (Plowden, p. 69). If the focus of the public's and Parliament's concern was more with safety and law enforcement, laced with more than a touch of interclass antipathy, the resulting Royal Commission on Motor Cars of 1906 did examine the wider issues, and on these it made a number of radical proposals. Largely adopting proposals submitted by the (Royal) Automobile Club, the Commission recommended the establishment of a central department to collect motor taxation, and to distribute grants to local authorities for repairs and improvements aimed at producing better, less dusty surfaces. After a lengthy period of debate, in August 1909 the government put forward the Development and Road Improvement Funds Bill, which contained the more radical proposal that a new Road Board should have powers not only to help local authorities but also to built new roads reserved for motor traffic and without speed limits. The Bill was considerably diluted during its passage through Parliament. A Road Board was constituted in 1910, but during its ten-year life it sponsored few major improvements and no new routes; over 90 per cent of its grants went towards small-scale improvements of road surfaces. 'Local authorities were handsomely subsidised, without losing any of their independence to the Treasury, and without being obliged to incur the much heavier cost of new construction or major improvement' (Plowden, p. 95).

Despite the considerable increase in motor traffic during the inter-war years, little progress was made. Successive Acts of Parliament introduced a greater element of central government responsibility for the nation's principal roads, and a corresponding increase in Treasury grants. While calls for a national road plan and for the construction of special-purpose motorways were more frequently and more explicitly made, not until the end of the period did the government begin to take positive steps in these directions. Throughout the period, the initiative for proposing new construction remained with the county councils and other local highway authorities; the Treasury was reluctant to provide supplementary funds for road projects, which understandably during the depression it saw as having low priority; and the quick succession of junior ministers of transport were generally unable to wrench transport policy away from the dominant influences of unemployment, the severe problems of the railway interests, and safety on the roads.

The Ministry of Transport had been established in 1919 and in the following year took over the functions of the Road Board. Preliminary work on reforming road administration had been started by the Board immediately before the First World War, when it began the threefold classification of roads (Bagwell, 1974, p. 271). The Ministry continued this work, and used the resulting categories to allocate varying levels of grant to the local highway authorities. The mileage of roads attracting grants, and the proportions of total costs that these represent, have been increased on many occasions since this time. By 1929, the mileage of classified roads in Great Britain had expanded to 66,800, or nearly a third of the total. A further change came with the Trunk Roads Act of 1936, under which 4,500 miles of major roads were transferred from the county authorities to national administration. Although the Ministry of Transport drew up ambitious plans for new construction costing £131 million, of which £106 million had been accepted in principle by March 1939, the war came and the plans were shelved (Dyos and Aldcroft, 1971, p. 369). Despite the administrative and financial changes, as Walker (1947, p. 31) has shown, throughout the 1930s total road expenditure remained below the level of 1930. Throughout the inter-war period, the bulk of expenditure continued to be for improving road surfaces, although during the 1930s several inter-city roads were completely reconstructed and many separate schemes for by-passes and ring-roads, many of them in and around major cities, were completed. These schemes hardly scratched at the growing congestion problem within the cities. Indeed the emphasis on impriving radial routes in the less densely

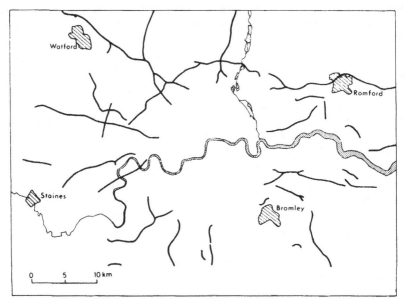

Figure 8.1 *New arterial road construction around London, 1900–39* (Sources: *Ordnance Survey maps; Greater London Council, 1970, various figures*)

developed outer suburbs had the effect of improving the accessibility of town and city centres by car compared to other forms of transport, but not even traffic management measures were being adopted on a large scale to facilitate the internal distribution of road vehicles. The clearly radial nature of twentieth-century road construction around London to 1939 can be seen in figure 8.1.

During the war and the following period of austerity, although announcements continued to be made about the progress of plans for new motor roads, very little was achieved. From 1939 to 1952 practically no major improvements or new road construction were carried out, and worse, from 1948 to 1953 expenditure on maintenance and minor improvements was only 66 per cent, and new construction 21 per cent, of the average expenditure for the years 1936–9 (Bagwell, 1974, p. 368; Savage, 1966, p. 199). In the transport field, the post-war Labour government's principal interests as reflected in the 1947 Transport Act were related to the public ownership, coordination and modernization of the national passenger and freight undertakings. The succeeding Conservative government shared these interests, but its Act of 1953 moved in the directly opposite direction of reducing public control and of greater independence for each transport sector. Even when the combination of greater national wealth, and rapidly growing car ownership and urban traffic congestion, encouraged the government to react more postively to the growing calls for increased road expenditure, the priority it adopted was the construction of the rural sections of inter-urban routes. These had the merit of relatively low costs of land acquisition, compensation and construction, and moreover were more spectacular in the transformations they produced in travel time and average speeds.

Not until the 1960s did motorway construction begin in clearly urban settings, and even then the projects undertaken were determined by their importance for the national network rather than for meeting the transport problems of the host city. Meanwhile, little change was made to the pre-war system of road administration and financing. In the early 1960s the levels of Ministry grants to local authorities were 75 per cent of the cost of any maintenance, improvement or new construction of Class I roads, 60 per cent for Class II, 50 per cent for Class III; and none for unclassified roads. All costs of motorway and trunk roads were paid for by the Ministry of Transport. This whole system was roundly criticized by Foster (1963, pp. 173–6): 'It is a muddle, and a failure, if, as seems likely, it ever were meant to be just [in its] attempt to divide costs fairly between national and local authorities . . . . In short, it is illogical, inconsistent and arbitrary.' Symptomatic of the low priority given to roads by central government was the fact that in the mid-1950s only three of the sixteen Under Secretaries in the Ministry of Transport dealt with roads or road transport, and symptomatic of the low status given overall to transport was the fact that from the beginning of the Second World War until 1957, the Minister did not have Cabinet rank (Plowden, 1971, p. 349). As late as 1959, official policy towards the urban road problem still rested on trying to push more and more traffic through existing streets. A London Traffic Management Unit was set up within the Ministry with the principal objective of speeding traffic flows. At about the same time, however, other influences which had been germinating ever since the war years, were strengthening, and in remarkably few years led to considerable changes in public policy. To a very great extent, the evolution of policies to deal with motor vehicles in urban areas since that time underlies the whole shape of current urban transport policies, and therefore requires close attention.

The debate on traffic in towns

Because of the pressing problems of the national passenger and freight undertakings, the post-war Labour government's transport policies gave little attention to urban problems. Yet Herbert Morrison, a leading member of that government, had already shown his interest in and command of the London passenger transport problem of the 1920s. It was another enthusiastically pursued aspect of the government's policy, that of town and country planning, that was eventually to lead to a reappraisal of urban transport policy. Under the 1947 Act, county and county borough planning authorities were required to produce development plan maps. As with the major conurbation planning studies, such as those of London, Glasgow and Hull by Sir Patrick Abercrombie, and with the contemporaneous New Town plans, roads were seen as an integral part of the physical plan. While the roads and transport-planning elements of Town Development Plans cannot be said to have quickly produced new thinking or expertise–road planning was still determined largely by a combination of route selections based on simple estimates of flow and congestion and of carriageway design based on engineering criteria – the creation of local authority planning departments and the resulting expansion of the planning profession was in the longer term to be important. Until the middle of the 1960s, urban transport policies were strongly guided by the attitudes and concerns of highway and traffic engineers, who took on and often effectively met the problems of new road construction and urban traffic management. More recently, however, other groups with both differing and wider interests in the city, including many with experience of town planning problems, have challenged and contained the sectionalist approach that was being pursued. Two fallacies were exposed: that the congestion problem could be solved by building more and more roads, and that the whole population would in the matter of a few decades have access, let alone independent access, to private cars. The simply impossible costs of road schemes recommended on the basis of traffic demand criteria was revealed, and most widely and populistically the social and environmental effects of roads solutions to urban problems were highlighted and rejected. In many accounts, these changes in thinking and attitudes are attributed to undoubtedly important individuals: Tripp, Buchanan, Smeed, Beesley, Foster and Hillman are names that immediately spring to mind; but many others clearly were thinking along the same lines and ready to encourage and implement policy and practice changes.

It was (Sir) H. Alker Tripp, an Assistant Commissioner at Scotland Yard, who, in two influential volumes (1938, 1942), first emphasized the importance of integrating land-use and communications planning. The second, *Town Planning and Road Traffic*, illustrates well the state of progressive thinking at the time. Tripp saw the problem in terms of road casualties, traffic delays and congestion, and inappropriate road design. The problem had to be contained in the short term by restrictive laws and traffic controls, but 'constructive work represents the path to complete solution of the whole problem, restrictive measures are only the provisional safeguards which in the meantime are indispensable' (Tripp, 1942, p. 10). In coming to this conclusion, the lack of an understanding of the potential scale of the traffic problem in towns was revealed, as well as ignorance of the now commonplace understanding that road improvements generate traffic, and that in larger towns not enough road space can physically be provided to cater for the latent demand. Only towards the end of the 1950s, when the limited ability of traffic management measures and piecemeal construction improvements to deal with the growing urban traffic volumes became apparent, did this attitude change. Even as late as 1965, a digest of the accumulated knowledge of the Department of Scientific and Industrial Research's Road Research Laboratory entitled *Research on Road Traffic*, can fairly be said to reflect Tripp's views and concerns: in nearly 500 pages there are just thirteen on economics, and none on road or transport planning.

In Britain Colin Buchanan with his two volumes (Buchanan, 1958; Ministry of Transport, 1963b) provided a much more detailed analysis of the impact of the motor vehicle in urban areas and, of equal importance, stimulated enormous public interest and reappraisal. The first volume (1958) was written when Buchanan was working in the Ministry of Housing and Local Government as a planner, and was offered as the first 'general survey . . . of the impact of the motor vehicle on our society'. It challenged the assumption that existing policies and remedies would eventually solve the problem, because a saturation level of car ownership was being approached. It also again challenged the still prevalent view that the main problems raised by cars were in rural areas, and with this reiteration, the meassage that dealing with road traffic in towns raised major issues of urban planning was followed by policy studies. In 1960 the Ministry of Transport set up a Study Group to study the long-term problems of traffic in urban areas. The Steering Group was headed by Sir Geoffrey Crowther, and the Working Group by Buchanan, and their reports were published together in 1963. The Working Group's conclusions were that 'very few of the statutory development plans really face up to the future problems of traffic and transport' (Ministry of Transport, 1963b, p. 192). While advocating much greater investment in towns to achieve the accommodation of higher traffic volumes through 'the canalisation of longer movements on to properly designed networks serving areas within which, by appropriate measures, environments suitable for a civilised urban life can be developed', it was also recognized that some degree of reliance on mass transport is unavoidable. 'If movements have to be contrived by more than one means of transport, then clearly some planned co-ordination between transport systems is necessary' (pp. 191–3). This could be achieved by the supplementation of development by 'transportation' plans, which would be part of the statutory submission. A fundamental element of any such plan would be the decision as to what level of road traffic to accommodate: this would determined the scale and cost of the road proposals, and also be the principal factor in governing the size of the residual task to be handled by public transport. These summary remarks in no way do justice to the number and range of the conclusions made by the Study Group, and the interested reader should regard the report as required reading. It is worth noting, however, that the rather tentative and skeletal suggestions by Buchanan's group about policy and legislative changes were strengthened by the prefatory recommendations of the Steering Group.

In their concluding paragraphs on 'Ways and Means', Sir Geoffrey Crowther's team argue that the then-existing arrangements for co-ordination between the local planning authorities, the Ministry of Transport, the Board of Trade and the Ministry of Housing and Local Government, 'effective though they may be for the purposes for which they were designed, do not lend themselves to the taking of prompt initiatives on a very large scale and embracing the fields, now administratively separate, of town planning, transport, housing and industry (Ministry of Transport, 1973b, Section 44, unpaginated). Instead, they went on to suggest in some detail, there should be a fully integrated hierarchical planning system, with coordinated policies and responsibilities at national, regional, and local levels. With unusual temerity for such a report, they went on to suggest that the 'new machinery should take the form of a number of Regional Development Agencies, one for each recognisable "urban region"' (Section 49). The mandate of the Agency should be to oversee all aspects of urban modernization, but not to take over functions which were, or could be, effectively performed by the existing (local) authorities. To carry out this task, the Agency should be the channel through which all grants for development purposes were directed. Although there were caveats explaining that the groups had not considered all the details and implications of their suggestions, they were forcibly and confidently expressed, and, above all, it was firmly maintained that the necessary tasks 'cannot be done by any existing agency or by any joint body formed from existing agencies' (Section 52).

With hindsight it can now be seen that some of the assumptions of both the Steering and Working Groups proved misplaced. In 1963 it was widely believed that the relatively rapid rates of population and economic growth of the previous fifteen years would be maintained, and this climate clearly influenced the optimistic estimate of the amount of resources that could be moved into urban modernization, and in particular urban road construction. It is also now argued that the report adopted a too generous view of the utility and potential availability of the private car as a means for satisfying personal transport requirements, and correspondingly devoted too little attention to the needs of large groups of the urban population, such as the young and elderly, housewives and other semi-dependent adults, and those who cannot afford or do not wish to own cars. Finally, the almost exclusively architectural and engineering background of the Working Group was reflected in the dominantly physical and aesthetic interpretation of the environmental problems generated by roads and traffic. In more recent years we have heard contrasting opinions, often vociferously expressed, which have variously stressed the impact of modernization and, specifically, transport investment on social, community and other aspects of the quality of life in urban areas.

Several weaknesses of the reports' analyses of urban traffic, future car ownership and future urban land-use distributions, and of its suggested criteria for transport investment were subject to critical debate, particularly on the part of transport economists (Foster, 1964, Beesley and Kain, 1964, reprinted in Beesley, 1973, pp. 189–222). The crudity of many of the Working Group's assumptions had been admitted, and the report had itself argued that a great deal of detailed research and thinking was necessary to validate or modify its suggestions, but even so, the team was undoubtedly weak in the areas of economics and investment planning. On the other hand, quite often the report's critics were equally guilty of over-emphasizing their own interests. For example, Beesley and Kain agree with Buchanan that there was a need for a change in the traditional evaluating criterion for a road investment, i.e. that a reduction in the total operating costs for vehicle users should result, but take issue with him for arguing that 'within any urban area as it stands, the establishment of environmental standards automatically determines accessibility, but the latter can be increased according to the amount of money that can be spent on physical alteration' (Ministry of

Transport, 1936b, p. 45). They argue that 'this puts the decision in the form: decide on the standard of environment as an absolute; increase investment to increase accessibility' (Beesley, 1973, p. 192). Instead they suggest that a more comprehensive analysis of the distribution of benefits and costs of changed accessibility and changed environment should be made. While these calculations could not be precise, a useful estimate of the main environmental losers could be made, and these should be compensated by payments added to the cost of road improvements. Interestingly, however, Beesley and Kain see environmental characteristics in much the same way as the *Traffic in Towns* team, namely in terms of townscape and property values. Largely because more explicit and well-defined investment criteria can best be applied to relatively small-scale and short-term projects, but also for the good reason that programmed and flexible plans are likely to be much more satisfactory than grandiose long shots, Beesley and Kain concluded that the 'piecemeal approach to investment which the Report abhors seems to us to be the right one' (Beesley, 1973, p. 216). In many respects their argument convinces, and is a useful corrective to the inevitable fancies of less-disciplined thinking, but in going on to suggest four types of improvement that would result (all of which were road schemes of a cosmetic character), they highlighted both the limitations of a gradualist approach to a complex problem, and the value of the much wider conceptions and strategy for dealing with urban transport which Buchanan and his team had exemplified.

If *Traffic in Towns* can be said to have somewhat overstated the problem, and, in the last resort, not predicted all the hurdles in the way of a solution, these were the inevitable blemishes on a valuable and very influential study. Many of its recommendations in terms of policy, grand design and detailed layout have proved of enduring value and have been incorporated into planning practice. Its impact on government attitudes was rapid and substantial. In early 1964, a joint circular (1/64; cited by Beesley and Kain, 1964) from the Ministry of Housing and Local Government and the MoT stated that 'the Government accept Buchanan's analysis of the situation.... They agree also with the main planning concepts and techniques set out in the Report.' Even the Regional Development Agencies have appeared, albeit in much less substantial form, as both conurbation Passenger Transport Authorities and more recently among the functions of Metropolitan Counties. The report was not the only or the last influential study, and other have since brought contrasting viewpoints and skills to the analysis of the urban transport problem. The economists, Foster and Beesley (see e.g., Coburn *et al.*, 1960; Foster and Beesley, 1963; Foster, 1964), developed and applied the techniques of cost-benefit analysis to the evaluation of major proposals; the traffic scientist, Smeed (1961; MoT, Smeed Committee, 1964) examined the technical and financial possibilities of restraining traffic by pricing systems;

Thomson (1969) and Hillman (1973, 1975) articulately summarized and extended the voices dissenting from the roads solution to the urban transport problem. Before embarking on an assessment of the most recent contributions to this continuing debate, and attempting to foresee the future development of policy, we turn from considering the climate of opinion in which policy has its origins, to the details of policy and legislation since the mid-1950s.

Urban policy in the 1968 Transport Act

In his review of transport policy written in 1963, Gwilliam found few explicit statements dealing with the urban transport problem. The 1962 Transport Act was in lineal descent from its predecessors of 1947 and 1953, and provided for the reorganization of the nationalized transport undertakings, in part to give the constituent industries greater commercial freedom. The Act was 'notably silent on the question of relationship between the various forms of transport', even outside the towns (Gwilliam, 1964, p. 231). Even the Labour opposition at that time uncharacteristically had done little thinking on this subject. They showed a greater disposition to see transport problems as part of multi-dimensional regional problems, and to argue that 'an essential step towards the efficient use of resources is that all transport investment decisions are made on the same basis and perhaps by the same body.... To this end it had been suggested that regional and central planning agencies should be responsible for both road and rail investment.' Little thought had apparently been given to the provision of urban road space and the control of the ways in which it was to be used (Gwilliam, 1964, p. 240). In the following year, the Labour party was elected into office, and in December 1965 Mrs Barbara Castle was appointed as Minister of Transport. Briskly, and no doubt sometimes abrasively, the Ministry was stirred into publishing a series of six White Papers which contained encouragingly fresh thinking on urban, as well as other aspects of, transport (MoT, 1966, 1967a, e, f, g, 1968a). These were the product of teams of experts, including many of those named in the previous section, who were appointed by the Labour ministers to aid the permanent staff.

The Ministry's diagnosis of the urban transport problem was clearly set out in the first White Paper, *Transport Policy* (1966, p. 14) and is worth quoting at length:

> Those responsible for traffic management and parking are not directly affected by the consequences of their policies for the operation of public transport undertakings. Public transport operators, for their part, must try to pay their way without having any control over many of the factors which affect their power to do so. Again, because local authorities have a financial responsibility for their own bus undertakings but not for local

railway services, their attitude to fare levels is different in the two cases. The lack of a carefully weighed local transport plan is often exacerbated by the fact that highway development is grant aided while rail or bus development is not. Local authorities therefore have little incentive to explore the development of rail or other rapid transit systems which might better serve local needs.

This analysis emphasizes the interdependence of the various sectors or modes of urban transport, and clearly takes the view that public transport operators had been working under increasingly difficult conditions, in part as a result of the measures taken to accommodate the private car. Among the legislative and financial needs visualized at this early stage were the extension of highway capital grants to certain types of expenditure on traffic management, the widening of the powers of local authorities in respect of traffic management and parking control, and the creation of a new responsibility for local authorities to prepare traffic and transport plans for their areas. These ideas were considerably developed in the White Paper on *Public Transport and Traffic* (1976e), in which the principal points of urban policy were (see p. 2):

1. The planning of public transport, particularly in our cities, must be a responsibility of local rather than central government.
2. Transport planning must be done over wider areas than those covered by any existing local authorities outside London.
3. Operation of the basic network of urban passenger transport services is best carried out by publicly owned bodies rather than by private companies responsible to shareholders.
4. The provision of transport can no longer be considered in isolation from other developments.
5. Investment in local public transport must be grant aided by central government just as investment in the principal road network of our cities and towns receives capital grants of 75 per cent from the exchequer.

More generally, it was argued that all urban transport studies had suggested that our major towns and cities can only be made to work more effectively, and provided with an improved living environment, by giving a new dynamic role to public transport as well as by expanding facilities for private cars. Unless we recognize this, we shall pull down the centres of our towns in an attempt to get rid of congestion, but the result will be that congestion will not be removed, and the character of our towns will be destroyed.

The form of the government's proposals took clearer shape in the specific suggestions for the strengthening of the transport planning and operating responsibilities for the Greater London Council, by the creation of a Greater London Transport Planning Group with wide representation, and the transfer of the London Transport Board's functions to the GLC (MoT,

1968a, p. 9). With only detailed modifications, the urban aspects of the government's policy were incorporated in the 1968 Transport Act which, 'with its 161 clauses and 18 schedules, established an all-time record for length in a measure of this kind' (Begwell, 1974, p. 351). The principal provisions of Parts II and III of the Act were that within the major provincial conurbations Passenger Transport Authorities (PTAs) were to be established under broad local authority control to be responsible for all forms of passenger transport within their areas. There was a particular responsibility to integrate road and rail services. PTAs were proposed for the conurbations of South East Lancashire and North East Cheshire, Merseyside, the West Midlands and Tyneside. Their boundaries were eventually fixed after consultation with local authorities and the appraisal of information on journey-to-work patterns from the 1966 census, and the authorities formally established on 1 April 1969. The PTAs controlled policy and finance and were responsible to the local authorities in the area concerned as well as the Minister. As Gwilliam and Mackie (1975, p. 374) have observed, 'the existence of the PTA as an independent unit rather than as part of a centralised transport and land use planning function within the county could generate ... conflicts of policy', not least among the constituent local authorities. The day-to-day management of the (mainly bus) undertakings taken over by the Authorities, and the technical aspects of drawing up the transport plans, became the concern of professional executives (PTEs and the LTE) appointed by the controlling authority. The PTEs were responsible for entering into agreements with the newly created National Bus Company for the coordination of overlapping and adjoining services, and with British Rail to agree upon the role that suburban railway services were to play in the comprehensive transport plan for the area.

For the rest of the country outside London, it was argued that the problems were somewhat less acute, and the question of establishing PTAs should wait until the future shape of local government became clearer with the findings of the Royal Commission on Local Government. For London, a separate Transport (London) Act of 1969 carried out the proposals of the White Paper, and created the London Transport Executive with a chairman and other members appointed by the GLC. In the event, the Royal Commission's recommendations were broadly consistent with the view of the Transport Policy White Paper. The Local Government Act of 1972 departed from the Maud Commission's recommendations by retaining two-tier local government in all areas and, in response to various political pressures, by defining the areas of metropolitan counties more restrictively and therefore less appropriately for transport planning and operating purposes. On the other hand, the retention of both land-use planning and transport responsibilities in the higher-tier authorities will enable the comprehensive planning envisaged by the 1963 Steering Group, the 1965 White Paper and the 1968 Act to proceed.

The 1968 Planning Act and urban transport planning

Legislation concerning the PTAs paralleled much more general changes in the objectives, structures and finance of local transport planning at the end of the 1960s. These were partly brought about by the reform of the local planning system specified by the 1968 Town and Country Planning Act, and by changes in central government organizations culminating in the creation of the Department of the Environment (DoE), but equally important were changes in the attitudes and methodology of planning and planners towards what is now known as the systems approach to planning. These reforms have been continued and reinforced during the early 1970s by the reorganization of local government and by radical changes in the system of distributing public expenditure on transport. At the time of writing, there are many indications that further changes are imminent in government transport policy and, in particular, in policies affecting urban areas. While it will be many years before the full effects and the success of the reforms can be reliably evaluated, the likely implications for urban policies can be described and tentatively assessed, at least in terms of their claimed advantages and initial prolems.

In order to judge the nature and importance of the reform of local planning, it is useful to review briefly the limitations of the system which has been replaced. The roots of British town planning lie in the housing and sanitary legislation and the modernized local government system of the late nineteenth century – a period already referred to in connection with associated reforms in highway administration. The 1909 Housing, Town Planning, etc., Act represented a direct continuation of already established sanitary and building regulations, and did little to inaugurate positive planning. A contemporary observer concluded that the Act would not lead to the 'remodelling of the existing town, the replanning of badly planned areas, [or] the driving of new roads through old parts of a town – all these were beyond the scope of the new town planning powers' (Aldridge, 1915, cited by Cullingworth, 1972, p. 20). Nor did the 1919 Housing and Town Planning Act strengthen town planning *per se:* its principal effect was to increase local authorities' powers to provide housing. Cullingworth argues that the linked and accelerating trends of suburbanization and transport improvement during the inter-war period outpaced the growth of the planning system, despite partial measures such as the 1932 Town and Country Planning Act, which extended planning powers to almost any type of land, and the 1935 Restriction of Ribbon Development Act.

These were, then, only insubstantial precursors to the achievements of that remarkable period of visionary social and economic reconstruction during and after the Second World War. The 1947 Town and Country Planning Act established the first comprehensive and effective local planning machinery. While its principal concern was with the planning of land uses, elements of transport planning were certainly envisaged as part of its responsibilities. This is clear from the 1944 White Paper on the Control of Land Use, which included among its statments of aims the following points:

> Provision for the right use of land, in accordance with a considered policy, is an essential requirement of the Government's programme of post-war reconstruction . . . the new layout of areas devastated by enemy action or blighted by reason of age or bad living conditions; . . . a new and safer highway system better adapted to modern industrial and other needs; the proper provision of airfields – all these related parts of a single reconstruction programmed involve the use of land, and it is essential that their various claims on land should be so harmonised as to ensure . . . the greatest possible measure of individual well being and prosperity.

Planning responsibilities in these fields were placed on the major local authorities, the counties and county boroughs, which reduced the number of planning authorities in England and Wales from 1,441 to 145. This rationalization was marred from the point of view of urban transport planning by the arbitrary nature of the administrative boundaries in question, as they were based more frequently on historic areas than on contemporary functional urban units, which in fact they often fragmented. As one example, responsibility for the Greater Manchester area was divided between three counties – Cheshire, Derbyshire and Lancashire – and no less than seven county boroughs – Bolton, Bury, Rochdale, Oldham, Stockport, Salford and Manchester. While consultation between individual planning authorities and central government, and adherence to national plans, in for example road building, was built into the planning system, the machinery for inter-authority consultation and coordination was much weaker. It appears to have often been the case that, even within a single authority, transport planning was poorly coordinated. Many local authorities 'had more than one committee dealing with transport matters (quite apart from land use planning and redevelopment): those running bus fleets were almost always managing them through a separate public transport or trading committee. On the wider issue of bringing together transport and land-use planning at committee level . . . in only two out of twenty-three county and county borough councils surveyed was this in fact done' (Starkie, 1973a, p. 341). The two principal instruments of the 1947 planning system were the Development Plan Map and the largely negative system of development control to prevent proposals not conforming to the Map (for a detailed description of the planning system, see Cullingworth, 1972, chapter 4). Nevertheless, plans for the transport development of an area were an important and explicit subject for the surveys carried out preparatory to drawing up a development map.

While the detailed critical review of the whole system in 1965 by the Planning Advisory Group argued that the resulting map was often little more than a backward-looking inventory of existing land uses (Department of Housing 1965, cited by Donnison, 1975, pp. 263–4), and was incapable of clarifying the strategic issues facing a rapidly evolving economy, the thousands of planners engaged in producing the maps were inevitably instilled with the inseparability of transport and land-use planning, perhaps to a greater extent than the technically more sophisticated but more specialized staff of the land-use–transportation surveys that practised in the 1960s. Legislative effect to the Planning Advisory Group's criticisms of the 1947 system, and to their alternative proposals, was given after the 1967 White Paper on Town and Country Planning in the 1968 Planning Act. In essence, this replaced the development plan with the structure and local plans. The structure plan is primarily a written statement of policy dealing broadly with questions of land use, the management of traffic and the improvement of the physical environment. Emphasis is laid on major economic and social forces and on policies or strategies for large areas. It is under continuous review, unlike the intended five-yearly reviews of the development plans, which generally proved impractical. The local or 'action' plans were to deal in greater detail with limited areas. Both were intended, and have in fact, concerned themselves with a wide range of urban transport issues.

The complementary reforms of central government responsibilities

Changes in the structures and objectives of planning at the local authority level made invitable, but were not the only reason for, a number of substantial reorganizations of central government departments at the end of the 1960s. During the 1950s town and country planning was the responsibility of the Ministry of Housing and Local Government, and transport planning by central government the responsibility of the Ministry of Transport. Although the latter had effective control over most facets of transport except civil aviation and airport planning, its crucial weakness 'was the comparatively poor relationship it struck up with land-use planning; the planning process was effectively split in two' (Starkie, 1973a, p. 335). From April 1965, certain town and country planning functions were transferred to other ministries and departments, sometimes for only short periods. In October 1969, the Labour government attempted a more radical restructuring of departments, and created the position of the Secretary of State for Local Government and Regional Planning, who was given federal powers over Housing and Local Government and over Transport, together with direct responsibility for the Regional Planning Councils and Boards. This reorganization had a life of only one year before the 1970 Conservative government carried the process one stage further, and created the DoE, with its functions outlined as follows:

> It will cover the planning of land – where people live, work, move and enjoy themselves. It will be responsible for the construction industries, including the housing programme, and for the transport industries, including public programmes of support and development for the means of transport . . . . Local authorities are profoundly involved in these fields and the new Department will, therefore, carry responsibility at the centre for the structure and functioning of local government as well as for regional affairs. (Cabinet Office, 1970, cited by Cullingworth, 1972, pp. 48–9)

In Scotland similar changes had been made in 1962, when town and country planning and environmental services were transferred from the Department of Health to a new Scottish Development Department, which at the same time took over all the local government, roads and industry functions of the Scottish Home Department. In Wales, increasing responsibilities over a wide field have in recent years been transferred to the Welsh Office, which like the Scottish Development Department, was unaffected by the English reorganization (Cullingworth, 1972, p. 103).

The DoE began with three functional parts: Local Government and Development, Housing and Construction, and Transport Industries, each with a separate Minister. Particularly significant was the formation within the DoE in late 1971 of a single group of planning directorates concerned with all aspects of urban policy. As may be seen in table 8.1 this group had within it the Urban Policy Directorate combining under a single Under Secretary divisions responsible, among other things, for urban planning, motorways, transport policy, conservation and amenity. 'This was a far cry from the MoT of the mid 1960s, when the responsibility of one Under Secretary was restricted to urban transport policy, and an urban transport policy which failed to include effectively even motorways' (Starkie, 1973a, p. 336).

The administrative logic of these reorganizations as a response to the redefinition and expansion of the role of local planning is clear, but they were also stimulated by the attitudinal and policy shift towards structure and strategic planning. Linked with the creation of the DoE was the devolution of some of the powers of its ministerial ancestors to the local authorities. Only the structure plans generated by the local planning authorities have to be submitted for approval to the Minister: the local plans, which are detailed elaborations of the proposals sketched in the structure plan, do not normally have to be approved. While the plans must conform to the structure plan, the proposals they contain must be 'adequately publicised' before inclusion, and

Table 8.1 Planning directorates of the department of the environment in January 1972

Directorate	Divisions
1 Development Plan System	Statutory Planning System
	Statutory Planning Techniques
	Planning Methodology
	Cartographic Services
2 Land Use Policy	Land Compensation Policy
	Land Commission
	Development Control
	Specialist Casework
3 Urban Project Appraisal	Transportation Policy
	Urban Roads
	Urban Economics
4 Urban Policy	Urban Development
	Urban Planning
	Urban Property
	Urban Motorways
	Urban Transport Policy
	Urban Conservation
	Urban Amenity
5 Urban and Passenger Transport	Passenger Transport Industry and Investment
	Passenger Transport Operations
	Traffic and Parking

Source: Starkie (1973a) table 1.4, p. 340.

are subject to the normal procedures of making objections and inquiries; all of these stages would normally not involve the central government department (Cullingworth, 1972, p. 103). Moreover, a lower percentage of planning appeals are now dealt with by the Minister, except in a formal way on the recommendation of a DoE inspector (Cullingworth, pp. 56–7). The corollary of these changes is that the Secretary of State for the Environment and his staff can devote a greater proportion of their energies to national strategic and financial policy, as exemplified by the much delayed consultation document on *Transport Policy* which appeared in April 1976, and which will be discussed towards the end of the chapter.

Local government reorganization and a new system of local transport planning

Under the 1972 Local Government Act, on 1 April 1974 the number of local authorities in England and Wales was substantially reduced, and a new two-tier system of counties and constituent districts was created. Six metropolitan counties – Greater Manchester, Merseyside, South Yorkshire, Tyne-Wear, West Midlands, and West Yorkshire – were differentiated, mainly by the greater range of responsibilities given to the metropolitan districts as compared to the 'shire' districts. All the counties, however, were given responsibility for public transport policy and structure planning, thus producing a more integrated framework for environmental planning at the local level to match the earlier central government reforms. Non-metropolitan counties retained their functions as highway authorities, but in the metropolitan authorities these functions have been vested with the districts.

As well as being responsible for producing the structure plans, which inevitably will need to consider the interaction between land use and transport projections and proposals, under the 1972 Act the new county councils have also been charged with the duty of developing policies 'that promote the provision of efficient and co-ordinated systems of public transport' (DoE, 1973b). As the DoE pointed out, however, the former 'multiplicity of transport grants, at different rates, some payable to operators, some to local authorities, was not suited to the comprehensive approach that is now needed. In addition, the present specific grants involve the Departments in a considerable amount of detailed control of individual projects.' To replace all these, under the 1974 Local Government Finance Act a new block grant system took effect from April 1975. The grant is based on the estimated eligible expenditure for all transport, and is distributed partly through the continuing Rate Support Grant and partly by Transport Supplementary Grants, which will cover 70.4 per cent of expenditure over a prescribed threshold level. The threshold is based upon road maintenance and lighting expenditure plus a *per capita* sum, with adjustments for committed capital projects, and is subject to annual revision (Bayliss, 1975). At first it appeared that the county councils would have considerable discretion over the composition of their intended expenditure, so long as they justified their intentions by annually submitting a Transport Policy and Programme (TPP). This must set out policy objectives for 10–15 years, the means by which the county intends to achieve them, and details of forecast expenditures over a five-year period. In addition, more detailed estimates of expenditures in the first year of the quinquennial programme, and a review and assessment of past spending and development, must be provided. In devising this new system, the government hoped that they would promote the development and execution of comprehensive transport plans, eliminate bias towards capital or current expenditure and towards particular forms of transport, distribute exchequer funds in a way reflecting local needs, and reduce the degree of detailed supervision by central government.

It is too early to tell whether the good intentions and the widely approved theory underlying these fundamental changes has been supported by a practical and efficient system which will produce the desired results. But already a number of difficulties and limitations of TPPs and the new grants have been identified and voiced. Much government transport expenditure remains outside the scope of the TPP, such as trunk road spending, grants to public transport operators for new vehicles, and grants to British Rail. It has been argued that the TPP system will be unable to finance large once-and-for-all projects such as underground railways, and that the allocation of the supplementary grants on the basis of population size ignores the alleged greater needs of metropolitan relative to shire counties (Hellewell and Ledson, 1975). Gwilliam (1976) has pointed out many problems of accommodating the intended greater degree of public participation in the formulation of the TPPs and discussed the difficulties of harmonizing the contributions of elected representatives, county officials, the constituent districts, transport operators, central government and other interested parties. He also suggests that there is as yet insufficient expertise or will in all the counties to respond to the aims of the TPP system, and that in a few of the shire counties the statements are too dominated by highway engineers and their concerns. Writing late in 1975, Bruton was also pessimistic on studying the second round of TPPs: 'Little attempt seems to have been made to relate transport policies to land use and community objectives; the same isolated and limited road improvement schemes are put forward; the same subsidies for public transport are proposed; the same expenditure on parking places is provided.'

A more fundamental concern has been aired by Bayliss (1975), Chief Transportation Planner of the GLC. He argues that while in principle the TPP system does not seem to obstruct properly integrated land-use transport planning, in practice 'the danger is that, by focussing attention and resources on local transport planning, other activities may be under-served and some distortion of the relationship between transport and other sections may result'. He adds that, 'although TPPs are framed within structure plans, their frequent updating and financial content may make them the dominant instrument of local transport planning. Should this become the case, TPPs and structure and local plans may need to be linked more formally in a unified planning system which would replace the present informally linked system.'

Given the acute economic problems of the country in the mid-1970s, it has not been a propitious time to introduce a revised and extended role for local government in local transport planning. Additional funds have not become available, and although they may not be condition of the new system's success, their appearance would be the surest way of generating enthusiasm for TPPs within local authorities. Most of the present grants remain earmarked for committed investment, or have been consumed by the inflated costs of existing policies, particularly public transport fare subsidies and concessions. It remains to be seen how the TPP process can be integrated with the wider system of structure and local planning, not least in terms of public participation. Finally, and as with many other important local government issues, there is considerable uncertainty and difficulty surrounding the respective roles of the two tiers of authorities in the field of planning. It is to be hoped that transport planners and interested politicians will not be diverted for too long by such organizational problems and debates from the deepening problems of public transport and lagging road investment in our urban areas.

Current policy objectives

Since the series of White Papers in the late-1960s, there were few comprehensive statements of the government's transport policy until the 1976 consultation document. Instead, policy changes were revealed by decisions on specific projects, changes in grant-giving practice, and circulars and statements (as in connection with local government reorganization). In other words, change did not occur dramatically but through a gradual evolution and refinement of policy. It may be useful to stress that the distinction between policy formulation and administrative practice is not hard and fast: shifts in the former often emerge from a modification of the latter. The furore created by a major public planning inquiry, for example, can be most effective in encouraging the DoE to alter its policies, sometimes in quite fundamental directions. Before 1976, one of the most recent and fullest statements of government policy was given in the evidence submitted to the Parliamentary Expenditure Committee (House of Commons, 1972–3) during its investigation of urban transport policy. This has been summarized by Self (1972) as follows:

> There are no correct general principles for urban transport and much depends upon local circumstances. The basic aim should be to achieve the best possible 'modal split' between private and public forms of transport, which will vary for each conurbation or city. Much importance has therefore been attached to the transportation–land-use studies which (emulating American methods) have now been done for the principal urban areas.
>
> Despite this scepticism about principles, the Department accepts the

need to restrain car use and to improve public transport. The former aim depends at present primarily upon parking controls, and the 'most sensitive' additional method that might be contemplated is said to be road pricing (sensitive because unlike, for example, supplementary licences the charge paid depends upon what actual use is made of roads). For public transport we are told that 'positive promotion has two aspects – the improvement of the services themselves in terms of reliability, speed, comfort, and frequency, and making the cost at least comparable with the perceived costs of private motoring'. The first task is primarily up to the transport operators, helped 'to some extent' by central and local government; the second requires appropriate injections of public money.

Further developments in the DoE's thinking, and in the likely directions of change of transport policy, became clear in April 1976 with the publication of the 'orange' paper on Transport Policy (DoE, 1976a). Put forward as a consultation document with an invitation for any interested parties to submit comments and recommendations, it nevertheless is usually quite categorical about the changes that the Department regards as desirable. Its analysis of the present situation, and most of the recommendations it makes, are understandably concerned with the long-standing national problem of integrating or coordinating investment in different forms of transport. In essence, in the mid-1970s this becomes a question of determining the level of government transport expenditure, and its allocation between road transport, mainly through road construction or improvement and subsidies to bus service undertakings, and rail transport, mainly through the size of the subventions required to finance British Rail's deficit. The preeminent constraint upon the government in its consideration of this question is the decision to halt the rapid growth of public expenditure on transport since the late-1960s. This requires first a painful reappraisal of the priority to be accorded to transport as against the competing claims of housing, education and the social services, and secondly a ruthless reexamination of transport expenditure to ensure that it is actually (which many critics doubt) achieving the social and economic ends in view (DoE, 1976a, p. 2). Other proclaimed objectives include ensuring that those without cars can remain reasonably mobile at reasonable cost, attempting to improve the residential and urban environment and the quality of life, and a retreat from the use of blunt, unselective subsidies aimed at vehicles or facilities, but the wider use of more selective assistance, aimed at the populations most in need because of low incomes, age or inaccessible location.

As regards urban transport policy, the then government's view was that a great deal had already been achieved in devising arrangements for the execution of integrated and effective policies, at least in the metropolitan counties. 'Urban transport planning is now a comprehensive process that considers movement as a whole', is the ambitious claim, although the substantial volume and substitutable nature of movement on foot is barely recognized in the document (DoE, 1976a, p. 40). It does appear that the DoE urban transport policy is still a matter of arriving at policies for dealing with the various transport forms to cater for a given level of mechanically assisted transport determined by economic growth and car ownership. Only in a small way has it moved towards the more difficult, more embracing and longer-term possibility of influencing the need, and the demand, for movements for various purposes and with various characteristics. Such a criticism is easily made, and it is not intended as a condemnation of the document, but it is surely not now unreasonable to expect an awareness of the fact that increased car ownership produces a replacement of longer for walking trips, as well as generating entirely new journeys. The statement that the car predominates for all lengths of journeys is not true, unless a journey only takes place in, or on, a vehicle.

A number of short-term, institutional problems are identified. The untidy situation of the National Bus Company's (NBC) operations in metropolitan county and PTE areas, and the lack of coincidence between NBC area boundaries and the shire counties, are discussed, and a number of options for reorganization are outlined. The likelihood of continuing exchequer and rate support for the bus industry underlies the view that 'one of the most pressing needs is to involve locally elected members more widely with the running of the bus industry'. This will be particularly difficult to arrange in medium-sized urban areas served by both district and county councils, and the proposal to create new PTAs covering the areas of counties or even whole regions could be counterproductive in this respect. Urban bus services would probably be required (certainly, where small municipal undertakings would be extinguished) to provide greater cross-subsidies for rural services, and the politically elected body would be more distant. The problem of coordinating and economizing on competing bus and rail services, particularly as has been mentioned in London, is another which is likely to lead to further institutional reorganization (DoE, 1976a, pp. 45–8, 57).

The substantive content of the urban policy section of *Transport Policy* represents a radical reordering of priorities and intentions from those held only a few years earlier. The principal statements are not about the desirability of large-scale urban road-building programmes, but outline a range of measures including traffic restraint, parking controls, bus priority schemes and comprehensive traffic management, which is seen not simply as increasing road capacity but as giving priority to particular places. A considerable rundown of urban road expenditure is envisaged, both on construction and maintenance, and the emphasis of expenditure will be on safety, and on making the best use of existing roads. New construction and improvement schemes will generally be linked to traffic management, the improvement of public transport or environmental gains (DoE, 1976a, pp. 40-4).

As a result of these changed priorities, or at least partially so, the government introduced a number of departmental changes affecting the administration of transport policies. In September 1976 the Department of Transport was created, with William Rodgers MP as Minister with Cabinet rank. Although representing an elevation of transport in the affairs of the government, many observers were concerned about the separation of land-use and transport planning. As an editorial in *Traffic Engineering and Control* (October 1976, p. 385) commented, the chief criticism of the discussion paper on transport policy had been its omission of any reference to the interrelation of transport and land use . . . 'and it is paradoxical that the main result of the reshuffling of responsibilities at the DoE could be a weakening of the already loose link between them'.

From the outset the two departments were to have joint responsibility for major road proposals, and for the appointment of inspectors to conduct public enquiries and jointly to decide on the final verdict, but subsequently there has been little clarification of the steps that are being taken to achieve more general liaison. It is not felt, however, that the divorce need strongly deflect the long-term trend towards a greater harmony of the two branches of physical planning. The government continues to recognize the value of such integration, for in its 1977 White Paper on Transport Policy, it stated its intention to require English and Welsh non-metropolitan counties to prepare and publish county transport plans within the context of the TPPs. As it concluded, 'more responsibility for planning transport to meet local needs should be devolved to local government since the most practical and democratic approach to co-ordination is local'. The restructuring is not, therefore, inconsistent with the prospect of more effective and integrated transport policies and planning.

Within the guidelines and financial controls imposed by the DoE, there is clearly room under the TPP system for a greater variety of policies to emerge among local authorities. This is not inevitable, however, even given the explicit aim of allowing and encouraging local policies to reflect local needs, because the DoE is retaining the power of approval over the TPP. Probably, the intention is that cities of different sizes, and with varying quantities and qualities of roads, suburban passenger rail networks, and bus services, would need to develop policies of markedly different character. It is also probable that, in the event, local transport policies will reflect the party politics of the dominant group on the local authority. While the scope for party clashes in the transport field is almost certainly less than in, say, education or housing, nonetheless one can see the clear possibility of trials of strength between a local authority controlled by one major party, and a central government of opposite colour. At one time or other it has been argued that the best or most rational programmes for virtually every area of government policy are designed when party ideologies or manifestoes are put to one side, and it is tempting to repeat calls that urban transport policies should be similarly ideologically unsullied. Such pleas are, however, naive as well as unrealistic: it is quite clear, for example, that no government would remain passive when local councils pursue policies directly contrary to government policy. In the urban transport field, policies regarding fare subsidies, road pricing, and rate precepts are clear candidates for major political dissension. Rather than seek the impossible, those concerned to improve urban transport must encourage increasingly well-informed decisions. It is acknowledged that considerable strides have been made in this direction during the last decade, but much more can be done to provide the politicians and their closest advisors with clear, well-argued proposals. These should reflect the understanding of the advantages, implications and disadvantages of alternative policies as provided by specialist transportation studies, economic analysis, environmental impact studies, and as reflected by public opinion through the democratic system and the enquiry procedures. We take the view that the quality of proposals and decisions will also be improved by taking the trouble to acquire the widest possible appreciation of the purposes and characteristics or travel, and the role of mobility in satisfying and enriching urban ways of life.

Conclusion

In this chapter we have seen that from slowly-growing and diverse roots have grown the climate of opinion, the professional expertise and the political will to develop and execute an urban transport policy. The recentness of rapid movements towards this end mean that except in broadest outline, the urban transport policy that we will pursue is still undecided. The areas of wide consensus are still limited, but at least there is clear agreement that the present situation can and should be improved. If nothing is done, then rising car usage and the worsening financial situation of public transport operators will inevitably lower the quality of urban transport and even of urban living; and this could occur very soon and in a short time. Immediately one tries to specify the nature or directions of improvement that should be pursued, the field of continuing debate is entered. As with all resource decisions, questions of choice or priority are inherent, and these can only be solved in the political forums of the country. In the urban transport field, there are clear issues of trade-offs between, for example, investment in roads, which will primarily give short-term benefits to private car users, and investment in public transport; between attempting to improve the facilities that cater for present and future mobility demands, and attempting by land-use controls and changes to restrain or even reduce the demand for mobility; and between present and future benefits. Perhaps the most fundamental and difficult choice is to determine the proportion of our resources, and particularly of public expenditure, that should be devoted to transport as opposed to the other many claimants: health, housing, social services, defence, employment policy, and so on.

There is every sign that recent reforms have gone a long way towards creating the administrative, financial and executive structures to facilitate a much-improved and more ambitious urban transport policy. It remains to be seen whether the new structure has the innate capability to carry out what is intended, and more importantly, we have yet to learn whether there will be sufficient finance to increase the net benefits of urban movement, or sufficient political consensus to apply necessarily long-term policies with this aim.

Postscript

After the completion of the final draft of this chapter, there was a change of government (May 1979).

At the end of 1979 the transport planning policies of the new Conservative administration were by no means clear, but their first actions made apparent two priorities: the reduction of public expenditure and the encouragement of competition in part by the de-nationalization of transport undertakings. December's White Paper on Public Expenditure for 1980/1 shows that expenditure on roads and transport will be £2,914 million, £204 million (7 per cent) below the 1979/80 level. This will mainly bear upon capital expenditure on new roads and local transport, although Transport Supplementary Grants will total £350 million, about 29 per cent of estimated expenditure. Under the Transport Bill published 15 November 1979, the existing provisions relating to the licensing of public service vehicles in Great Britain will be altered to enable more innovation and competition in routes and services, and the assets of the National Freight Corporation will be transferred to a limited company. The principal urban implication of these proposals is that the Minister of Transport will be able to designate trial areas in which no road service licences are required – how this would work in practice is difficult to forsee, but it could result in the very antithesis of coordination of public transport services. The extent to which these two revealed priorities will deflect the secular trends towards coordination and towards the accommodation of more diverse travel needs remains a completely open question.

9
Management of urban travel demands

Transportation studies permit alternative responses to the control and management of urban travel to be monitored and evaluated in the light of evidence taken at a point in time about existing travel behaviour. The time scale involved is usually at least 20 years, but the approval of a strategy based on a time period of this kind does not easily resolve the serious problems which confront movement in cities now rather than in a decade or two. The solutions suggested by a sophisticated analysis of existing problems may take large amounts of money and time to implement and, indeed, goals may never be achieved because of changes which take place between the present and the forecast year. There are a large number of short-term problems to be solved, and much time and effort has been devoted to finding ways of overcoming them. Probably the most crucial problem concerns the conflict between towns and road traffic. It has become apparent that the physical form of most British towns is not suited to the effects of uncontrolled growth of car ownership and use. Saturation level for car ownership has yet to be reached but there is already insufficient room to accommodate flows of private vehicles at peak hours or to provide space where vehicles can be parked conveniently near to the destinations that their users are attempting to reach. Private vehicles offer flexibility of choice, but this very flexibility is choking urban arteries, particularly those which lead to, and provide circulation within, the heart of the city, the central business district, which still contains a large proportion of all the jobs, shopping and other trip attractions.

Congestion

The most tangible short-term consequence of uncontrolled movement of vehicles and people is congestion, which can be defined as 'waiting for other people to be served' (Thomson, 1974, p. 72). This is an important characteristic of the supply of services as compared with goods and, because the demand for any service fluctuates and it is not economic to provide sufficient capacity to handle the highest levels of demand, transport is particularly prone to the effects of congestion. Traffic congestion is inevitable at certain times of the day or at particular points, irrespective of time, in a city's transportation network. Vehicle congestion or the 'delay imposed by one vehicle on another' occurs on roads where there is insufficient space to accommodate the flow of vehicles, while on the railways or underground where the movement of rolling stock is scheduled, person congestion is the main result of fluctuating demand (Thomson, 1974, p. 76). Both types of congestion occur on buses which, although scheduled during peak hours, suffer from the congestion caused by other road vehicles and become 'bunched' as well as experiencing person congestion during peak hours. Congestion on urban roads in Britain is said to occur when average traffic speeds fall below 40km per hour, but the significance of this value in real terms is difficult to comprehend because most traffic in urban areas has never reached this average speed. It is used to calculate the money cost of congestion but, even if by showing that costs are increasing because average speeds are falling even further below the chosen limit and therefore are able to justify more investment in transport infrastructure, the increases in the volume of traffic on roads, existing or new, ensures that the amount of congestion increases. It is notable, however, that the effect of congestion as

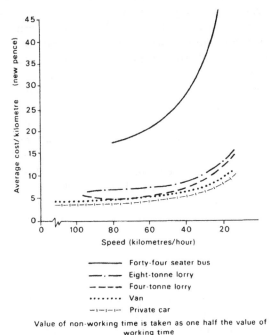

Figure 9.1 *Effect of congestion on vehicle operating costs* (Source: Thomson, 1974, fig. 19, p. 81)

defined by average operating costs per kilometre for various types of vehicle (figure 9.1) is least serious for private cars. Operating costs increase much more slowly for private cars as average journey speeds decrease, while for buses the operating costs rise sharply for any unit decrease in average speed. In this situation, congestion is less likely to act as a disincentive to the car user than it is to the bus operator concerned to provide a satisfactory service at a price which allows operating costs to be covered.

Effects of congestion on public transport

Traffic congestion is undoubtedly contributing to the demise of road-based public transport in cities. It has been shown that when the traffic using a stretch of road is at 25 per cent of its maximum capacity, the effect on journey times per mile is to increase them by approximately 12 per cent (Smeed, 1968). When traffic is at 50 per cent of maximum, the journey time is 40 per cent longer than in light traffic conditions. The nearer the level of use reaches the maximum capacity of the system, small changes in the level will lead to large increases in journey times. When traffic is at 98 per cent of its possible level, the average journey time per mile is seven times higher than under light traffic conditions. Unfortunately, such conditions are most likely to occur in central business districts during the two peak periods associated

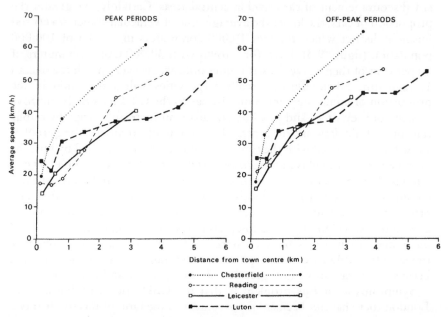

Figure 9.2 *Average journey speeds at peak and off-peak periods in relation to distance from the city centre* (Source: *Vaughan* et al., Traffic Engineering and Control, **14**, *1972, p. 227*)

with the journey to work, when public transport use is usually at its highest level. Elsewhere in cities, average journey speeds increase with distance from the centre, but they are always least dissimilar in central business districts (figure 9.2). Buses must stop en route to pick up and drop off passengers, it is then necessary for them to get back into traffic streams and this, combined with the less direct routes used by buses, creates a considerable difference between the trip times of buses and cars. This happens despite the fact that buses occupy about three times the road space of one car but can carry the occupants of 60-70 cars, assuming one person per car. Hence car commuters, who are major contributors to congestion, pass on time costs to other road users and, in particular, public transport passengers. Public transport trips therefore appear to offer inferior service, irrespective of any cost differences, real or perceived. The journey times of both bus and car users increase as the proportion of car travellers increases, and although the rate of change is low in small towns it increases sharply with increases in the size of central business district. The vertical difference between the curves in figure 9.2 is more or less constant, so that the differences between the travel times of bus and car travellers seem to be more or less independent of the proportion of car commuters, but these figures exclude walking and waiting times, which are clearly much more important for bus than car travellers.

A dramatic increase in public transport trips would not necessarily follow any discouragement of car travel in central areas. Certainly, the greater the proportion of cars, the lower the average time of vehicular travel in a central business district which receives 10,000 commuters in a town of 100,000 population (figure 9.3). Even for a town with 30,000 CBD commuters, if 80 per cent of them travel by car it does not increase very much the overall average time taken. Only in really large cities will a reduction in the proportion of commuters travelling by car to the CBD materially improve vehicle journey times and possibly stimulate an increase in trips by public transport. Calculations for central London road commuters (figure 9.3) show that, compared with the time taken if everybody travelled by bus, the average time of travel is not much altered if up to 20 per cent travel by car, but it is about 80 per cent longer if 40 per cent travel by car. Clearly, the bus commuter is at a consistent disadvantage whatever the mix of public and private transport; and even more so in situations where the proportion of commuters travelling by car is low, rather less so in larger cities. The small difference in trip times between buses and cars when a large percentage travel to the CBD by car really places car commuters in a superior position because they can also enjoy greater comfort, privacy and convenience.

Symptomatic of this problem is that for CBD-orientated journeys in London there has not been an improvement in the ratio of bus to car travel times. Referring to the relative costs of car and bus travel at *off peak* periods in central London in 1972 along 25 different routes, Buckles (1973) found

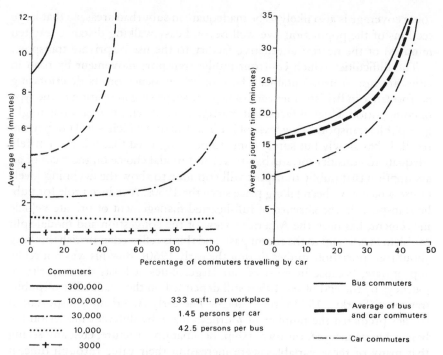

Figure 9.3 *Average trip times for car and bus travel within central areas of cities according to number of commuters and the proportion travelling by car* (Source: *Smeed, 1968, figs 5, 6, pp. 17–19)*

that for journeys of less than 0.68 miles it was quicker to walk than to catch a bus and that the average journey took 2.5 times as long by bus as it did by car, compared with an earlier survey in 1963 when the ratio was 2.0. Indeed, for some journeys, bus times were four times as long as those by car. Changes in average direct speed from 4.9 miles per hour in 1963 to 4.4 miles per hour in 1972 for buses was partially responsible for this, along with an increase from 10.2 miles per hour to 11.2 miles per hour for the average speed of cars. It would seem reasonable to infer that these characteristics of internal trips at an off-peak period will be exaggerated for external journeys during the peak hours, so that the comparative advantage of cars will be even greater for CBD-orientated travel.

Transportation studies cannot easily resolve problems of this kind. Short-term solutions are needed urgently to ease the congestion problems which invariably are particularly acute in central areas, but which also exist at some suburban locations. Here the problem is not so much congestion but inadequate levels of service, especially on routes between suburbs not linked by radial routes into the centre of the city. The density of public transport route coverage is also likely to be inadequate in suburban areas, so that large sections of the population live well beyond easy walking distance (say ten minutes) of the nearest alternative facility to the use of private transport.

The difficulties which confront public transport movement by road in cities are not insurmountable. The research evidence on its deteriorating performance in British cities cannot be very surprising in relation to the type of competition which it faces, but management of central-area travel might give public transport a new lease of life in a situation which is still not beyond recall. It has slowly but surely come to be recognized that it is not entirely adequate to consider intra-urban travel and modal choice on the basis of an assumption that public transport will continue to show the declining levels of use which have been taking place since the 1950s. A reduced role for public transport, in the absence of full-hearted management of private vehicle movement, has been the American view of the congestion and modal-split problem in cities up to the recent past, and this is reflected in British thinking about transportation studies during the early 1960s. But this was not really appropriate, because in most of our larger cities at least, the majority of journeys by the end of the 1950s still depended on the availability of public transport (Starkie, 1973b, pp. 370–3). The early American transportation studies predicted the number of journeys made by different modes on the bases of zone income, car ownership, occupation structure, etc., and seeing that many of these variables were increasing their value through time, it seemed reasonable to assume that public transport could assume a less important role in urban travel. This was irrespective of whether attempts were made to improve the level of service, and the access of urban residents to public transport facilities, or whether control on the movements of private transport were introduced. Such assumptions began to be modified in British transportation studies from the mid-1960s onwards when public transport accessibility indices, of the kind used by Wilson (1967) or in the *London Traffic Survey* (Greater London Council, 1966) and the *Leicester Traffic Plan* (City of Leicester, 1964), were incorporated in modal split models. Some later studies went into more detail and tried to establish how households which owned cars actually made decisions about whether to use public and private transport (see, e.g., Traffic Research Corporation, 1969a). This improved information about potential demand and whether it would be worthwhile managing urban travel demands more closely with a view to reducing congestion and other costs, but there was little indication as to how the future contribution of public transport could be enhanced.

Improving the role of public transport

On economic, social and environmental grounds it is accepted that it has become necessary to improve public transport services in urban areas. But there are many difficulties to be overcome, the most serious being the need to resolve: firstly, behavioural problems, i.e. the attitudes of users towards public transport; secondly, infrastructural problems in the transport system; and thirdly, an outdated organizational framework whereby rail and road-based public transport in cities is the responsibility of different bodies. This is now being rectified by the establishment of Passenger Transport Executives which attempt to coordinate conurbation transport in a way which has not been possible in the past. The behavioural problem is a product of the perceived disadvantages of public transport which have already been discussed in the context of modal choice. Many of the difficulties are real enough but it is no solution to assume that present trends towards higher car ownership lead to demand for more highways and that public transport, with the exception of peak-hour work journeys to CBDs, will remain in an inferior position *ad infinitum*. More positive action designed to influence people's behaviour in relation to public transport use is needed by way of making road and rail public transport more attractive. More frequent services, better route coverage in order to make more households accessible to services, and more reliable, competitive journey times are the kinds of change needed. Regulations which control the movement of cars in favour of buses will help to resolve the latter, and the Ministry of Transport (1968b, p. 8) made a number of suggestions, some of which are now being tested in experimental schemes. These include provision for buses to circulate in the heart of the town centre as freely as possible; giving buses priority over other traffic; allowing bus routes through areas from which other traffic is discouraged; reviewing the siting of bus stops and the facilities available there. Traffic restraint of this kind is one possibility, but before discussing it in more detail let us briefly discuss the attempts which could be made to make public transport attractive by using inducements.

Free public transport

One of the alternatives to restraint is to introduce free public transport (Greater London Council, 1975). Advocates of this approach/solution consider that there are important social benefits to be derived by the captive users, many of whom are amongst the most disadvantaged economic groups in urban society. Secondly, they believe that the introduction of fares-free public transport would be so effective that it would help to reduce traffic congestion in cities and, in the long run, make public transport even more attractive. The effect on the level of public transport patronage of abolishing fares will depend on the direct price elasticity of demand, which is revealed by the response of passengers to fare changes (which usually increase) that have been made in the past. Studies in Germany, in which public transport demand was seen as a function of the density of private vehicles, length of public transport routes, bus and tram kilometres, time, and fares, have shown that fare increases have the greatest effect on demand. They produce price elasticities which reduce demand by roughly 50 per cent of the percentage increase in fares, e.g. a fare increase of 16.7 per cent in Duisberg induced a demand reduction of 8.8 per cent (Baum, 1973). It also seems likely that the introduction of free public transport will cause demand changes related to trip purpose. Work trips are likely to be least affected, partly because public transport is already extensively used in the larger cities and partly because the rolling stock is already so heavily used at peak hours that any additional demand could not be accommodated without considerable investment needing justification vis-à-vis off-peak use. Shopping trips will probably increase, but it is difficult to know to what extent the increased demand would reflect transfer from other transport modes rather than a simple increase in shopping frequency because public transport was free. Free public transport could help to redirect urban shopping trips towards the CBD because most public transport services terminate there and this would allow the CBD, at least in British cities, to retain its traditional functions and status at the top of the urban retail hierarchy. Social/recreational trips will not be drastically altered by free travel facilities because they will only provide access to a few of the recreational alternatives available and these, in turn, only correspond to a small part of the overall range of recreational pursuits followed by urban dwellers. One difficulty with free public transport is that it is likely to generate 'unnecessary' journeys by children and others, which could have the effect of reducing the marginal utility of public transport for those with private alternatives offering more comfort and other advantages at slightly higher cost.

For free public transport to have a significant impact on congestion it will clearly be necessary to induce the transfer of motorists. Modal cross-price elasticities are low, however, and Starkie (1967a) takes the view that the best strategy is to improve journey times and level of service as the most persuasive way of producing a shift in modal split (see also Beesley, 1973, pp. 318–22). Although only hypothetical questions were answered, one American study has shown that even if public transport were free, only 13 per cent of motorists would transfer their allegiances to it for the journey to work. To persuade 24 per cent to do so it would be necessary to 'pay' 10 cents to each motorist (Moses and Williamson, 1963). Similarly, other studies in Germany have shown that almost 50 per cent of motorists would still prefer private transport for the journey to work even if public transport were modern, comfortable, clean and free.

After interviewing 400 regular commuters into central London in 1965, Williams (1969) found that 43 per cent of respondents claimed that they had worked out in detail the cost of bringing their cars to work, but only a third considered the total cost to be relevant (i.e. costs other than fuel and oil). Just over 50 per cent said they would transfer to public transport if it were free. Most would transfer to British Rail and the underground rather than the buses, but this might change if faster buses were introduced as a result of lower traffic levels induced by low fares or free public transport. Williams estimates that a 44-per-cent decrease in the number of regular car commuters into central London resulting from the introduction of a standard fare of 5p (1965 prices) would generate a reduction of 35 per cent in peak-hour car traffic. This might, however, induce people already using public transport to revert to private transport again if road conditions improved to such a marked extent in the absence of other kinds of disincentive to car movement. Quarmby (1967) has estimated that a reduction in public transport fares to zero would produce a shift in the balance of private/public transport from 69/31 to 52/48. An increase in parking charges by 15p per day would shift the ratio further to 41/59. In practice, many motorists do not view fares as the major incentive to transfer to public transport. They remain prepared to absorb the higher costs of using private transport, although many do not consider this to be the case, in exchange for greater flexibility (Lansing and Hendricks, 1967; see also Goodwin, 1974). Field experiments to assess response to free travel facilities have been disillusioning. Baum (1973, p. 10) cites the experience of Hanover, which experimented with a park and free-ride system for shopping trips into the city during Saturdays in the Christmas period. Despite extensive public relations activity only 2,615 cars were parked during four successive Saturdays, and 6,000 persons actually used the free bus service (there are well over 200,000 motorists living in Hanover).

The effects of free or reduced-fare public transport therefore remain uncertain. Opinion surveys provide only very general guidelines to what travellers would actually do, in that results from such surveys vary widely and, in most cases, represent highly hypothetical situations. Perhaps, on average, a 15- to 20-per cent diversion can be achieved in the most favourable circumstances, although nobody is sure what these circumstances actually are, but actual changes will depend on existing levels of public transport use (which is usually linked to the size of cities), the degree of centralization of the urban area and, most important, an acceptance by the community at large that subsidy of public transport by local or central government via tax revenues is socially justifiable. Many local authorities, as well as central government, are already subsidizing public transport services; to ask them totally to support such services the expense of other community services which also make a call upon their resources is perhaps unreasonable.

Fares-free public transport will also have consequences of note for land-use patterns in cities (Batty *et al.*, 1974). In a study of its impact on the West Midlands conurbation, using a gravity model to simulate public transport use for journeys to work, it was shown that fares-free public transport produces the now well-documented pattern of decentralization from inner city (inner Birmingham) to outer suburbs and beyond. The model predicts that zones within 2 miles of the centre of Birmingham would lose up to 45 per cent of their population, and increases of over 75 per cent would occur in some of the outer suburbs between 8 and 14 miles from the centre. In total, about 61,000 of 590,000 workers using public transport would relocate their households because of the fares-free situation. This is based on an assumed 49-per-cent reduction in travel cost which therefore leads to a relocation of 10 per cent of economic activity, i.e. the workforce using public transport (or about 5 per cent of the total population). It is assumed that land-use policies, e.g. the use of Green Belt controls to restrict development would permit such changes; the model does not take constraints of this kind into account.

Alternatives to free public transport

While free public transport is intuitively an attractive, radical solution to congestion caused by private transport, there are many economic and political factors which make its introduction on a large scale unlikely for many years ahead. It is therefore necessary to look for other solutions which, with reference to buses in particular, increase the average speed of public transport and lead to improvements in the level of service by increased frequencies and more comprehensive networks. The alternatives can be classified according to, firstly, whether they are measures relating to urban roads and the way they are utilized; secondly, whether improvements are introduced to the conventional operation of public transport undertakings; and thirdly, whether operational improvements result from the establishment of new services. It is unlikely, however, that each of these alternatives will provide a complete solution independently of any action taken in relation to the other elements. Hence the following discussion does not attempt to consider each element in turn but illustrates the possibilities with schemes which, by their characteristics, invariably incorporate all three alternatives.

The demand for a role for buses is not simply a product of the need to ease congestion but also reflects their continued importance, particularly for journey-to-work travel, in all the major British cities except London, where 70 per cent of the peak-hour journeys are by train. It has been shown that buses are highly competitive with all other modes where traffic volume along a corridor exceeds 10,000 passengers per hour, and only on the most heavily used routes into major city centres would rail investment be worthwhile (Meyer *et al.*, 1965, pp. 182–3). During peak hours, rail facilities can carry as many as 55–60,000 passengers per hour, which is considerably more than central bus services, but the latter would carry more passengers if they were operated along the same principles as an intra-urban rail service, for example with frequent picking-up and setting-down points but uninterrupted headway allowing higher average speeds between these points. A bus consumes about three times as much road space as a car but, on average, it is eight or nine times as efficient in its use of that space. The average bus in central London during the rush hour, for example, is transporting 39 passengers compared with the 1.5 passengers in the average car. The potential savings in road space are therefore considerable.

Express bus services

There are a number of possibilities, the first of which is to designate certain bus services as 'express' routes. There is no attempt made to separate buses from other traffic, but routes which begin in the outer suburbs pick up passengers over a limited length of the first part of the route and then proceed without stopping to the city centre. For the reverse journey there will be one or two embarkation points in the city centre and then the bus proceeds without stopping to the first suburban stop, from which it operates in the conventional way to its terminus. Frequency of stopping is reduced but the actual improvement in trip times remains heavily dependent on the degree of conflict between the bus and other traffic through delays caused at junctions, etc. Express bus services will obtain the best operating advantages where the effects of congestion will be limited to as low a level as possible. This is easiest to achieve on motorways which provide access to city centres, provided that serious congestion does not occur where the motorway ends and enters the conventional street system of city centres. Express services between Heswall/Irby on the Wirral and Central Liverpool use the M53 motorway for more than half of the journey, and this has led to journey-time improvements of up to 5 minutes on a 30-minute journey and a high level of utilization by passengers during the peak hours. But this service operates along a motorway which is still relatively under-used because it is not yet linked to the national motorway network. Traffic densities are therefore much lower than they will be in future, so that any benefits to public transport users and operators will be short-term rather than long-term gains. For a true express service to operate using a motorway standard route for much of its length, it is necessary to reserve a lane for exclusive use of buses. An example is the Shirley Highway Corridor Scheme in Washington D.C., which has an 11-mile route servicing the Pentagon and downtown Washington from the southern suburbs (Hall, 1974; Morton, 1971). Some 5 miles of the Highway incorporates two central bus lanes to which buses can gain access to three points. Journey times have been reduced to 20 minutes compared with 40–60 minutes for rush-hour travel by car. The bus companies providing the service are making an operational profit on the service. There are no similar examples in Britain and there are few cities with urban motorways which could easily be modified to create a system similar to that on the Shirley Highway.

Stevenage superbus experiment

One of the most successful express bus services using standard urban roads is the service in Stevenage which started as the experimental Blue Arrow service in December 1969. The decision to improve the role of public transport in Stevenage was based on a cost-benefit study which concluded that if a sufficient proportion of the working population of Stevenage who used cars for their journey to work could be attracted to bus travel, than a community benefit would be gained by avoiding the need to provide increased road capacity at intersections (Lichfield and Associates, 1969). These improvements would have a total community cost of £51m at 1967 prices, while bus improvements, which could be implemented much more quickly, would involve costs of £42m. The report suggested that if the bus scheme were put into operation, a reduction of 5 per cent in the number of workers driving cars to work would be needed. The actual response achieved in experiments designed to test the validity of the assumptions in the cost-benefit studies is illustrated in figure 9.4, which shows the ridership between the Chells neighbourhood (population 11,000), the town centre and an industrial estate to the east of it. There have been a number of later experiments, each of which has produced subsequent improvements in

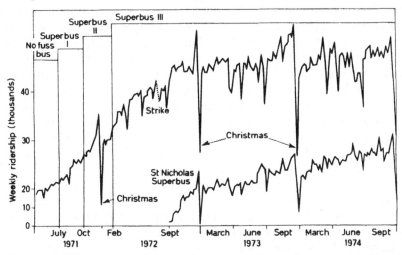

Figure 9.4 *Patronage of Chells and St Nicholas 'Superbus', Stevenage, 1971–4* (Source: *information provided by Stevenage Development Corporation, Engineers' Department, June 1975*)

patronage of the system. The five phases shown in figure 9.4 essentially involved the creation of a more direct route, an express length, an improvement in frequency to 5-minute-intervals throughout the day and a reduction in fares from variable rates to a flat 6p in phase three (4p for children and 3p for old-age pensioners) to 4p in phase five, in order to compete with perceived motoring costs (fares for children and OAPs reduced to 3p and 2p respectively). Single-deck buses are now used on the service with a pay-as-you-enter system. In May 1971, when the experiment commenced, the weekly ridership of the whole Superbus catchment was 17,300 or 4.3 trips per household. In the period of four weeks ending 2 November 1974, the average weekly total was 48,152 or 12.0 trips per household, an increase of 180 per cent (DoE and Stevenage Development Corporation, 1974, p. 414). The effect of the Superbus on modes of travel used during peak hours and for journeys to work has been to reduce the proportion of private transport journeys (mainly by car) from 42 to 32 per cent for trips to the town centre or the industrial area, but this is accompanied by an increase from 46 to 51 per cent for those working elsewhere. Hence the improvement since 1971 is only 3 per cent overall (table 9.1). It seems that the Superbus has been more successful at diverting car passengers than drivers, and the modal change forecast in the cost-benefit study has not been achieved as yet. But a passenger survey in 1973 did show that in a situation with no traffic restraint a substantial number of car drivers, with serviceable cars available, were using the Superbus because they preferred it. Hence, it may only need modest traffic restraint to attain the goal indicated in the cost-benefit study. The number of Chells residents switching from car to bus is equivalent to about 2.8 per cent of the active population of the district. The demand for the Superbus from Chells does seem to be

Table 9.1 Modes of travel to work: Stevenage Superbus experiment, 1971 and 1974

Mode	Working in town centre or industrial area		Working elsewhere		All workers	
	1971	*1974*	*1971*	*1974*	*1971*	*1974*
Car driver	42	32	46	51	43	40
Car passenger	24	14	15	12	20	13
Superbus	20	48	6	13	14	33
Other	14	6	33	24	23	14
Total	100	100	100	100	100	100

Source: Stevenage Superbus Survey, surveys in Chells neighbourhood. 1971 and 1974.

smoothing out, a phenomenon also apparent in the curve for a confirmatory experiment in the St Nicholas neighbourhood which started in September 1972.

Bus lanes

There has been a growing interest in bus priority schemes on non-motorway routes in a wide range of British cities including Reading, Sheffield, Liverpool, Leeds, Newcastle, London and many others. The concept of priority for public transport goes back much further than the present surge of interest, however, as exemplified by the tram systems in European cities which have been assisted by reserved lanes and traffic signals which give them priority over other traffic on the roads (Collins and Pharoah, 1974, p. 414). Priority to buses has also been given much earlier in Paris, Milan or Marseille than in British cities, which have only begun to implement such schemes during the last decade. There were only 45 reserved lanes approved or under discussion in London at the end of 1972, for example. Lanes reserved for buses are of two types: those in which a lane is reserved in the same direction as the general traffic flow; contra-flow or against-the-flow lanes which save long detours and allow the locational advantages of certain bus stops to be retained. Lanes which run with the flow of traffic are being used in Manchester, Sheffield, London and Dublin, but the main problem is that it is difficult to avoid other traffic 'straying' into them and so holding up the buses, while it is also important to permit access to shops and other premises along bus lanes which need access for service vehicles as well as customers' vehicles, e.g. petrol stations. Hence a with-flow bus lane of 950 metres along Vauxhall Bridge Road in London operates on a no-waiting basis for all but bus traffic between 08.00–0.900 hours, and from 09.00–16.30 hours goods deliveries and parking by other vehicles is permitted. Contra-lanes are operating in Reading, Derby, Manchester and Teesside (Redcar), and while some of the weakness of the with-flow lanes are reduced, access to business premises remains a problem. A good bus priority scheme should maintain the total carrying capacity of passengers and goods on a network, unless there is some other objectives, and it should also be self-enforcing although providing for adequate servicing of buildings. Safety levels for pedestrians and other should also be maintained or even improved.

The real test of a bus lane, however, is whether it can create trip-time and other improvements which make bus transport more attractive to users of other vehicles competing for more road space. It has been estimated that reserved lanes can move 25–30,000 seated passengers per hour at 15–45 miles per hour. A study of the impact of the Reading contra-flow scheme showed that time savings of 14 per cent were achieved in one direction and 10 per cent in the other (Bendixson, 1970). When the scheme was eventually extended to cover some 2 miles of route the services using it attracted 4 per cent more revenue and passengers during 1969–70 and 6 per cent more in 1970–1. A later study of bus lanes in the European cities, including London, Manchester, Reading and Dublin, reported that average journey times were reduced by 2 minutes (Westler, 1972). This suggests that bus lanes only introduce marginal improvements for the public transport user and are probably more effective in generating operating economies for the bus companies.

The impact of bus lanes in British cities is possibly muted because few of them are more than 914 metres in length, and most are considerably shorter. Research on bus priority schemes and alternative strategies for inner London suggests, however, that short bus lanes positioned so that they enable buses to 'jump' queues at known congestion points along routes do go a considerable way towards freeing buses from serious congestion. This can be achieved with only limited coverage of the road network, because delays caused by congestion are only concentrated at a limited number of points; e.g. in a survey of central London, 50 per cent of the total delays were caused on 10 per cent of the intersection approaches (Buchanan *et al.*, 1974). It has been estimated that if 1 per cent of the bus network in inner London were operated in bus lanes during the morning peak it would cure 10 per cent of the total problem at that time of day; 2–3 per-cent coverage could cure over 30 per cent of the problem. Where congestion is less serious there are diminishing returns obtained from the introduction of bus lanes. Continuous bus lanes were also considered but only 50 per cent of the bus routes used roads which were wide enough to incorporate bus lanes without seriously curtailing the flow of other traffic; elsewhere there is sufficient width but little congestion, so that bus lanes would simply interfere with stopping arrangements, etc., rather than improve travel characteristics. With some 60 per cent of the various bus lane schemes examined for inner London along shopping frontages, it is important to maximize the benefits to buses but to reduce the negative consequences for other activities. In addition, in streets such as Oxford Street, pedestrian and general-access problems also tend to make bus lanes an unattractive proposition, while any assessment of bus lanes must establish the ripple effects caused by the reactions of other traffic which has less road space and may try to use adjacent streets to avoid congestion. Another procedure for allowing easier movement of buses might by separation of buses and private transport at intersections by means of different crossing levels. This is not likely to be very acceptable on cost grounds, however, and much more likely is that priority is given to bus-only lanes at intersections by signals which detect buses in motion as they approach junctions and allow them to proceed without stopping, or at least with minimum delay. This method is already being used in Leicester (Richbell and van Averbeke, 1972).

One of the major objectives of bus priority schemes should be to reduce the behavioural cost of using buses for the journey to work (Lesley, 1975; see also Huddert and Allen, 1972). It is necessary to reduce passenger riding time, bus stop time and delay time to the lowest possible level. The values of these components depend on the type of bus operated; one-man buses without fare boxes may be slower than buses with a conductor at peak hours; the average running speed on urban bus routes is 24km per hour. Assuming that the average maximum speed for town buses will be 50km per hour, then the average running speed can only be about 35km per hour. Lesley estimates that improving the three elements could improve the first component of behavioural cost by so much as 31 per cent. The reduction of passenger excess time, walking and waiting, can be brought about by siting of bus stops, but most important will be improvement in service reliability, and without some kind of traffic restraint or bus priority scheme this is the most difficult element to improve. Changes in passenger fares would also contribute to the behavioural cost but it would be necessary, in many cases, to pay the bus passenger for each journey undertaken!

Despite these desirable objectives of bus priority schemes, it remains difficult to know to what extent the measures used are successful in achieving transfer from car to bus. There have been few attempts to detect whether any transfer has in fact occurred, so that the benefits are usually assigned to too few bus passengers and the disbenefits to too many car users. It is also necessary to assess the benefits not just to users of the bus priority scheme but also to passengers outside the scheme who may benefit from more punctual arrival of buses transferred from one route to another during the peak, for example. Benefits to bus passengers arising from superior waiting conditions in cases where traffic flows have been reduced are also ignored. If these conditions are improved, then the value of excess time might be reduced so that total behavioural costs would also be lowered. Bus priority schemes are often introduced too late, i.e. when the transfer to car-based trips is well-established. An attempt should be made to anticipate deteriorating modal-split conditions which are not in favour of buses, and take preventative rather than curative action. Bus priority schemes will not be very helpful in community terms, however, if the general traffic conditions improve in a way which continues to favour car users. If an effective modal split is to be achieved it must be undertaken in conjunction with a reduction in road capacity for other traffic.

In order to achieve the sort of modal transfers sought after it may be necessary to produce bus routes using extensive networks which are totally independent of all other forms of traffic. Hall has noted that there are a number of large British cities with major arterial roads containing former tram reservations which could be converted into bus-only routes, e.g. Princess Road in Manchester, Speke Boulevard and Queens Drive in

Liverpool. This would in many cases be at the expense of street 'furniture' such as trees, shrubs and lawns which now cover former tramway lines, and objections to such a change of use are likely to be strong. Much better-placed are cities such as Nottingham, Edinburgh and parts of inner London, which possess a network of abandoned rail routes which could be converted to 'busways' operating along the same principles as train services and completely segregated. In Nottingham there are over 30 miles of former rail routes connecting the suburbs to a major new shopping complex in central Nottingham, the Victoria Centre. Some 10 miles of former rail track has already been purchased with a view to conversion.

Busways

Probably the best opportunities for creating busways exist in the New Towns, and an example in Britain is provided at Runcorn New Town (figure 9.5). The busway is not, however, confined to the new residential neighbourhoods such as Castlefields or Halton, but will also be extended to incorporate the 'old' town which had a population of approximately 28,000 when the new town was designated in 1964. In the formulation of the master plan for the New Town, two problems, which had not been well-considered in the past, were seen as important (Mercer, 1971). The first was how to create an environment which, while allowing complete freedom for growth in car ownership, would provide an opportunity to reduce the undesirable side effects. Secondly, there was a problem of how to establish a reliable, efficient and attractive public transport system which would not be subject to the spiral of increasing fares and a reduction in the number of passengers and passenger/miles. The response to these two problems has been to provide a busway which furnishes the framework upon which the town's neighbourhoods are based, rather than the road system as is so typical of most of the other British new towns. The structure of each neighbourhood (see figure 9.5, inset) demonstrates the emphasis on providing rapid pedestrian access to the busway, with every resident no more than 5 minutes' walk or 500 yards from a busway stop. Most of the busway halts also coincide with neighbourhood shopping, social, educational and cultural facilities and these, in turn, are linked by the busway to the new town centre, where there is a completely enclosed bus station segregated from all other types of traffic. The busway also provides direct links with Runcorn's industrial estates in an effort to encourage workers to travel by public rather than private transport. The routing of the busway is so direct that there might be as much as a 50-per-cent saving on overall route length when compared with bus routes in towns with a similar population size but having a spider-web road pattern. In addition the busway has helped to save space in that an extra 20 acres in industrial areas and 60 acres in residential areas would be required for

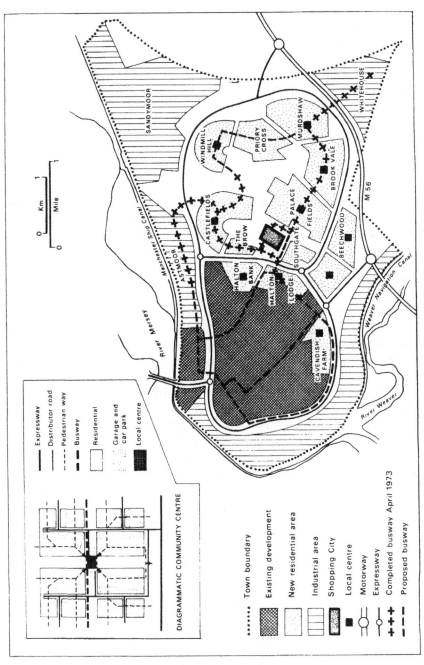

Figure 9.5 *Busway network in Runcorn New Town* (Source: *Runcorn Development Corporation 1975*)

Table 9.2 Patronage of Runcorn Busway, 1972–74

Year	Total mileage operated	Total patronage	Approximate average patronage per work day	Cost per mile (p)	Total revenue (£)	Revenue per mile (p)
1972	334,059	475,000	1,600	18.23	39,099	11.70
1973	383,922	885,000	3,000	20.53	65,192	16.98
1974	623,678	1,600,000	5,300	25.05	122,137	19.58

Source: Runcorn Development Corporation (1975).

parking space if the busway had not been provided. The busway also removes the second tier in Runcorn's road hierarchy, the district distributor, and this imparts considerable environmental benefits.

The buses run to tight schedules, which they have no difficulty in keeping to because they are given priority where the busway crosses conventional roads. By 1 April 1975, some 8 miles of the busway was operational and a further 1.88 miles was under construction. The objective is a 50:50 split between public and private transport trips and although it is a little early to assess whether Runcorn has achieved the right blend of accessibility and attractiveness in the system, all the signs are encouraging (table 9.2). A survey by the TRRL in 1973 showed, however, that few car users were using the busway despite the high level of service provided. This may reflect the general lack of traffic congestion and the fact that 45 per cent of Runcorn's employed residents work outside the town, compared with the Master Plan target of 30 per cent. The Runcorn busway is primarily designed to provide efficient circulation rather than the benefits of an expressway system of the kind which has been proposed for the area at present served by the North Tyne Loop Railway (Costinett *et al.*, 1971). Redditch New Town is also introducing a busway, of which 3 miles is operational out of a total of 15, on new and converted roads. Clearly, not every town will have the opportunities which exist in Nottingham or the New Towns but, where they can be properly integrated into an urban transportation network, there seems to be good reason for promoting them. But more data about their impact on travel behaviour is vital.

Dial-a-ride buses

All the bus services discussed so far operate along fixed routes and at frequencies which bus operators estimate will meet the demand. Hence, services are more frequent at peak than off-peak times, and the Victorian concept of scheduled buses running over fixed routes with the minimum of dynamic control is still very much in evidence in our cities (Oxley, 1970). If public transport expects to compete effectively with private transport it should perhaps be attempting to provide more demand-responsive services, and a number of dial-a-ride or dial-a-bus services are now in operation. Owing to their size and manoeuvrability, conventional buses are not suitable for demand responsive systems, so that 10–20-person minibuses are usually used.

This approach to public transport provision has its origins in the United States, where in some towns such as Peoria or Decatur, employees in large factories were picked up at home at prearranged times, conveyed to work and then transported home in the evenings at a prearranged time (Oxley, 1970, p. 146). Such a service is relatively easy to provide because route planning can be optimized in a situation where origins and destinations are known in advance. The British experiments, and many others in the United States, attempt to provide a facility which is flexible and can provide access to a wide range of destinations. A service which started in Harlow in August 1974, has shown a steady increase in patronage, and since November 1974 some 4,000 passengers have been using it each week (Slevin and Ochajna, 1974). The transit minibuses which are used are at least 50-per-cent full throughout the day and 75-per-cent during peak hours. Some 30 per cent of the trips are for personal/shopping purposes, 16 per cent for the journey to work, and 32 per cent for social/entertainment purposes. Over 50 per cent of the users in Novermber 1974 were from one-car households, and 26 per cent of the passengers had transferred from private transport modes. Over 75 per cent of the trips were pre-booked pick-ups. London Transport's dial-a-bus service in Hampstead Garden Suburb is also still operational following a review of its performance in April 1975 when it was carrying over 700 passengers a day.

In a dial-a-ride system, the buses are dynamically routed and scheduled on the basis of a quick response to a telephone request for service. Control is exercised from a central point, using a computer. The system provides door-to-door service for diffuse trip patterns, and by serving a number of passengers at the same time it is possible to keep the fare at a reasonable level with most systems utilizing a flat-rate for all trips. Each bus has a two-way radio and a digital-printer communications device, and when a request is made for a journey the computer selects that bus which is best able to serve the passenger concerned in relation to the existing number of passengers on each bus, their destinations, and each bus's position on the network. The computer algorithm used to route buses operates within the framework of waiting and travel-time constraints which are necessary to guarantee passengers a certain total travel and waiting time (for a more complete description of operating principles, see Oxley, 1970, pp. 146–8). Kissling (1973) has shown how the details of dial-a-ride operations vary widely

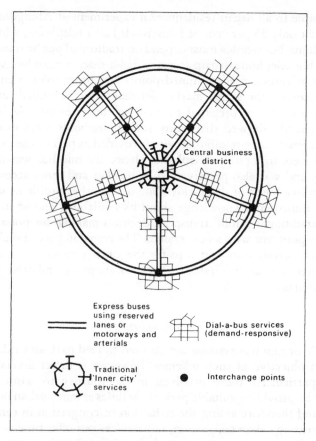

Figure 9.6 *Idealized bus system for a medium-to-large British city* (Source: Hall, 1974, p. 800)

between cities because size and shape of operating areas vary, as do socio-economic characteristics. Some offer routes from 'many to one' location, from 'many to few', and from 'many to many' locations.

It would appear that dial-a-ride is most desirable for short-haul trips of all kinds within urban areas, perhaps serving suburban interchanges for express buses or rail/underground services to the city centre. If they attempt to provide longer-haul services, they will suffer similar problems to those faced by conventional buses, particularly in and around the city centre. This has led Hall to suggest a bus transport hierarchy of the kind shown in figure 9.6. It is assumed that there will be a radical restructuring of public transport systems in most British cities, but it may be necessary in the interests of reducing the level of private transport travel. Before such a pattern could become viable, however, we should recognize that dial-a-ride systems are not accessible to all urban residents. An experiment at Abingdon in 1972 showed that only 55 per cent of households had a telephone, which means that remaining households must depend on traditional public transport, can use the public telephone to gain access to dial-a-ride, or must hail-stop a dial-a-ride bus at certain predetermined points and times (Slevin and Cooper 1973). Therefore the actual market for dial-a-ride is limited unless non-telephone owners are prepared to walk to their nearest telephone. The Abingdon study showed that many such households were nearer to an alternative fixed-route minibus service (provided as part of the experiment) than they were to a public telephone. Hence the minibus was used more frequently and was also perceived to be quicker and more accessible than normal-service buses or dial-a-ride. One of the other problems with dia-a-bus is the diffusion of knowledge about its availability and an in-built bias towards traditional public transport decision-making by potential users asked to experiment with a new system. The actual impact of dial-a-ride on peak-period private transport trips for the journey to work is unclear, and the system may be most suitable for off-peak shopping and other trips from residential areas.

Park-and-ride

Dial-a-ride or minibus systems can be used to feed park-and-ride facilities. The main objective of such schemes, however, is direct diversion of car drivers, particularly those involved in the journey to work, to pubic transport by providing suitable parking facilities at inner-urban or suburban stations and therefore aiding the reduction of congestion in central areas. With increasing urban congestion, rail services can offer time savings over other modes, particularly at peak periods, while beyond a certain threshold level the operating cost of rail as opposed to bus facilities becomes lower. Additional benefits from park-and-ride facilities include the possible elimination or diversion of some existing bus services, with consequent cost savings and a reduction of the pressure being experienced by local authorities to build new highways and all-day parking facilities. The Merseyside Passenger Transport Executives are experimenting with bus-to-rail and car-to-rail facilities and have found that patronage has nearly doubled at car parks included in the experiment (Millward *et al.*, 1973; Choudhury, 1971). Reducing the scale of charges at car parks has a considerable effect but making the parking free dramatically increases car-park use; in one case from 50 to 200 cars per week. In the case of bus–rail facilities an increase in feeder services resulted in an increase in bus–rail users from 100 to nearly 600 return journeys per week. Interchange systems seem particularly suited to peak-hour diversion of car-borne commuters but may be less attractive to other users in the case of car–rail facilities, because the car parks are often full unless space is reserved for users who arrive later in the day. Off-peak users, such as shoppers, of bus–rail interchanges may well prefer to use stage bus services direct rather than transfer from one mode to another during a trip. It is also worth noting that an increase in the number of driving licences among wives permits an increase in 'kiss-and-ride' mode choices, and such a demand must also be allowed for at suburban stations.

A variation on this approach to encouraging higher levels of public transport, particularly buses, was attempted in Nottingham during 1975–6 (figure 9.7) (Trench and Slack, 1973, 1974, 1978; Milne, 1974). Instead of

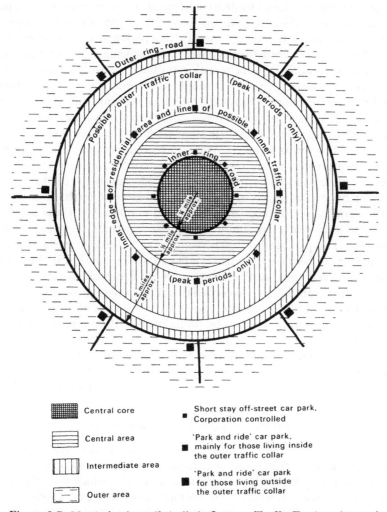

Figure 9.7 *Nottingham's traffic 'collar'* (Source: **Traffic Engineering and Control, 14,** *1972, p. 348)*

using conventional methods of bus priorities a circle of 'intercept' points were selected about 2.5–3.0 miles from the centre but which were inside, though near, the outer ring road. Motorists arriving at these points during the peak were given a threefold choice: park-and-ride; circulate around the city (using the outer ring road); queue at the 'collar' for entry by car using traffic-light controls which stayed on red for up to 10 minutes followed by a very short period on green. Buses and permitted commercial vehicles were allowed through the collar, which only operated restrictively during the peak period. It was therefore left to the motorist to decide how much value he placed on his time; it was hoped that the length of queues or the extra journey length introduced by diversions would cause motorists to use park-and-ride services, thereby reducing congestion and parking demands in central Nottingham. It may be, however, that park-and-ride facilities should be further from the centre than planned in this scheme if they are to be really attractive to motorists, while a major problem is likely to be motorists undertaking alternative routes through residential streets in an effort to avoid the 'collar'. Nottingham's morphology prevents this on the western and northern sides of the city but not in the other two sectors. It was even suggested that barriers be placed in some residential streets likely to be most vulnerable, but this is likely to interfere, in an unreasonable way, with the vehicle movements of local residents. It is also difficult to know how many commercial vehicles to permit and how to identify between through and locally-generated commercial traffic. There was considerable resistance to Nottingham's 'zone and collar' scheme, and in mid-1976 it was decided to neutralize the 'collar' of traffic lights barring the city centre to most motorists. The park-and-ride scheme, costing £120,000 per annum, was also phased out because there had not been a noticeable shift in the mode of travel from car to bus. A Transport and Road Research Laboratory report on the scheme subsequently recorded that it had a negligible effect on traffic congestion, reduced bus-journey times by less than a minute on average, increased private-journey times by no more than $1\frac{1}{2}$ minutes and produced no significant changes in travel habits. Nottingham's traffic problem may not be serious enough to merit a scheme of this kind, and an effective and appropriate solution will to some extent be controlled by the size of cities; some of the alternatives have been suggested in the *Leicester Traffic Plan* (City of Leicester, 1964, p. 61).

Traffic restraint and road pricing

However sophisticated the attempts to make public transport more attractive, the success in terms of reduced congestion is likely to be limited unless efforts are also made to control other vehicles which use urban roads. These are generally referred to as restraint procedures, of which there are many kinds, and which are sometimes described as 'congestion pricing' or 'road pricing'. They are based on 'broad considerations of the optimum attraction of resources as between transport and other sections, and within the transport sector' (Beesley, 1973). One of the characteristics of urban road systems is that only relatively small parts of them are heavily used and any response designed to match supply and demand at these points is both slow and usually only a temporary palliative for the problem. Elsewhere on the road system, excess supply is the rule, and excess demand at certain points on the network, such as on some roads into the CBD, arises because no 'rent' is sought for the use of scarce road space as it is for development land, for example. In his decision to use a road an individual does not suffer the total delay he causes, only the amount suffered by every individual using the same stretch of road. In other words, individuals ignore the opportunity costs of their decisions, and the idea of congestion pricing is that each individual should pay a charge that reflects the delay imposed on other travellers and that this should bring the marginal private cost to the individual up to the social cost suffered by all individuals collectively.

The major difficulty is how to select a system of pricing which leads to equtable distribution of its effects on different types of road users. Because there are wide variations in the way in which people perceive costs, according to their value of time, the vehicle they use, their income, etc., this means that some individuals would be prepared to find alternatives. A road tax of some kind would increase speeds for those prepared to pay the tax, assuming that the levy would be more than some individuals would be willing to pay. Hence, traffic volumes would be reduced, speeds increased and the real costs to all vehicles using the road system at congested times would be reduced, and those paying tax would be 'paying' for those 'not paying' it to use some alternative mode. Vehicles continuing to use the road system at congested times would be paying more out-of-pocket costs by the amount of tax less the reduction in travel costs. Most of the tax paid would be transfer payment, i.e. it would not involve any consumption of resources but a transfer of money from those who continued to use congested roads, to the government or to the local authority. The real benefit would be the reduction of congestion.

The desirability of road pricing in British cities began to be seriously considered in the early 1960s. A report by the Ministry of Transport (1964; see also Walters, 1961; Beckman *et al.*, 1956) listed six net benefits to the community, as well as some disbenefits, resulting from the higher traffic speeds consequent upon pricing: savings in fuel and other vehicle running costs (1); greater productivity from buses and other commercial vehicles in that the same number of journeys can be made using fewer vehicles, meaning less capital investment in the short and long term (2); savings in the paid working time of persons who travel in working hours, including crews of commercial vehicles and buses (3); time saved travelling to and from work

(4); changes in the costs caused by accidents (5); losses to people who as a result of the price changes refrain from making journeys which they would otherwise have made (6); gains and losses to other road users such as pedestrians and cyclists (7); changes in the costs imposed by road users on the rest of the community by way of noise, fumes, etc., derived from motor traffic (8). It is estimated in the Smeed Report that savings in paid working time (3) would amount to at least 40 per cent; in other time savings (4), more than 63 per cent; vehicle running costs (1), in excess of 7 per cent; capital savings (2), more than 10 per cent; losses (6), -20 per cent. The classes of net benefit would yield a total, allowing for the value of time savings, of £100–£150m a year under traffic conditions prevailing at the time of the Report. As congestion increases more rapidly than vehicle mileage, it is likely that the potential benefits from pricing will rise at a steeper rate.

It is important to appreciate that road pricing will not, as with the managing of public transport mentioned earlier, completely eliminate congestion; it will only reduce the level and make public transport more viable. Indeed, above 20 miles per hour, the Smeed Committee estimates that road pricing would be unnecessary and that the benefits are greatest in cases of severe congestion. Meters were nominated by the Smeed Report as the best way to effect road pricing. The meters would be attached to all vehicles and actuated electronically as they crossed pricing points and entered areas where pricing operated, and vice-versa. A number of meter systems have been tested and developed but have yet to enter general use. The concept is simple but its application poses a wide range of problems, such as the extent to which it will be necessary to vary the location of pricing points at different time periods during the day or week and at different locations according to changes in the distribution of congestion. (The Smeed Committee gives nine important and eight desirable objectives for the operation of a road-pricing scheme. It would be difficult to satisfy all the criteria and, in general, it remains unclear whether such schemes would be practicable.) It will also be necessary to vary the charges according to vehicle size: higher for large vehicles such as lorries and lower for small cars. The need to produce equality in the effects of road pricing is also a major consideration, and it may will end up that the rich will pay and the less wealthy will be squeezed out. Methods of reading meters and collecting taxes levied also pose a wide range of problems for administrators of such a scheme. 'Black boxes' which identify vehicles using toll roads are the subject of experiments in New York and San Francisco. Charges are billed by computer without vehicles being required to stop, but at present such schemes Are exceptional.

Other methods of road pricing

There are a wide range of other possible methods for charging for the use of urban roads (figure 9.8). The fundamental distinction to note is that between direct and indirect methods, with the latter having more weaknesses than the former. Purchase tax or annual licences for vehicle ownership may be easy to collect but they do not discriminate between the different journey purposes for which the same vehicle may be used, while fuel or tyre taxes are not

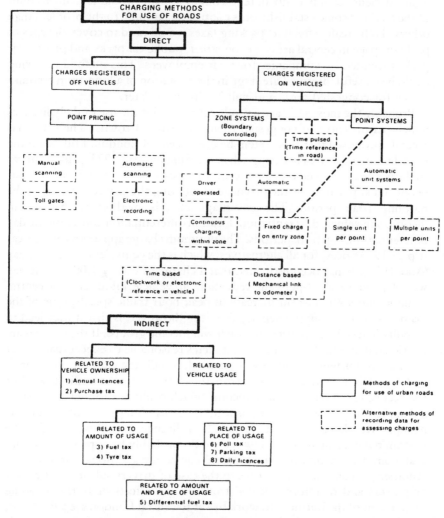

Figure 9.8 *Alternative methods of charging for the use of urban roads* (Source: Beesley, 1973, fig. 9.2, p. 227)

specific as to the location of congestion. A differential fuel tax related to amount and place of vehicle usage might be administratively simple, but it would be difficult to produce differentials between areas which would be low enough not to encourage special trips to low-price fuel areas to obtain supplies or the purchase of fuel in non-congested areas en route to a more highly taxed congested area. A poll tax would involve taxing employees working in congested areas or a tax on floor space in properties such as offices or warehouses. This assumes that employees in a particular establishment actually contribute to congestion and would take no account of the fact that some establishments contribute far more to the problem than others. To be really effective, parking taxes would need to cover all types of parking space in central areas, i.e. on-street, public car parks and private car parks provided by companies for their employees. This method of taxing could be a useful intermediate stage in the evolution of a long-term restraint policy for congested areas and will be discussed later.

Probably the most satisfactory indirect method is the use of daily licences which permit entry to congested towns or parts of towns. The GLC has recently proposed that this system be use in central London. Following the introduction of an area licensing scheme in Singapore in 1975, peak-hour car traffic has been reduced by 75 per cent. Private cars carrying fewer than four people must pay 60p for the privilege of entering the city centre during the morning peak period. The advantage of this method is that it can incorporate vehicle differentiation, location differentiation and a time-of-day differentiation, e.g. peak periods. The optimum charge appears to be 60p/car trip (at 1973 prices) for all alternatives, which is the equivalent of 75p/car day (May,1975). It has been estimated that a charge of 60p–£1.00 per car/day would produce a 37-per-cent reduction in vehicle kilometres in central London and a 40-per-cent increase in peak-hour traffic speeds. One of the main criticisms of supplementary licensing, however, is that it may lead to inequity in its effects, particularly on lower income groups. If the licences are restricted in use to the central area, it seems reasonable to expect that higher income groups will benefit most because they will stay with car transport if possible and they will also benefit from the improvement in traffic flows. But the licence fees would produce income which could be channelled towards an improvement of public transport, thus benefiting the lower income groups. Residents of areas included in licence zones would also be discriminated against, and a low fee would probably be appropriate rather than complete exemption which would encourage additional car journeys. Another problem of equity arises in the case of drivers subsidized by their employers and therefore shielded from the full effects of restraint. Some 40 per cent of the journeys to work by car to central London are subsidized, usually in the form of fringe benefits and not because the car journey is important to the employer. Interviews with employers during the study upon which the proposals are based, showed that most would be reluctant to pay their employees' supplementary licensing fees, so that sensitivity to licensing may be more elastic than is commonly assumed. Supplementary licensing would also affect industry and commerce in central areas of cities, and in central London exemption of commercial vehicles would produce a 35-per-cent increase in the flow entering the area. The additional costs and benefits for individual land uses also vary, with offices experiencing insignificant additional costs. For shops, places of entertainment or hotels the additional costs would also be small compared to existing ones, but there might be loss of custom. Industry would be worse-affected because its traffic-generation rates are higher than offices and its other costs, rents and salaries for example, lower. Public reaction to the scheme for central London was wide-ranging, and it has been decided not to implement supplementary licensing for the time being. Revenue was calculated to be £56m at 1975 prices, and with buses and taxis exempted (as well as other essential vehicles) it is estimated that public transport would gain an extra £24m in fares. The most important difficulty is deciding when a suitable set of zonal charges are being levied for the same town or for various levels of congestion in different towns. Although in some ways similar to a parking tax, the daily licence would be required by all vehicles using the congested area, irrespective of whether they were simply passing through rather than having destinations in the licensed zone.

Indirect pricing is therefore generally considered inferior from the marginalist point of view, and there has been considerable interest in direct methods such as meters of various kinds. The alternatives are outlined in figure 9.8 and involve a basic distinction between on-and-off-vehicle registration of charges, using a variety of electronic devices to record duration or distance travelled. As already mentioned there has been considerable testing of alternative devices at the Road Research Laboratory but meters have not yet been used to operate a congestion-pricing scheme.

As the objective of these pricing procedures is to ease congestion, it is hoped that car drivers will be persuaded to transfer to public transport or that they will make their journeys to the limits of restricted zones before proceeding on an alternative mode. If pricing is effective it should leave more space on congested streets for buses, which will have to accommodate the increased demand from users of cars/cycles, from the underground railway stations (in central London, Liverpool, Manchester, Glasgow) and from 'new' travellers attracted to the traffic-restraint area by the improved speed and services of public transport. Using data produced by Thomson for central London, Beesley calculates that of the extra passengers travelling on the buses, 22 per cent will be transfers from cars/cycles, 23 per cent from underground services and 55 per cent will be new customers. Most of these transfers will take place during peak hours and will therefore mostly relate to

Table 9.3 Comparative effects of daily licenses and parking tax on peak-hour traffic and passenger flows in central London

Flow (veh./h)*	Without restraint	With 30p daily licence	With 3.7p hourly parking tax
Cars	1,201	633	854
	(6,720)†	(3,540)	(4,800)
Taxis	208	298	283
	(748)	(1,072)	(1,020)
Cycles	432	261	267
	(1,896)	(1,148)	(1,176)
Commercial vehicles	369	418	398
Buses	139	193	175
	(15,040)	(20,800)	(18,880)
Total	2,349	1,803	1977
	(24,404)	(26,560)	(25,876)
Average Speed (mph)	9.1	11.4	10.4

Notes: * Flows on main survey roads of central London
† Passenger flows.
Source: Thomson (1967), table 19, p.366.

the journey to work. There will, of course, also be some transfer from cars/cycles to the underground and the surface railways, so that, although on the whole it will be easier for buses to operate, the effects of the improvement on their operating reserves will not be great.

The level of restraint used will also affect the flow of traffic and the flow of passengers (table 9.3) (Thomson, 1967). The effect of a 30p daily licence is to reduce the flow of cars per hour in central London by approximately 50 per cent (during the peak) while a 3.7p hourly parking tax does not have the equivalent effect even though the total cost per day would exceed the daily licence fee (although most cars probably spend less than 8 hours per day parked). The effects on passenger flows in cars are similar. As expected, the flow of buses increases to compensate; an increase in the number of buses of the order of 30 per cent occurs if daily licences are used, with a rather smaller increase as a result of parking taxes. Traffic restraint means that investment will be necessary in additional buses, while in order to maximize the benefits to public transport it will be important to reorganize bus routes, make wider use of flat-rate fares, and a better mix of stopping and express services tailored to satisfy the traffic requirements of different types of travellers at different times of the day. As Beesley (1973, pp. 269–70) suggests, traffic restraint provides public transport with the opportunity to increase its share of urban travel; the response depends very much on the enterprise of the operators.

Parking controls as a means of restraint

It remains uncertain, however, whether any of the above methods of restraint will be used in practice. On social and economic ground they make good sense but, politically, they may prove difficult to implement. Hence, considerable attention has also been devoted to parking control (as well as public transport) as the most promising method of restraint for the short term (Ministry of Transport, 1967d; Bouton, 1973). Not that the demand for parking is a new phenomenon: special off-street facilities were provided in Rome to accommodate chariots, and vehicles were not allowed to enter the central areas of large Roman cities because of traffic congestion (Brierley, 1962, p. 1). The objective of every parking policy has usually been to satisfy free demand but, more recently, it has become clear that the supply of road and parking space cannot, in most cities, match free demand. There has therefore been increasing emphasis during the 1960s on restraining parking demand in central areas and important suburban centres in order to allow traffic to flow freely and safely and trade and commercial activity to operate more efficiently. The demand for parking spaces comes from several sources (figure 9.9), and the problem is how to reconcile the different needs of users. Total parking demand can be expressed as a function of total shopping floor space, total office floor space and public transport accessibility of the defined area (Creyke, 1971). Controlled parking schemes ensure that moving traffic can move safely and freely; that kerbside space for loading and unloading goods and for picking up and setting down passengers is available where it is needed; places where vehicles may or may not park are clearly indicated; priority is given, where necessary to the short-term parker (Ministry of Transport, 1963a, pp. 9–10). The car commuter is the main target of most parking controls and the level of restraint depends on the extent of the problem to be solved; for the central area of London it has been estimated that a 35-per-cent reduction in peak-period private car traffic entering the area could lead to the necessary improvements in public transport and the environment, without adversely affecting mobility.

The flow of traffic can be improved by introducing parking controls which discriminate between areas where no parking of any kind is allowed, areas where waiting is permitted for loading or unloading only, and areas where any parking is permitted but which is limited in duration and/or

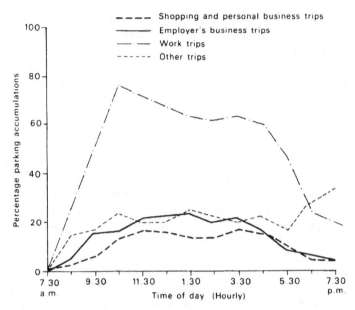

Figure 9.9 *Parking demand by trip purpose, Croydon* (Source: *Creykie, 1971, p. 303*)

controlled by price (e.g. parking meters: parking meters were first introduced in London's Mayfair in 1958 under powers contained in the Road Traffic Act, 1956; they first appeared in Bristol, Manchester, Birmingham, Southend, Newcastle and other major urban areas during the early 1960s – see Lambe, 1972). These are on-street controls administered by local authorities and usually 'policed' by traffic wardens who ensure that the regulations are observed, issue fines, etc. Flow-orientated parking controls discriminate against long-term parkers, particularly commuters, who will be forced to seek on-street facilities further from their trip destination or must attempt to utilize off-street parking facilities where long-term parking is possible, although this can still be priced in such a way as to encourage or discourage commuters (table 9.4). The table shows how uncontrolled parking has been reduced severely in central London, especially between 1963–70, while the number of meters (representing controlled parking) has doubled since 1961. Apart from space supplied in multi-storey car parks, the volume and location of off-street parking places tends to fluctuate considerably in the inner areas of older cities as plots of land are cleared for development and used as temporary car parks until new development begins. Land cleared for redevelopment also often provides opportunities for uncontrolled parking which may undercut overall local authority parking policies, at least in the short term, and temporarily help to

Table 9.4 Structure of parking accommodation: central London, 1961–81

Sector	1961	1963	1970	1981
Off-street				
Public	14,780	25,360	29,000	50,370
Private	36,670	46,660	53,000	59,180
Residential	4,550	8,330	17,000	26,680
On-street				
Uncontrolled	53,000	30,000	2,000	—
Metered	10,500	14,500	21,000	20,000
Residents	775	1,550	7,000	9,000
% off-street	46.5	69.7	76.7	82.4

Source: GLC *Business Traffic Generation Study* (1973).

reduce the role of public transport. In July 1975, Liverpool District Council decided to seal off a number of car parks of this kind in and around the central area in order that its parking policies should not be prejudiced.

While it may be possible to restrict the availability of public parking facilities, either on- or off-street, and to integrate them into some cohesive parking strategy, it is much more difficult to deal with the large proportion of private parking spaces in the central areas of our cities (table 9.4). Private (non-residential) parking spaces exceeded public facilities by approximately 2:1 in 1970, and the balance has changed little since 1961. They account for 40 per cent of the central-area parking stock in London. Parking facilities in this group are attached to offices, shops and various public buildings, and employees and visitors are either not levied a charge for their use, or only charged a nominal sum which is well below the price of public parking in the same area. This clearly induces car owners to drive into the central area and to contribute to the congestion problem. It is difficult to restrict use of such facilities in existing development but the future supply can be limited by restricting the non-operational parking spaces attached to new office development, for example. Alternatively, it may be necessary for local authorities to acquire the larger private garages in order to make them available to the public at a suitable price (Ministry of Transport, 1967b). It may also be feasible to charge a tax for each existing parking space in private ownership, but this will only indirectly affect the supply of parking in that the funds produced could be used to provide more controlled public facilities. The GLC has suggested a tax of £6 per week for each private office car space, or an annual tax of £312, on firms keeping private parking spaces in central London (*The Times*, 21 February 1976). It is estimated that this would reduce peak-period traffic by between 15 and 22 per cent. The cost of bringing private parking facilities in city centres under public control is likely to be very expensive and, unless funds and/or legislation are forthcoming from central government, it is unlikely that local government will be able to spend money levied from the rates on acquiring a facility which will benefit non-residents as much as the residents of the local authority area; with the former not having to make a contribution to purchase costs either. Compensation to firms losing their private parking spaces could amount to at least £60m in central London alone. While there have been numerous efforts to limit non-operational space in private development and a number of attempts to increase residential parking on on-street space, it is only comparatively recently that the supply and control of parking has been seen as a way of regulating travel patterns (May, 1972).

Integrated parking strategies

In any parking strategy it is important to distinguish between the needs of long- and short-term stayers. It is unlikely that increases in the cost of parking will completely deter some commuters, while excessively high costs would keep out the shoppers and others who do not wish to stay very long, perhaps two or three hours (figure 9.10). The kinds of solution which could be adopted can be illustrated by reference to Leeds City Council (1969), which during the course of preparing its central-area traffic plan for 1981

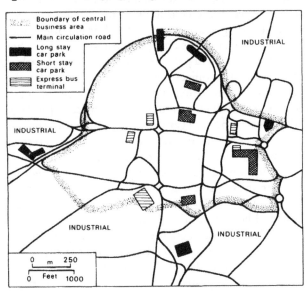

Figure 9.10 *Central area traffic control: Leeds, 1981* (Source: Rose (ed) 1973, figure 2.9, p. 2.35)

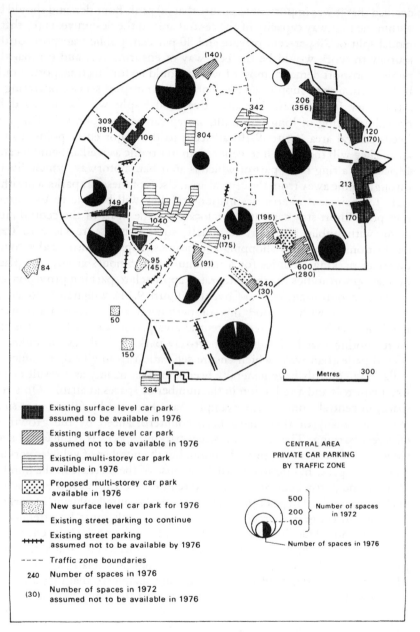

Figure 9.11 *Parking in central Coventry, 1972–6* (Source: *City of Coventry, 1972, fig. 2.9, p. 2.35*)

decided that it was important to ensure that (a) peak-hour demand should be within the highway capacity of the central area in the design year; (b) that a modal split of 20-per-cent private and 80-per-cent public transport for the journey to work would be the best way of ensuring (a); and (c) that the public transport demand, most of which would be for bus transport, would be large enough to support an efficient public transport service operating at satisfactory frequencies. The objective modal split will be achieved by providing a ring of long-stay multi-storey car parks around the fringe of Leeds central area (figure 9.10) adjacent to the major entry points to the area. They will therefore intercept commuter traffic as well as being located adjacent to a ring road, part of which is an urban motorway, which diverts through traffic away from the central area. Users of these car parks will either walk to their central-area destinations or they will be able to use public transport. Short-stay car parks are located in the heart of the central area, along with public transport terminals, so that users are close to their destinations. Short-stay shopping and business trips will also be able to use on-street parking facilities for a maximum of 2.5 hours at any one place. Similar opportunities for this integrated approach to parking provision and public transport usage exist at Coventry (figure 9.11), which in 1976 will be well-provided with a good mix of peripheral and core-area car parks combined with efforts to reduce central-area private car parking. It is also worth noting that in Leeds the short-stay parking facilities are linked to special pedestrian ways which ensure separation from all other traffic.

Parking controls have generally been effective, mainly as a result of on-street controls and a reduction in the number of spaces available. On-street spaces in central London, for example, have been cut by 60 per cent since 1963, but many of them have been counterbalanced by the growth in off-street parking. The use of on-street parking spaces is reduced by such tactics, by 30 per cent in central London, but the number of vehicles parking on private spaces has doubled; partly because of the increase in the number of space and partly because they are more intensively used. If the reduction of at least one-third in peak-period traffic is the favoured solution of parking problems in congested areas, then parking controls must be extended to cover all kinds of facilities and, in addition, they will only be fully effective if combined with controls on through traffic. As a method for traffic restraint, parking controls have the advantage that they are relatively simple to administer and understand and are reasonably flexible, so that they are likely to have a dominant place in transport planning for many years to come.

Staggering of work hours

The traditional approach to the resolution of central-city congestion has been to invest either in new roads or railway systems, or to invest in traffic-management schemes which bring about a shift from a more to less space-consuming mode of transport. Everything is geared to handling peak-hour demand, and the availability of central government funds and grants, as well as the application of various policies over the years, has reinforced these responses. Many of the projects involved are large-scale and may be an unsatisfactory way of dealing with peak-hour congestion, for several reasons. The time lag between the initiation of an idea for a scheme such as the Victoria Line in London or the Piccadilly/Victoria Line in Manchester, the decision to go ahead and the implementation of the plan is usually very long; 10–20 years is not exceptional and it can be much longer. Substantial changes in urban transport and travel can take place in the meantime, and the original stimuli for a new project may be overtaken by subsequent events which may undermine the benefits foreseen when the project was first mooted. Secondly, large new facilities such as urban motorways or comprehensive traffic control/public transport provision schemes are expensive both in fiscal terms and, in some cases, in their environmental consequences (see chapter 10). Thirdly, new facilities will generate traffic which will eventually induce peak-hour congestion similar to that which existed before, because traffic tends to increase to meet maximum capacity. The new construction attracted to sites along new transport routes also militates against an overall reduction of congestion because it inevitably attracts 'new' traffic. Opposition to urban highways is also becoming more effective and delaying new projects even longer than usual, and this makes long-term investment plans both uneconomic and difficult to justify.

Some of these disadvantages could be alleviated to some degree by staggering the hours of arrival and departure of workers in city-centre workplaces. This is a low-cost method of traffic control which involves spreading the peak-hour travel demand, most of which is a product of the journey to work. The idea of staggered work hours has been examined on and off for more than 50 years, particularly with reference to central London, with initial experience centred on industrial jobs in London's outer suburbs where, by the end of the last war, 600,000 industrial workers were participating in staggered hours which fitted the capacity of the transport system (Collins and Pharoah, 1974, pp. 176–85). This scheme had largely lost its impetus by 1971, but since 1955 there has been considerable interest in applying the same principles to central London (Ministry of Transport, 1958). A committee for Staggering of Working Hours in Central London was set up in 1956, but its success never matched that of its predecessors and when it was wound up in 1964 some 60,000 workers were working adjusted hours (Collins and Pharoah, 1974, pp. 176–85). This is very small in relation to the number of commuters entering and leaving central London, which exceed one million during the peak hours. About 70 per cent of the firms asked to participate in the project declined on the grounds that it would lead to a loss in business efficiency; of contacts with customers; the need to maximize hours for intercontinental communication; the need to maintain business hours; and staffing problems arising from the unattractive working hours introduced by the scheme.

A successful programme of staggered hours depends on the cooperation of a wide range of business and other activities, and the interdependence of different functions makes it difficult for many firms to consider starting and finishing at different times because intercommunication could be prejudiced. Against this, firms must consider the increased worker-efficiency following a less arduous work trip and the possibility of lower staff-turnover rates. Individual workers will experience a reduction in commuting times, especially for access by road, and this could be linked to a public transport policy. There should also be reductions in the air pollution and noise which are associated with stop–go driving under congested conditions. Staggered hours would also spread lunch-hour demand for restaurant and other facilities and reduce further the tendency for a minor peak in the demand for internal trips during the period. Clearly, to have any effect, several firms must participate, and not just individual firms scattered throughout a central area. The latter would create transport benefits for individual workers but the community benefits would be marginal. Other difficulties with the introduction of staggered hours include the need to coordinate it carefully with public transport timetables so that employees and undertakings can share as much of the benefit as possible; it is important to try to control any shift of workers from public transport to cars because congestion is reduced. Staggering may reduce the possibilities for operating car pools, which in themselves help to ease congestion and lead to more efficient use of resources; it is important to analyse the effects of staggering hours on all parts of a city's transport system, including stations outside the central area or suburban bus routes, because new peaks may be created elsewhere where transport facilities are inadequate (Goodman, 1972; O'Malley and Selinger, 1972). In addition, staggered hours may have adverse effects on non-work trips to the central area, because of a greater potential for conflict between shoppers and others who often wish to avoid the rush associated with work travel or want to use off-peak concessionary fares. Finally, the autonomy of firms will always make it difficult to obtain cooperation for any staggered-hours project.

An alternative way of spreading peak demand for travel is 'flexitime', in which workers can choose their own working hours between a longer, specified period, say 07.30–19.00 hours. This method is used by some offices in London and elsewhere, but it is difficult to predict its effect on congestion levels because the likely pattern of work will vary according to individual preferences, which may in turn then vary from day to day or week to week. Social, recreational and other trips connected with dental or medical appointments can be more easily accommodated without consulting employers. On the other hand, inter-office communication becomes even more difficult than with staggered hours.

A third possibility for spreading the peak is to introduce three- or four-day weeks, which experiments suggest lead to higher productivity and lower unit costs, reduced absenteeism and turnover, and improved morale. It leads to an increase in the amount of leisure time available to individuals, a decrease in the number of commuting trips and hence a reduction in the cost and time spent travelling. The peak load can be eased by companies working different four-day periods, possibly on the basis of a Monday-to-Saturday week. A three- or four-day week may, however, lead to more recreational travel during the extended weekends. Mid-week peak-hour congestion would not be reduced very much and, since it would be necessary to work ten-hour days (assuming a forty-hour week is still necessary), a major rescheduling of public transport would also be necessary. The lower number of work trips would also reduce public transport services even further, at a time when many systems are finding it very hard to exist, let alone to operate profitably.

Public transport and land-use distribution

The relationship between land use and trip generation has already been illustrated. It can also be shown that the location of development in relation to transport facilities will be reflected in the balance of trips using public or private transport. These relationships are well-illustrated by office and other commercial development, which generates much of the peak-hour travel demand in city centres and at major suburban nodes. The most obvious opportunities occur at railway stations, which are either termini or are located along the radial routes serving the city centre and which are also nodes for bus routes. Commercial development proposals for stations in London have excited great interest since the start of the boom in office development following the lifting of building licence restrictions in 1954. The main transport benefits of commercial development at stations have been summarized by Collins and Pharoah (1974, p. 222; see also Wacher, 1974) as follows; it will produce site and rent become for the transport operator which can be invested in new or improved travel facilities; it will generate revenue for the transport operator arising from additional travel along the line(s) serving the development and its interconnecting bus services; it will induce some transfers from private to public transport; by diverting travellers to buildings above or adjacent to stations, the provision of distribution facilities for passengers arriving at stations can be reduced; the peak demand will be staggered to some extent because workers will not have to allow for distribution time at the terminal station in their journey to work. The last three benefits will only occur if commercial land use is redistributed within cities, and this is a process which is occurring, partly as a response to congestion in city centres and partly as a result of land-use controls which make it difficult or impossible to expand establishments in congested areas (Daniels, 1974).

Experience in London has not been encouraging, however, because of a consistent failure to produce an overall policy for commercial development at interchanges (Collins and Pharoah, 1974, pp. 250–3). Opportunities, such as the redevelopment of Euston, to include several million square feet of office floor space have been missed, as the various authorities involved such as the Department of the Environment, Greater London Council, British Rail, the London Transport Board and the London Borough of Camden disagreed on the costs and benefits of the scheme. The same has been true of proposals for office development at suburban railway stations. It is clear that many of the conflicting interests involved could be better-informed if a suitable procedure was available for assessing the benefits of interchange developments; the property and financial benefits; social transport benefits; and the social land-use benefits. The evidence available does suggest that a properly coordinated policy would produce benefits for the public transport system of our larger cities (Parker, 1967; Daniels, 1975a).

Cost-benefit of public transport improvements

Until comparatively recently the general attitude towards public transport in Britain was that it should operate at a profit. This reflects its former dependence upon a myriad of private operators actively competing against each other for business, but as the structure of public transport operations has been rationalized and enlarged whilst levels of use have been declining it has proved difficult to satisfy this criterion. This has also inhibited the level of investment necessary to produce the kind of role envisaged for public transport as outlined in the preceding discussion. It is now recognized that subsidies from central government are necessary both to retain the existing role of public transport as a social service in many cities and to expand its clientele through improved organization, efficiency, and new vehicles and rolling stock. This growing involvement of public funds in transport undertakings has also created demand for accountability for investment decisions and more conscious attempts to assess the costs and benefits of projects, particularly large-scale infrastructure investments, the effects of which may be felt throughout a city's transport system. Hence, cost-benefit analysis has emerged as a 'practical way of assessing the desirability of projects, where it is important to take a long view . . . and a wide view . . ., i.e. it implies the enumeration and evaluation of all the relevant costs and benefits' (Prest and Turvey, 1965, p. 683). Put another way, 'cost-benefit analysis is a method of assessing alternative investment projects by techniques based on economic criteria' (Independent Commission on Transport, 1974, para. 7.47; see also Barrell and Hills, 1972). The technique is not only applicable to large-scale projects such as London's Ringway system or the Third London Airport, but is just as applicable to the economic assessment of the introduction of bus lanes or of supplementary licensing.

The purpose is to make economic policy rational or to increase the efficiency of public intervention in transport. A good example of the use of cost-benefit analysis in the context of urban transport is provided by the case of the Victoria Line, linking north-east with central London, which was the subject of such a study in 1963. With existing (early 1960s) underground facilities heavily congested, especially in and around the central area, and the situation equally bad on the roads, the Victoria Line was a sorely needed addition to London's underground network. But assuming that fares remained the same on the buses and the underground, plus the need for profitable operation, London Transport concluded in 1962 that the line could not be provided. The investment involved would simply divert passengers from other London Transport services and would not therefore produce any gains; these would simply accrue to existing passengers through less overcrowding; to new passengers via the improved accessibility provided by the Victoria Line; and to the existing road users through the reduction of traffic congestion. There would also be general environmental benefits because of reduced traffic levels. But Foster and Beesley (1963; see also Beesley and Foster, 1965) calculated that the construction of the line could lead to a social rate of return of 11 per cent. The values of the various first-year costs and benefits estimated for the line are shown in table 9.5, and they comprise a number of cost savings to traffic diverted to the line and to traffic on routes relieved by it. There are also benefits experienced by traffic remaining on relieved routes, and to new or 'generated' traffic. An effort was made to put a value on comfort by estimating the probability of finding a seat on an underground train, although the figures produced are still extremely rough estimates. The outcome of the social-benefit calculations depends on the assumptions about the value placed on the gains and losses, and the discounting methods used to estimate future costs and benefits from present values. Nevertheless cost-benefit analysis applied to transport projects has become an important component in many studies, such as the *London Transportation Study* or the search for a third London Airport (Freeman Fox *et al.*, 1968b; Commission on the Third London Airport, 1970; other examples of the use of cost-benefit analysis in relation to transport include: Buchanan and Partners, 1970; Lichfield and Chapman, 1968; Lichfield and Associates, 1969).

There are some deficiencies in the potential of cost-benefit analysis which

Table 9.5 First-year costs and benefits from the Victoria Line

Item	£ (m)
Costs	
Capital cost (at completion date)	53.5
Current costs (annual)	1.4
Benefits	
To diverted traffic	0.4
underground: time savings	0.4
underground: comfort and convenience	0.3
railways: time savings	0.2
buses: time savings	0.6
other motor vehicles: time savings	0.2
other motor vehicles: savings of vehicle costs	0.4
pedestrians: time savings	0.2
To common traffic (running on relieved routes)	
underground: cost savings	0.1
underground: comfort and convenience	0.5
buses: cost savings	0.6
other road users: time savings	1.9
other road users: savings on vehicle costs	0.8
To generated traffic	0.8
Total:	6.8
Net benefits:	5.4

Source: Thomson (1974), table 12, p. 207.

are implicit in some of the above observations (Thomson, 1969, pp. 155–7). It is, firstly, extremely difficult to evaluate the direct and indirect costs and benefits of the environmental consequences of transport proposals, such as the urban motorways proposed in the *London Transportation Study*. In analyses which concern themselves with a particular component of the urban transport system it is vital that the effects on other forms of transport should also be evaluated. Hence, cost-benefit analysis of London's motorway proposals makes little sense unless the impact on the public transport system as a whole and its users is incorporated. Thirdly, it remains difficult to evaluate the suffering caused by accidents, particularly road accidents. If elements of this kind are not adequately included, then it must be recognized by transport decision-makers that cost-benefit analysis is really only a partial evaluation, and an allowance must be made for 'intangibles'.

The use of cost-benefit analysis also creates problems in cases where it is used to test alternative schemes which must satisfy different statutory requirements. London Transport or British Rail, for example, must cover their costs from revenue (hence the failure perhaps to invest in rail or underground improvements), while there is no such requirement for road schemes. If such conditions were a prerequisite for new-roads schemes in urban area, Thomson estimates that if London's proposed ringways were used to capacity, a charge of about 3p per ride would be needed to cover their cost, but such a charge would reduce their use and a higher charge would be needed and the volume of traffic would again fall. In such circumstances public transport would almost certainly make a profit. Another criticism of cost-benefit analysis is that the calculations ignore important areas of public concern, while there is also a belief that quantitative assessment cannot be a substitute for experience and judgement. The main problem is when to invoke these two elements in the decision-making process. Despite its inherent limitations, cost-benefit analysis is an 'important component of the social process of decision-making' (Independent Commission on Transport, 1974, para. 7.64). But it is unlikely that cost-benefit techniques will become so sophisticated as to be able to dispense with broad judgements on aesthetic and environmental effects.

10
Some topical issues in urban transport decision making

The transport policy and management problems which confront British cities are currently being resolved by using a mixture of short- and longer-term solutions of the kind outlined in chapters 8 and 9. But it would be misleading to suggest that there is unanimity about the 'correct' solutions: in practice, given the dynamic characteristics of movement in cities, it is unlikely that these can ever be identified. The problems which are demanding attention vary considerably between urban areas; the policy for a commuter centre in Surrey is unlikely to be appropriate to the needs of a more self-contained market town in Cumbria. Short-term policies and traffic-management schemes of the kind outlined in the previous chapter seem to vary, for example, according to the size and function of urban areas (see *Traffic Engineering and Control*, vol. 14, 1972, pp. 87–9), while many of our towns and cities still have very indeterminate policies on a wide range of detailed local problems arising from parking demand, public transport services or traffic congestion. Equally, while progress is being made in respect of improved coordination of urban transport and the legislation governing it, there is still some way to go (see chapter 8), especially in metropolitan areas where the transport problems are usually most pressing. In view of this it is rather hard to present broad generalizations about the trends apparent in the policy and management responses of British cities to the demands for movement with which they are all confronted to some degree. This task is inherently difficult because of, firstly, the constant shifts in the complexion and character of movement in each city and, secondly, the emergence of more general issues during the last ten years which are relevant at all levels in the urban hierarchy. These are the questions increasingly asked about the relationship between transport and the urban environment; the related consequences of utilizing new and improved transport technology on urban form, development and the environment; and the social consequences of contemporary attitudes towards transport provision in cities. It therefore seems pertinent to identify some of these issues in this chapter in order to demonstrate their importance for contemporary and future urban transport decision-making.

Cognizance of the effects of movement in cities on most aspects of urban life only dates from the early 1960s. Provided that road surfaces were adequate or there was enough space for road traffic or for passengers on public transport, all was considered to be well (see chapter 8). Any problems of congestion, especially those created by traffic, were solved by providing yet more road space in the form of new highways or by undertaking extensive road-widening schemes. Much interest also focused on the need to rationalize the administrative structure of competing transport undertakings in order to provide better transport planning, improved levels of service or an integrated urban-highway programme. The costs of achieving these goals, in terms of environmental decay, reductions in housing stock, changed land-use patterns, or human suffering, were never properly built in, however, to the investment calculations for a new road or railway in an urban area. Pressure to provide new rail routes in urban areas or to modernize existing routes was far less than that for the provision of new roads. This dichotomy was an inevitable consequence of the rapid increase in the number of private-car registrations since the last war. The massive investment which this generated for the creation of more road space in urban areas eventually stimulated a number of informed observers to stand back a little from the 'benefits' of the 'new mobility' to consider the wider implications for urban areas and their residents.

The transport innovations of the nineteenth century released urban dwellers from the severe constraints on personal mobility which had characterized earlier centuries and seems to have encouraged the view, endorsed by the attitudes to mobility during the first half of this century, that all increases in mobility are desirable whatever the cost. But, paradoxically, reality has not supported this assumption, because in a densely settled country with a large proportion of urban residents at comparatively high densities the advantages of more mobility have, at best, generated marginal real improvements and greater levels of disutility as the century has progressed. If mobility is treated just like any other commodity such as housing or durable goods, then it must be purchased at the expense of some of these. The opportunity cost of mobility to both the state and the individual has steadily increased while, at the same time, it has come to be realized that the cost can be adjusted or controlled to meet other priorities or to cover the costs of purchasing other 'commodities' such as the improved environmental standards discussed later in this chapter. Improvements in mobility are also self-defeating because the more varied activities and life styles which they permit, erode and eventually destroy the benefits they have originally conferred. Such benefits as have arisen have also been selective in that more of them have gone to certain groups of urban residents, particularly car owners and the higher socio-economic groups. There has therefore been a growing interest in more egalitarian approaches to urban mobility which are based on equity. Some of these new emphases have been encouraged by the adoption of more rational and/or quantitative assessment of existing and future investment and infrastructure proposals for urban transport of the kind prompted by the longstanding private versus public transport debate (see, e.g., Nash, 1976).

There can be no doubt that the major responsibility for this change in attitude towards urban transport rests with Buchanan and the ideas which he introduced initially in *Mixed Blessing* (1958) and which are articulated at greater length in *Traffic in Towns* (Ministry of Transport, 1963b). Urban traffic is described in the latter (p. 7) as 'a problem which must surely be one of the most extraordinary facing modern society . . . because nothing less is involved than a threat to the whole familiar physical form of towns'. The Buchanan study stressed the need to approach the problems caused by traffic in towns in a much more comprehensive fashion than that merited by the fragmented public attention already given to parking problems, road accidents or traffic jams. Of particular importance to some of the arguments advanced was the fact that 'the deterioration of our urban surroundings under the growing weight of traffic has passed almost unnoticed' (p. 19). Environmental damage arises from the functional conflict between the use of roads to canalize traffic movements and their use as urban shopping centres, pedestrian arteries, points of access for business and other vehicles to adjacent land uses, or as part of industrial areas. As the frustration of using cars in cities has increased with the rising volume of traffic, their usefulness has decreased and the main environmental issues have become clearer. Conservation of world energy resources, particularly oil, has also prompted more interest in alternative or energy-saving forms of transport which also confer environmental benefits.

Main environmental issues in urban areas

There are many environmental problems arising from urban traffic, but three central issues can be identified for the purpose of this discussion. Safety, with reference to the pedestrian/traffic interface and the circulation of traffic through an urban road network, is the first of the major problems. Accidents involving personal injury on the public highway (including footpaths) in which vehicles have been involved have shown a small decrease since the early 1960s, even though traffic has been increasing at some 5 per cent per annum since 1949 (Department of the Environment, 1974; Baerwald, 1973). There was a decrease of 7 per cent on urban roads (30–40 miles per hour speed limit) between 1973 and 1974, at a time when the accident rate per 100 million vehicle miles only decreased by 3 per cent. But there is no room for complacency; motor accidents comprise some one-third of all accidents in advanced societies and a large proportion of them occur in urban areas. The cost to the community in terms of police and hospital expenses or lost production are enormous. One official estimate (Dowson, 1971) puts the annual cost at £850 million in Britain alone. Many road accidents are a product of the conflict between an outdated urban fabric and the patterns of demand shown by road traffic, some 80 per cent of which utilizes just 20 per cent of the urban road network. As traffic densities increase, the opportunities for accidents increase and, although measures for reducing such accidents are available, they can only be introduced using massive investments of the kind which Britain, and many other countries, cannot afford. Street lighting, the use of high skid-coefficient surfaces, pedestrian-phased traffic signals, or median barriers (Mackay, 1973), provide part of the answer but can only be located usually where it is felt that the benefits to road users and pedestrians will justify the investment involved. Total vehicle/pedestrian segregation, limited-access roads and grade separation would all help to reduce certain types of urban accidents, but at best, they can only be provided at a very limited number of locations in cities.

A second environmental problem arising from urban travel is vehicle noise (*Report of the Committee on the Problem of Noise*, Cmnd. 2056, 1963). There are five main kinds of noise produced by motor vehicles: brake-squeal, door-slamming, loose loads or bodies, horns and, by far the most important, propulsion noises. It is necessary to control the amount of noise emitted by vehicles, especially from heavy goods and diesel lorries, but unfortunately these are the most difficult groups with which to achieve noise-level reductions. The incidence of this noise depends, to some degree on the general condition of vehicles and standards of maintenance. Some control on the quality of the nation's vehicle stock can be exercised but would have minimal effect on other sources of noise in urban areas. These are related to the density of traffic and its flow characteristics. Noise from these sources can be minimized by smoothing the flow of traffic as far as possible, thus reducing noise generated by vehicles accelerating from traffic lights or roundabouts. Reducing the density of traffic at any given point means that a higher proportion of all urban roads are likely to suffer traffic noise to some degree but at lower levels than the positively harmful noise levels in homes and workplaces adjacent to major traffic arteries.

Atmospheric pollution caused by vehicle exhausts is the third major environmental issue (Sherwood and Bowers, 1970). Ventilation holes in fuel tanks and carburettors and 'breathers' in crank cases also help to pollute the atmosphere. It has been estimated that exhaust and evaporation from motor vehicles contribute 50 per cent of the hydrocarbons, 33 per cent of the oxides of nitrogen, 66 per cent of the carbon monoxide and 90 per cent of the lead-bearing particles in the atmosphere. This means that at least 60 per cent of all air pollutants in cities are attributable to motor vehicles.

Chemical reaction of these constituents under certain sunlight conditions produces photochemical smog, of the type well documented for Los Angeles and Tokyo, which irritates the eyes and throat. High buildings help to confine air pollutants, as do cuttings and road tunnels, while it is also suggested that the incidence of lung disease is increased. In common with noise, an initial difficulty is that no agreement has been reached on how to measure and interpret air pollution data in a way which properly separates traffic and non-traffic sources.

There are a large number of other environmental difficulties consequent upon the use of urban transport. Vehicles intrude visually upon the urban landscape, particularly in the centre of cities where buildings appear to rise from a plinth of cars crammed into every available square yard of space available for parking or movement. Similar, but perhaps less serious, intrusion occurs in older suburban residential areas where space was not originally provided for garaging facilities. Such vehicles also create street hazards for children, act as litter traps, and generally make street cleansing more difficult. The use of streets has therefore been effectively preempted by motor vehicles. Streets as a place for children's play, socializing, argument, relaxation and exchange no longer exist. Parks and squares have become less secluded because of their exposure to both traffic noise and fumes. Vistas of architectural or historical significance are interrupted by moving or parked vehicles as well as by the proliferation of street furniture such as bollards, light signals, railings and direction signs which marshall urban traffic and guide it through the street network. Finally, construction is an inevitable by-product of the demand for urban transport; new and very large multi-storey car parks and new highways, which may be elevated, are typical examples. Unfortunately, such development is often out of scale with the surrounding areas and often consists of dreary, formless structures which devalue the quality of urban architecture.

It is hardly surprising to read in *Traffic in Towns* (p. 23) that 'no person confronted with the issues, from accidents to visual intrusion and burdened with the responsibility of deciding upon them, could do other than conclude that they are indeed very serious matters'. Traffic has been systematically destroying the fabric, economic and social, of our towns and 'the over-riding context in which the problems of urban traffic have to be considered is the need to create or re-create towns which in the broadest sense of the term are worth living in, and this means much more than the freedom to use motor vehicles' (p. 32). This is a clear statement of intent which has undoubtedly helped to mould contemporary attitudes to transport management in urban areas. The utilization of traffic-management schemes at various scales, of the kind discussed in the previous chapter, has only become a major consideration since the early 1960s. The degree of public involvement through participation in transport decisions which have far-reaching implications for the urban community at large has also only become significant during the last decade, particularly since the provisions made in the Town and Country Planning Act, 1968. It is no longer possible for planners to relegate the environmental consequences of urban movement to a minor position in their list of priorities. There has been a tendency to do this because the environment and its quality is to some extent an intangible item which cannot be objectively assessed and included in a cost-benefit analysis of anything from a transport plan for a large city to the construction of a pedestrian road-bridge. Transport planners are now under far greater pressure to produce more than purely intuitive judgements about the environmental effects of their proposals and, while the ideal of objective evaluation procedures remains elusive, much more attention has been given to the relatively more tangible environmental by-products of urban traffic. More information is required about the way in which individuals affected by urban roads, railways or airports perceive their environment and how the community, as well as individuals, values the numerous elements which combine to make the urban environment. Some positive steps have been taken to ameliorate the environmental effects of road traffic, through the introduction of lorry routes, pedestrian precincts, or environmental areas, for example, but it is not clear whether these are the solutions which are most appropriate, how the community or individuals react to the changes which are generated, or how individuals or groups outside the areas so affected react to the ripple effects of extra traffic or pedestrians diverted through their local area. These are but a few of the possibilities.

The list of problems and possible solutions, with their different advantages and disadvantages, may be very long but it is crucial to identify those elements which are most vulnerable to positive modification and which will reap the greatest benefits for urban dwellers. Hammond (1973) has identified noise, atmospheric pollution and severance of pedestrians from usual lines of movement as the most significant impacts of motor traffic and, although the survey related to the attitudes of residents living adjacent to new roads, it is reasonable to infer that these conclusions apply to those living near long-established traffic routes. Unfortunately, this does not mean that these are the effects which can be most easily rectified. The problem of severance is less likely to be a long-term difficulty, because the passage of time does allow individuals to adjust their movement patterns as required, but noise and pollution, whilst open to reduction, are likely to be far more persistent. Equally long-term and less vulnerable to positive modification are the side-effects of new road structures such as elevated motorways, which cause visual intrusion, loss of sunlight and privacy (DoE, 1972, pp. 17–18).

Space will not permit a fuller discussion of the characteristics of, and the responses to, all the environmental effects of urban transport, but it may be useful to at least look more closely at problems which have been scrutinized in some detail during recent years; noise, the pedestrian/vehicle conflict, and urban motorways.

Traffic noise

According to the (Wilson) *Report of the Committee on the Problem of Noise* (Cmnd. 2056, 1963) 'road traffic ... is the predominant source of annoyance and no other single source is of comparable importance (see also

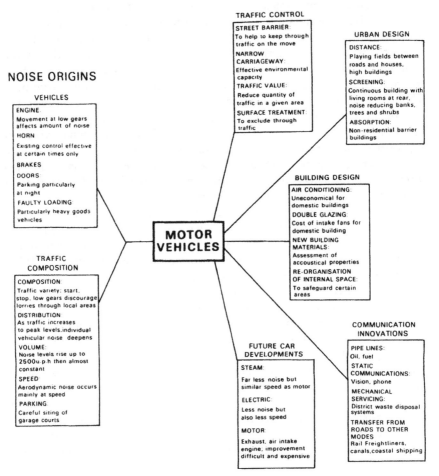

Figure 10.1 *Some relationships between the car, the origins of vehicle noise and its control* (Source: *Antoniou, 1969, fig. 1, p. 162*)

Transportation, Vol. 1, 1972–3, pp. 291–305), and the GLC have confirmed that at 80 per cent of the sites measured in a survey, traffic produced a higher level of noise than from any other source. The origins of noise generated by cars and other vehicles are summarized in figure 10.1. The intensity of noise from vehicle engines, traffic composition or traffic distribution patterns is not uniformly spread throughout an urban area and, in general, the noise level will decrease with distance and/or height from its origin. In order to improve methods of controlling noise (figure 10.1) it is necessary to measure it as accurately as possible, and the unit most commonly used, although by no means a conclusive method, for measuring traffic noise is known as L_{10}, which is a scale related to our perception of noise and defined as the level of noise in dB(A) for just 10 per cent of the time (Noise Advisory Council, 1973). The scale is such that an increase of 10 dB(A) represents a doubling of the perceived noise level. For the measurement and prediction of traffic noise the average of L_{10} values for each hour between 06.00 hours and 24.00 hours is the recommended measure. An L_{10} of 70 dB(A) corresponds to the noise 60 feet from the edge of a busy road through a residential area, with an average traffic speed of 30 miles per hour. In 1970 it was estimated that 9.7 million people in Britain were subject to road traffic noise of L_{10} above 70 dB(A), and a further 14.2 million experienced an L_{10} between 65 and 70 dB(A). There is every possibility that the number of people affected by these high levels of noise, particularly in urban areas, will increase. The urban population exposed to traffic noise greater than 70 dB(A) will, according to one report, be 65 per cent greater in 1980 than in 1970, with the number experiencing levels over 65 dB(A) likely to increase by 40per cent (Transport and Road Research Laboratory, 1970; Botton and Groome, 1969).

Various techniques of noise measurement using instruments are now well established, but it is much more difficult to measure the reaction of urban residents to noise. It must be recognized that there can be a wide range of human response to noise; what is highly intrusive to one individual, or in one area, may be acceptable to another individual in the same area, or at a different location (Bryan, 1973; Sexton, 1969; Starkie and Johnson, 1975, pp. 1–11; Walters, 1975, pp. 17–26). Residents in an area adjacent to a noisy factory might hardly notice any additional vehicle or other noise created by the introduction of a new road, but a quiet semi-rural residential suburb could well find such 'new' traffic noise totally unacceptable. Some people are mentally and emotionally better-equipped to tolerate noise than others, while people's attitude to noise will also vary according to its source, the time of day, or the activities in which they are engaged. The rumble of heavy vehicles past an office window may not offend an office worker during the day but the same sound during the night may cause extreme annoyance and discomfort for the same person. The range of variability in human response to traffic noise has made it difficult to specify maximum levels in the interests of environmental standards or the creation of normally acceptable levels of noise. The Noise Advisory Council has suggested that existing residential development should not be exposed to L_{10} levels exceeding 70 dB(A) unless some compensation is paid, or remedial action is taken, by the relevant local government department or central government agency. This L_{10} level is considered a maximum, and transport and highway engineers should be attempting to achieve lower values as a main priority. An external noise level of 70 dB(A) may not be acceptable as an internal condition and the Report (1972) recommends a maximum of 50 dB(A) for dwellings in busy urban areas and 45 dB(A) elsewhere (equivalent to 60 dB(A) externally), with 35 dB(A) as the nighttime standard for bedrooms in urban areas.

The problems of traffic noise and its control involve complex procedures and technical applications which discourage simple solutions (see figure 10.1). There are five main alternatives open to the transport planner, ranging from traffic restraint to improved building designs which tacitly recognize the need to minimize traffic noise. Communications innovations which, in the long term, will replace the need to travel by road or rail within and between urban areas (to complete business transactions or to get to work) could also be important. Various types of traffic control, such as phased traffic lights which help to keep traffic running smoothly, are already extensively used and are especially useful for limiting the effects of noise on existing urban development. It is more difficult to introduce screening or buffer zones between traffic and buildings in the older parts of cities without visual and other consequences for existing buildings and their occupants. But, clearly, improved urban design standards can be used most effectively in new or expanded towns as well as in extensions to established urban areas. There are limits, however, to the extent to which urban design can be utilized, because land for urban development is a particularly scarce resource in Britain, and attitudes towards residential development (the high-density versus low-density debate) will need to alter drastically if design criteria such as higher levels of open-space provision between traffic and buildings can continue to be used as buffer zones. Apart from reducing noise, such schemes also have other environmental benefits, such as pedestrian routes or leisure areas, and these must be taken into account in the cost-benefit balance sheet. Changes in building design can again be most effectively employed in new developments. Noise-reducing techniques such as double glazing or the use of sound-absorbing building materials remain only within reach of certain groups of urban residents, often the groups least affected by the incessant trundle of heavy goods vehicles and other traffic. In special circumstances, such as housing in the vicinity of London Heathrow airport, grants can be obtained for installing sound insulation, but these have not been generally extended to include the large number of properties seriously affected by road traffic noise. In the case of new urban roads the Wilson Committee stressed that householders affected by noise above the predicted level should receive aid from local authorities towards the cost of sound insulation. It would seem equally important to consider the case of other urban residents not directly affected by new roads but certainly not immune from traffic noise at unacceptable levels. Improvements in the design of vehicle propulsion units is an attractive method of noise control because it tackles the problem at source, but it is unlikely to completely resolve all the difficulties. Engine re-design may be helpful but the law of diminishing returns operates and produces a prohibitively expensive vehicle. Some of the alternatives to the combustion engine will be discussed in a later section of this chapter and some, such as electric-powered vehicles, offer real and lasting reductions in noise levels. It has been estimated that a decrease in car-induced noise to an average of 75 dB(A) should not be impossible by 1980, but goods vehicles and buses are a more intractable problem as many still do not meet the present 89 dB(A) standard (Noise Advisory Council, 1972).

Urban motorways

As levels of traffic congestion have increased in cities and less road space is available for each vehicle, one of the central problems has been how to decongest the traffic arteries. One of the main responses has been to provide yet more road space to filter off the existing excess demand, on the assumption that additional road space is the appropriate solution. This only provides a short-term answer because extra road space will eventually generate additional traffic and the system becomes congested again. When this happens it signals the opportunity to provide yet more new roads and to repeat the cycle of decongestion/congestion. But congestion is only serious for perhaps three to four hours each day and does not affect all parts of the road network. For much of the time, perhaps 70 per cent of each day, personal and goods movement in urban areas is relatively efficient, especially along high-standard roads. This suggests that investment of this kind is advantageous because of the improved journey times and lower costs created. Such benefits could be extended over a longer period each day by preventing the generation of private-car commuting, especially along routes to the city centre, which is the main cause of an eventual return to the *status quo* after new highway investment. Parking and pricing policies (see chapter 9) could be used effectively in this respect or in cases where the construction of a high-standard urban road causes a transfer of trips from public to private modes. More information is needed on the precise relationship between road-space provision and traffic generation, and the effects of traffic management policies, so that there may be more careful planning of new roads in the future.

There are a number of other reasons advanced in support of urban motorway construction. Probably the most significant of these is that motorways offer a cheaper solution to the provision of new road space than widening existing urban roads. Acquisition and demolition of buildings fronting commercial areas along existing major arterials is a very expensive affair, whereas the urban motorway allows several alternative routes to be considered which may involve minimal demolition and maximum use of derelict space or lower-value land, while overhead construction techniques can also be used and are cheaper than motorway construction at grade. Traffic separation on urban motorways is also much more effective and therefore reduces accident costs per vehicle mile, despite a capacity to handle far higher traffic volumes than conventional city streets. Such arguments in favour of urban motorways are persuasive and, until recently, have been sufficient to allow a number of schemes to proceed, e.g. the M1 extension into central Leeds or Westway in inner London. But this has been at the expense of understanding the overall and long-term effects of urban motorways, not only on city traffic but also on the urban environment. Demolition of 50 habitable dwellings and the displacement of families may be an unacceptable price to pay for a new section of road, and it is this kind of issue, as much as cost or changing government attitudes, which has led to the eventual abandonment of the proposals for the inner ring motorway around central London.

The impact of urban motorways on traffic and the environment, either directly or indirectly, can be divided into short-, medium- and long-term effects (Thomson, 1969, pp. 33–5). Short-term consequences are mainly related to distributional effects on traffic, such as the relief of congestion in streets immediately adjacent to new urban motorways. Diversion of trips from the old streets to the new will be supplemented by additional trips generated by the wider range of destinations made accessible by the motorway. Urban motorways will therefore encourage longer journeys, and this creates more traffic for the same number of trips. Some travellers may transfer from public to private transport because of the trip-time/distance advantages of using urban motorways, and the potential for environmental conflict is therefore increased. Medium-term effects are mainly confined to changes in the relative advantages of road versus rail travel in urban areas, so that car ownership is likely to increase. This will eventually add yet again to the congestion which is used to justify the demand for more roads. Redistribution of urban land use in a way which excacerbates the problems confronting public transport in cities is probably the major long-term product of extensive urban motorway programmes. The out-movement of residential uses is well-documented and is positively encouraged by new road construction. The indirect environmental costs through additional traffic and longer journeys may be spread from the short- to the long-term and are the considerations which have increasingly come to the attention of groups opposed to urban motorways.

Clearly the traffic effects of urban motorways will extend well beyond their immediate hinterlands, but there are also environmental conflicts arising from their construction and subsequent operational characteristics (Bor and Roberts, 1972; Cooper, 1975; Llewelyn-Davies *et al.*, 1971). Organized resistance to urban motorways has mushroomed during the last decade because, firstly, they have rarely been convincingly presented as part of a comprehensive transport package and, secondly, they have often been large-scale structures which are completely out of sympathy with adjacent buildings and have exposed many of them to unacceptable levels of traffic noise, air pollution, and glare from night traffic. Compensation for these effects has been consistently inadequate and has undoubtedly helped to fuel opposition during recent years.

Problems begin to emerge during he earliest stages of urban motorway planning. Areas likely to be affected by the proposals, which may include reference to alternative routes, will experience 'anticipatory' blight as property-owners, speculators and others become unceitain about appropriate investment decisions, such as whether to improve dwellings or to provide new office buildings. This will happen even though only one route will eventually be chosen and the chance of route x being selected may be as low as one in seven alternatives. Environmental decay at this stage must therefore be minimized by ensuring that public participation takes place as quickly as possible after proposals are published. Once a motorway route has been selected a new form of environmental decay is instituted. There will usually be uncertainties about when actual construction will begin because of financial constraints, for example. It may take 1–20 years to actually get started or, as in Greater London, proposed routes may be abandoned completely. For residents or companies directly affected by the motorway – i.e. their properties will be demolished – there is at least the prospect of selling at current market value as laid down in statutory regulations. But this does not apply to buildings outside the area of land directly along the line of the new motorway. Decay initiated by the original proposals for an urban motorway may therefore become endemic, with little hope of financial redress or improvement after the motorway has been completed. Noise, dirt, route diversions and dangers to residents, particularly children, follow the start of construction, which can last for more than five years. The loss of certain kinds of land can also cause problems. Particularly important in the urban fabric are parks, open spaces, gravel pits used for water sports or private open spaces, all of which may enhanced the quality of life and of the environment. The new road may well take all or part of an open space (construction costs will be lower) and even those parts not used will suffer the disutilities of, or environmental pollution from, the adjacent new road.

When it has been completed, the urban motorway continues to impose itself on the community and the environment in other ways, especially via the traffic which is using it. Operational side-effects of this kind include noise, which has already been discussed, and visual effects, which comprise the actual appearance of the motorway: its intrusiveness (i.e. is it at grade or elevated?), its contribution to changing vistas if at grade or elevated, the loss of sunlight or daylight as a result of any elevated structures, glare from headlights or motorway lighting, and the loss of privacy because the occupants of passing vehicles are sometimes able to overlook day-to-day household activities. The elevated sections of motorways are most likely to give rise to significant visual effects, but actual measurement of the way in which these effects vary according to the construction techniques used or the characteristics of individual urban residents, remains a largely unexplored area. Existence of significant operational effects cannot be doubted after the extensive demonstrations by local residents which followed the opening of Westway, the extension of the M4 into the heart of inner London. Some residential properties are no more than 20–30 feet from the hard shoulder of the elevated section of this motorway, and in addition to the noise and visual consequences, air pollution and vibration must also be endured. Exhaust fumes contain lead which can, in sufficient quantities, be harmful to the health of young children. There is insufficient evidence as yet to indicate whether children who live and play near new major or existing roads are more prone to health problems linked with lead but research is in progress. Vibration from traffic may be air-borne or ground-borne, with the former the most important, and it is caused by infra-sound (low frequency sound derived from vehicle exhaust pulsations, particularly from heavy vehicles in low gear) which is inaudible but causes nausea and headaches.

Interruption of 'usual' movements (journeys) across a new motorway, or severance, must also inevitably take place. Pedestrians are the group most affected in this way and it has been shown that changed or inadequate provision for pedestrian movement may in fact be one of the most unpopular side-effects of a new road (Urban Motorways Committee, 1972, p. 13). By conducting surveys it is not too difficult to establish the patterns of pedestrian and vehicular movement across the proposed line of a new urban motorway before it has been constructed. Such data help in the provision of adequate crossing and other facilities. But it is less easy to predict whether the demand for these trips will be the same once the motorway has been completed. Trip patterns will begin to change as construction proceeds. Pedestrians may start to use alternative retail outlets which were less popular before the motorway acted as a barrier and, given time to establish themselves during the construction phase, these changes may make comprehensive provision of crossing places inappropriate or poorly sited to deal with the demand which actually occurs. Alternatively, it may be considered essential to encourage people and traffic to cross major urban roads in order that they may benefit from new shopping facilities, may obtain access to schools for their children or use the local medical centre or any facility which cannot be easily duplicated on both sides of the 'barrier'. The major concern for pedestrians crossing these 'barriers' then becomes safety and shelter from the weather. Pedestrian underpasses or covered bridges provide most protection from traffic but not from vandalism and assault, while the gradients involved can discourage mothers with young children and prams. The dangers of light-controlled pedestrian crossings are equally daunting, and these facilities also impede the flow of traffic, which can in turn pass on the kind of environmental disbenefits which it is hoped to be able to minimize.

The proportion of an area's population and activities actually affected by any or all of these operational effects of an urban motorway will depend to some degree on distances from the new road (figure 10.2). Visual intrusion and dirt consistently affect the largest proportion of people over the greatest

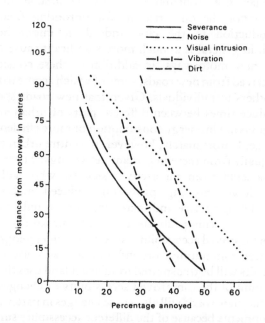

Figure 10.2 *Distance from an elevated urban motorway and the impact of noise* (Source: Bor and Roberts, 1972, fig. 15, p. 314)

distance; even at 100 metres, over 30 per cent will be annoyed by dirt. Vibration also decreases with distance from source less steeply than either noise or severance, and this underlines the importance of monitoring the effects of new motorways over a zone at least 150 metres wide on either side of them and possibly even further away depending on whether routes are at grade or elevated. Noise and severance decrease most rapidly with distance in terms of the percentage of the population apparently annoyed by them, but this should not be taken to mean that noise, in particular, does not have significant effects on health even at distances exceeding 100 metres from source. Time may also modify the response to urban motorways, particularly in respect of the level of visual intrusion. Hence, an 'aesthetically pleasing road would be one which the majority of the community considered attractive, or which (if nothing better could be managed) did not offend a significant number of people. It is quite possible that the majority opinion of a road's aesthetic quality could change over the road's lifetime' (Moser, 1972, p. 56; Joyce and Williams, 1972).

In some ways it is only too easy to overstate the substantial social and money costs of new road construction, especially in the inner areas of large cities. Noise pollution in inner cities is already high and urban motorways may not contribute to any noticeable difference. New roads are also not the only reasons for uprooting families or manufacturing production in the city. Indeed, it might be argued that urban renewal in general is a much more significant cause of housing loss in cities than new motorways. In addition to these considerations there are benefits derived from new roads, some of which accrue to the community as a whole, others to individuals. Given that new roads speed the flow of traffic and reduce times between different parts of urban areas, they must represent a net saving in energy consumption or more efficient use of time by the labour force. Unfortunately, however, the entire urban community will not benefit equally from these important net savings. There is likely to be a transfer of benefits from inner-city residents, who will suffer poorer environment, lower housing provision, reduced mobility/accessibility because of declining service and increasing cost of public transport, than higher-income, car-owning suburban and extra-urban residents. The latter will gain from improved accessibility, access to a wider geographical area for housing, reduced commuting costs and, in some cases, lower housing costs. Some net benefits will be transmitted to adjacent land uses through increased values on account of their improved accessibility or changes in retail prices and wage levels. New roads will also cause changes in patterns of investment in new developments because of the different accessiblity surfaces which are induced. Some of these benefits may be completely external to the road system of the city, such as the stimulation of new economic activity in one area and perhaps decline in another, or reduction in average costs in industries from which demand is diverted on to road transport. Thomson has also suggested that new urban roads may generate increases in taxation for the government through the extra expenditure on road travel which a new road encourages.

Evaluation of the environmental consequences of urban motorways against the construction costs and traffic benefits remains the most crucial part of urban motorway planning. Decision-makers need better guidance when selecting the best solution in any particular case. Three evaluation techniques were developed by the consultants to the Urban Motorways Committee (1972, pp. 33–7) in an effort to introduce quantitative techniques to the essentially value-ridden assessments which had already been made for some urban motorway schemes. One proposal was a method of assembling and analysing data about the costs of a range of types of motorway construction and the environmental factors pertinent in each case. In effect, once the cost data have been assembled, it is possible to calculate what it will cost to 'purchase' any particular set of environmental conditions. This is the most easily and cheaply implemented of the three evaluation techniques suggested and it can be applied equally well to small or very large projects. Cost-benefit techniques of evaluation were a second suggestion. A matrix is constructed which comprises the various costs and benefits against their distribution to the various groups which form the urban community: displaced households, road-users, local authorities or the government. A price is attached to each item to show what the community would accept as compensation for any negative effects of the motorway and what they would be prepared to pay to obtain any benefit from particular items. These valuations are mainly derived from individual responses or from people's revealed preferences. Both the cost-effectiveness and cost-benefit techniques were preferred to a third alternative, the Environmental Evaluation Index. In common with cost-benefit methods the Index is oriented towards measuring environmental factors on a common scale. A weighting is given to each effect of a motorway proposal according to a judgement of its importance to individuals. The value obtained is multiplied by the number of people affected, and results for all the effects are accrued to produce an index number. As in all these techniques, a considerable amount of subjective judgement of weightings is necessary because of the absence of suitable empirical data on most aspects of urban motorway impact. The advantage of cost-benefit analysis is that traffic benefits or environmental benefits can be taken into account in an identical way, although the allocation of values attached to environmental components by urban residents remains a major problem.

Movement of pedestrians

It has already been shown in Chapter 1 how, during the earliest stages of their development through to the comparatively recent past, the shape, size and morphological variation of cities was determined by maximum walking distances. Cities were planned around the needs of pedestrians with reference to both their social and business contacts, in addition to the basic need for comfort and convenience with protection from the weather provided by porticos, canopies and other enclosed routeways. This concern with the human qualities of planning for movement in cities was quickly abandoned once machine-propelled transport began to make its appearance. Trains, buses, cars all provide encapsulation from the elements, permit segregation from other travellers, and have made walking or the pedestrianized part of a journey primarily a distributional function rather than a means to an end. Most journeys in cities involve decisions about how the walking parts of a shopping or journey-to-work trip can be minimized so that 'inconvenience' is reduced to the lowest possible level. It is notable, for example, that although building densities have increased dramatically in the centre of cities as a result of the accessibility afforded by non-pedestrian transport, pavement widths have, until very recently, changed little to meet the increased volume of use which this implies (see, e.g., Schaeffer and Sclar, 1975, pp. 8–17; Fruin, 1971, pp. 1–11; Benepe, 1965).

Pedestrians have therefore become second-class citizens in urban movement patterns. Yet most journeys in cities require an element of walking and many, particularly by non-car-owning households, are still exclusively made on foot. In central business districts many interfunctional links are achieved via pedestrian-only journeys rather than vehicular modes. The facilities provided for pedestrian and vehicular movement in cities ensures that there will be constant competition for space between the two. They both use the same 'channels' between buildings or land parcels and are only separated by a three-inch-high kerb intended to keep vehicles 'in bounds'. In many ways it is true to say that the 'nature of urban geography dictates that pedestrians and vehicles are continually competing for space' (O'Flaherty and Parkinson, 1972, p. 34; see also Masing, 1972). This conflict is always most marked in city centres, where pedestrian densities are highest throughout most of the day in shopping streets and in office areas during the lunch hour and the morning and evening rush hours. In Coventry a survey showed that there was an average of 137 persons per 100 yards of street in the central area, decreasing to 43/100 yards further away from the shopping precinct and as low as 8/100 yards on the periphery of the centre (City of Coventry, 1972, pp. 12.50–12.51). For the larger suburban shopping centres the equivalent figures were in the range 15–20/100 yards declining to 7–15/100 yards in small suburban shopping areas (for a full discussion of the measurement of pedestrian density, see Fruin, 1971, pp. 37–69). The number of pedestrians in shopping and contiguous areas at various times of the day will of course vary, especially in the central area. But elsewhere in cities the numbers are more predictable and, according to the *Coventry Transportation Study*, pedestrian density can be related to seven factors: net retail and service floor space, population density in the area, employment within given distances, bus accessibility, school attendance within the shopping area and traffic volumes. In all cases the dominant variable appears to be net retail and service floor space, and small improvements in prediction can be achieved by adjusting this variable to give greater weight to more specialized shopping facilities, particularly in higher-order shopping areas. The number of buses stopping within shopping areas was the second most important variable, with the number of jobs available within a quarter of a mile making a further marginal contribution to the predictive value of the regression equation.

A solution to the pedestrian movement problem in cities must therefore focus attention on the areas of greatest pedestrian vulnerability, in the CBD and major suburban commercial centres, while still providing appropriate levels of accessibility by motorized transport. Pedestrian vulnerability can be measured using data on the number of pedestrians on pavements, the

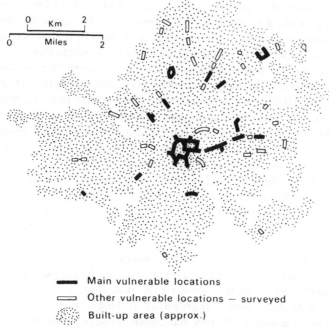

Figure 10.3 *Areas of pedestrian vulnerability, Coventry* (Source: *City of Coventry, 1972, fig. 12.12, p. 12.49*)

numbers crossing the road (classified by age), and the time taken by a sample of persons to cross roads (see, e.g., Ashworth, 1971; Cohen et al., 1955; also Copley and Maher, 1973). The most vulnerable locations can be mapped (figure 10.3) and used as a basis for planning to reduce the pedestrian/vehicular conflict. Greater separation is most likely to produce the quality of environment alluded to in *Traffic in Towns* where it is noted that 'the freedom with which a person can walk about and look around is a very useful guide to the civilized quality of an urban area. Judged against this standard many of our towns now seem to have a great deal to be desired' (Ministry of Transport, 1963b, p. 40). Separate route systems can easily be provided for people and vehicles in new urban areas such as the New Towns, e.g. the pedestrian-only routes in Runcorn, Skelmersdale, Stevenage or Cumbernauld, but it is more difficult to restructure existing towns and cities, where redevelopment is usually only gradual and piecemeal.

Pedestrian separation from vehicles can be achieved by controlling the number of hours for which the two may be in conflict or by horizontal and/or vertical segregation. Many city-centre commercial and retail activities depend on servicing by road hauliers and tradesmen, and to allow them to continue operations access may be permitted between 08.30 and 10.00 hours, for example, while for the remaining period the service roads are confined to pedestrians. The length of the restricted times can be adjusted according to the day of the week and the frequency with which individual streets, or parts of them, need to be used by commercial and other vehicles. Schemes of this kind are relatively easy to introduce, involving a minimum of street furnishings and related environmental cosmetics. Street configurations may, however, permit total exclusion of vehicles from certain areas (figure 10.4), which means that former streets must be repaved and street furniture and landscaping introduced to break up and make more interesting the large spaces between buildings which flank a pedestrianized zone. There are now many examples of horizontal segregation in Britain, such as those at Norwich, York, Glasgow, Leeds or Liverpool (see figure 10.4) (for some detailed examples, see Greater London Council, 1973; Transport and Road Research Laboratory, 1975). Few such schemes operate without supporting time-induced segregation and, in addition, adequate provision must be made to ensure that pedestrians have parking and public transport services close at hand. This sustains the economic viability of the retail and other activities in pedestrian zones and allays the fears of traders and others that business will be adversely affected by the inability of vehicles to transport customers or clients to the doorstep or at least within easy walking distance. Vertical segregation is best-introduced into schemes which involve a completely new start, such as the opportunities provided in the Barbican area of the City of London after wartime bombing. The renewal of this area has involved a mixture of residential and commercial

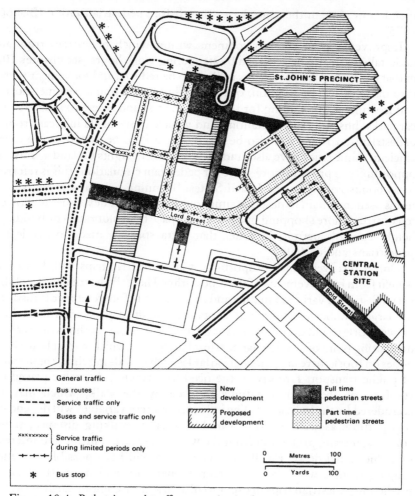

Figure 10.4 *Pedestrian and traffic segregation in the centre of Liverpool* (Source: Liverpool City Planning Department)

development linked by an extendable system of high-level walkways and spaces which will eventually be joined up to a comprehensive network of walkways planned for the one square mile of the City of London (Antoniou, 1968). Along Route 11, the commercial 'core' of the Barbican, six office towers are separated at ground level by a dual carriageway which aids smooth flow of traffic and can handle large traffic volumes, while the towers are linked by walkways some 20 feet above the road. Shops, restaurants and cafés are located on decks along the walkways. The reaction of users to segregation of this kind is usually favourable, and retail business may actually be improved because people come more often and are prepared to spend more time in the central area than before pedestrianization (Bishop, 1975).

Improvements in the environment within which pedestrian trips are undertaken can also be achieved by staggering work hours (see chapter 10), in that congestion on pavements is likely to be reduced and movement therefore made easier. In addition, queues for buses and trains become shorter and less time is lost waiting for transport. Egress and access to multi-storey car parks, bus and railway stations, and vehicle movements in areas around pedestrianized zones will also be eased and, it is to be hoped, made more efficient. Some cities have also traded development rights on land against an opportunity to provide more space for pedestrain circulation; this is referred to as bonus zoning in the Unied States. 'Planning gain' is the equivalent terms used in Britain to describe situations in which the developer is allowed to construct more shopping or office floor space than plot ratios or building-height controls permit, in order that extra space is made available for pedestrians and/or traffic.

Only a relatively small proportion of all pedestrian trips will, however, benefit from improvements of the type outlined here. Pedestrian trips exhibit very substantial variation in time and space, even before reference is made to the large number of trip variables which can affect pedestrian behaviour between origin and destination. Apart from the various methods of travel to the starting point of a pedestrian trip (excluding trips which start at household nodes on foot), there are personal variables such as walking time, the amount of public transport fare saved, the need for exercise, the value placed on comfort, or the level of fatigue induced by the walk, which may be considered by the individual. The influence of these variables will depend on trip purpose as well as 'path' variables such as walking distance, traffic conflicts, terrain, protection from the weather, security, image and interest, or traffic-signal delay. Finally, the pedestrian will have possible trip ends ranging from a workplace to a recreational area, and hence the task of predicting pedestrian trips in order to plan in a more comprehensive way for as many urban area residents as possible is likely to be a hazardous task. It is perhaps not surprising that most of the attempts at improving the facilities for pedestrian movement where conflict with traffic is a serious problem have focused on those areas in cities where pedestrian demand is greatest. This does not mean that the problem has been solved once our central and suburban shoping areas have been pedestrianized. Much remains to be done in residential areas or industrial estates, where pedestrian trips are often rendered unsafe and unpleasant because of parked cars, lorries and other vehicles as well as difficulties encountered when crossing busy roads or 'quiet' residential streets used as time-saving short cuts during the rush hour. Certainly, some pedestrian linkages are more easily planned for than others, in particular journey-to-work trips; it is much more difficult to cater for the lunchtime trips of office workers which may, for example, contain elements of personal business, exercise, and shopping purposes. Land uses also take on a dual role as pedestrian attractors in the morning and trip generators/attractors at the lunch hour, e.g. office buildings; retail stores act as attractors and generators throughout the day.

Environmental areas

There is little point in treating each of the symptoms of the traffic/environment conflict in cities as if they are mutually exclusive. As with so many other aspects of transport management and planning it is necessary to adopt a comprehensive approach. In this particular context the basic problem is that the attempts to sustain adequate levels of accessibility and to maintain suitable environmental standards are conflicting objectives, however desirable both of them may be to the creation of habitable, but functionally efficient, towns and cities. The conclusion reached in *Traffic in Towns* is that these two opposing objectives can in fact be adequately reconciled by producing urban designs which 'contrive the efficient distribution, or accessibility of large numbers of vehicles to large numbers of

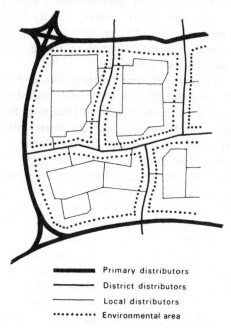

Figure 10.5 *Environmental areas and their relationship to primary distributor roads* (Source: *Ministry of Transport, 1936b, fig. 59, p. 44)*

buildings, and to do it in such a way that a satisfactory standard of environment is achieved' (Ministry of Transport, 1963b, p. 40).

Creating an appropriate balance between the two depends on the idea of 'environmental areas' (figure 10.5). These are 'urban rooms' where people can live, work, shop, look about, and walk about on foot in reasonable freedom from the hazards of motor traffic. Crucial to the operation of this concept is the need to channel traffic along 'urban corridors' which filter non-essential traffic out of environmental areas while at the same time providing access for traffic needing to enter any particular area. Hence, environmental areas are not completely free of traffic, because this would not be fully possible if the activities located within each area continued to be able to function adequately. But the design of environmental areas is such that the traffic using them is related in character and volume to a required set of environmental objectives. Increased noise and pollution are an inevitable consequence for boundary routes but this must be set against the improved environmental standards within each area. The product of this concept is a cellular structure consisting of a set of environmental areas set within an interlacing framework of distributing highways. It is important to appreciate that the organization of environmental areas is not confined to suburban residential zones where overall activity levels are low, but it includes the very active regions of a city, such as the central area, where there may be a large volume of extraneous traffic filtering through along various routes to get to destinations elsewhere.

The concept of hierarchy in traffic distribution and cellular structure in urban form is not new. Such systems are already well-developed in Britain's New Towns where the neighbourhoods and Radburn Street layouts, which form such a distinctive feature of various master-plan proposals either consciously or by accident, represent all that Buchanan envisages in his environmental areas. The problem is how to apply the concept to the existing urban fabric of older cities, which are so diverse in structure, form and connectivity according to the history of their growth. Each environmental area must be approached on its own merits, with the traffic problem considered with reference to three main variables: the level of accessibility, the standard of environment, and the cost that can be incurred as a result of physical alterations to improve one, or both, of the two preceding variables. Determination of environmental areas is therefore likely to be controlled largely by fiscal considerations rather than concern with minimizing social or economic damage to the areas affected. A rough and ready 'law' (see also chapter 9) is suggested in *Traffic in Towns* whereby 'within any urban area as it stands the establishment of environmental standards automatically determines the accessibility but the latter can be increased according to the amount of money that can be spent on physical alteration' (Ministry of Transport, 1963b, p. 45). It is clearly going to cost a great deal of money, estimated in 1964 as £18,000 million by Beesley and Kain (1964), to make the necessary alterations in the interests of traffic without taking account of the social and other costs to residents and activities contained within the environmental areas. Although pedestrianized zones in central areas probably reflect the influence of the environmental area concept to some degree, they are essentially located in unifunctional retail/ commercial areas with low levels of residential land use. Most people need to shop at some time or other, and pedestrian precincts therefore do not appear to have had adverse effect on retail trade, but unfortunately this does not mean that such ideas can be easily applied to other lower-density and functionally intermixed zones outside city centres.

Opportunities offered by new transport technology

New urban motorways, pedestrian precincts or controls on traffic noise will not provide complete answers to the amelioration of movement problems in cities. It is also necessary to tackle the problem at source by introducing new methods of transport or by improving existing transport modes. The evolution of both public and private transport modes has been a continuous process, with changes continuing to take place in response to new demands for movement and those anticipated in the future. Decentralization of urban residential and industrial land uses, lower densities of urban development, or new attitudes towards public transport are the kind of stimuli which have been operating. Indeed, it should be stressed that the evolution of new technology to meet urban transport problems is not just a characteristic of the period since the war but has been taking place for over a century (Richards, 1969). There are now a large number of new transport modes and systems in experimental use throughout the world, and some may eventually become viable in operational terms and help to change the shape and structure of contemporary cities. Most of these do not necessarily involve new technology of the kind heralded by the coming of the iron rail, for example, but involve modification or reorientation of the existing alternatives in a way which leads to more efficient use of resources such as energy, better use of urban space, less congestion, better pollution control, higher speeds for certain kinds of transport or reductions in the visual impact of transport. In other words, it is recognized that transport is indispensable to the various functions which cities perform, but more conscious efforts must be made to limit its detrimental effects on so many other aspects of urban life – a factor which has only been recently recognized. This chapter is therefore concluded with a selective review of some new transport technology which will have implications for movement in cities: auto-taxis, guided buses, moving pavements, monorails and automated highways have all been examined and it seems important not 'to exclude altogether the possibility of a major new form of transport emerging within ten to twenty years' (Thomson, 1969, p. 84).

Demand for new transport technology is also a response to the search by individual urban travellers for better and cheaper travel facilities. The problem facing urban society is how to promote technological development which meets this demand for progress but also enhances the environmental conditions now considered to be so important. The evolution of transport technology provides society with a threefold problem: firstly, which is the best mode to choose for individual trips in urban areas (private transport, personal public transport or mass public transport); secondly, how can suitable interchanges be provided between modes, given that many urban trips involve transfer from one mode to another; and thirdly, how can society develop both better modes and the best possible means of interchange in a way which promotes both technological and economic efficiency (Thomson, 1974, p. 252).

It is notable that many of the products of recent transport research are mainly personal or mass public transport systems. Contemporary attitudes towards public transport in cities have probably exerted some influence on this trend, which has been reinforced by the fact that the car cannot satisfy the travel demands of all urban residents at a time when dispersed patterns of employment and residences is the norm in many cities. This has made it imperative to find alternatives which bridge the gap between the level of service provided by the car and public transport. Such systems must be able to cater for diverse movement patterns, and a reduction in congestion at points of maximum movement convergence, and offer the kind of service flexibility and comfort which only car owners can at present enjoy. Hence, many of the new transport systems being tested, or suggested, are an attempt to provide personal transport via a public system.

Personal public transport involves small, family-sized, vehicles operating in an uncoupled mode which allows each unit to provide personal travel in privacy and security with good opportunities for the kind of comfort expected of the private car (Ross, 1971). Mass public transport, on the other hand, involves the use of large vehicles which convey 65–80 passengers coupled into trains carrying 200–1,000 passengers. This entails loss of privacy, a lower level of personal security and a relatively limited control over environment, e.g. non-smoking facilities may be provided but many non-smokers will inevitably find themselves in 'smoking' compartments. This type of public transport has achieved little success in attracting car drivers, while private rapid transit could well be competitive and even superior to the private car. Such superiority could be achieved because personal rapid transit can be operated non-stop without transfers en route, whereas mass public transport invariably involves frequent stops and interchange with other modes if trips are to be completed. Because mass rapid transit operates to a schedule, constraints are placed on when trips can be made, waiting time must be allowed for when planning a journey, and there are lower frequencies at off-peak times. Demand actuation is the additional key asset of personal rapid transit; individual passengers can request a service as and when required and it will not be necessary to reduce the service at off-peak times, because the system is far less dependent on operators such as drivers to run individual vehicles. From the point of view of the environmental lobby, personal public transport also has much to offer. The structures needed to guide vehicles can be much smaller because of reduced vehicle sizes, so that impact on urban vistas, etc., is reduced. Equally, the need for extensive demolition of the urban fabric is reduced when compared with attempts to fit new mass transit routes into existing built-up areas. Mass public transport usually involves large-scale structures which do not mesh well with the scale of most urban environments. On cost grounds, personal rapid transit is also advantageous because small lengths can be added economically as and when required, while large-scale projects are the only viable way of providing mass public transport if the installations are to function well as a system.

Duorail and Monorail systems

It is possible to divide new public transport systems into reserved-track and public-highway groups. The former has the advantage of speed and reliability which cannot be equalled by the latter because it is in competition with private transport and other modes. On the other hand, reserved-track systems lack the flexibility of response to changing demand patterns possessed by highway-oriented systems, while in the case of rail-based solutions, the costs of new facilities could well be prohibitive in a social as well as economic sense. It is, however, possible to utilize existing duorail facilities for new types of rolling stock such as small vehicles carrying small numbers of people at frequent intervals. Such systems can virtually provide personal transport, they may carry as few as four people, but they can never equal the capacity of more conventional high-capacity trains. The Westinghouse Transit Expressway, for example, comprises smaller units than the conventional underground and they can operate singly or in linked groups as demand varies during the day (Richards, 1969, p. 95). The units have rubber-tyred wheels which reduce noise levels and make overhead structures more acceptable to the environmental lobby as well as making it possible to provide new track at lower cost. Operating costs can be further reduced by introducing automatic control, and the possibilities in this respect are already demonstrated on London's Victoria Line even though on-board 'drivers' are still used at the present time.

Monorail systems have also been mooted and two, Alweg and Safege, have been fully developed. But the advantages over duorail systems are not great and they are not likely to be acceptable as methods of transport through city centres, for example. A 2.5-mile operational system on the east bank of the Rhone in Lyons, URBA, comprises cars with a capacity of 30 or more persons but, although it is not yet known whether the system is acceptable, it is claimed that installation and running costs are low enough to bring it within reach of relatively small cities of 250,000 population. The URBA system involves light supporting structures, and routes can therefore be rapidly extended and altered to suit changes in centres of traffic generation with far less interference with the local environment than would be the case with other monorails. It is basically a commuter system and if its overhead support installation can be suitably merged into an area it can be built and run cheaply. The most financially successful monorail is still operational; built between 1898 and 1903, it operates along a 9.3-mile route at Wuppertal, West Germany. A more recent example is an 8-mile-long line connecting Tokyo International Airport with a station in the central area of the city. Monorails do not, however, appear to offer significant advantages over duorail systems, and their general adoption is unlikely to occur for some time into the future.

Cabtrack and minitram

Cabtrack is an area personal rapid-transit system based on a fine-mesh network of segregated track, providing the opportunity for journeys to be made between any two points without intermediate stops (Russell, 1972; Langdon, 1971). During the journey, the control system, by taking account of the traffic situation and by a coordinated rerouting, can avoid temporary non-availability of links. It is demand-activated, i.e. it operates in response to a particular need. Cabtrack exploits dynamic merging-and-passing switching and has the ability to feed cabs into slots automatically prepared for them. Direction at a junction can also be selected by a mechanism mounted on the vehicle rather than by moving parts on a track. It also permits close headways and the coordination of flow, from which follow high levels of track utilization, capacity and service. In addition, it relies on the control of intake to avoid congestion by the indication of delays before the passenger is committed to the system, which also produces a high degree of vehicle utilization and the elimination of storage (parking) problems. This is significant for land-use patterns when it is remembered that half the cars parked in London are left for more than four hours. Car users do not seem to need ready availability of their vehicles for short trips. If there were an efficient inner-city distributing system such as Cabtrack, cars could be left further out and a great saving of valuable central-area space could be made.

Cabtrack is nowhere near as land-use intensive as monorail, for example, because it only needs a small structure on which to operate. Demand actuation of Cabtrack can be achieved by advising passengers to use a magnetically coded ticket obtained at the station of embarkation, and when a cab arrives it is automatically routed to the destination coded on the passenger's ticket. A study in central London has shown that Cabtrack could not provide sufficient capacity to carry all public transport demands and the underground and some bus services would still be needed. In smaller cities, however, it might be able to provide all the services required.

An alternative to Cabtrack is the Minitram, which combines the operating characteristics of the bus and the underground train (Grant and Russell, 1973, pp. 176-221). Some of the advantages of the shorter headway of buses are exploited in combination with the advantages of fixed-rail operation such as dependability and predictability. The trains, which run on a guideway similar to Cabtrack, can either run independently and provide space for up to 14 people or they can be linked in units of three. Headways can be as short as ten seconds and specially designed stations ensure that this flow of vehicles is maintained. The ability to split up the trains into single units means that the same level of service can be provided at off-peak hours but using less rolling stock. The feasibility of a Minitram system has been examined for Sheffield city centre. It would link the major focal points – the railway stations and the main shopping centres – along a route consisting of 2.5km. of double-track elevated guideway operating as a 'public demonstration project' (Mathew *et al.*, 1974; Bordass, 1975). Testing Minitram in public would help to resolve questions about its public acceptance, reliability, environmental impact, and vulnerability to vandalism, in a way which would not be possible on a private test track. The consultants in the Sheffield study concluded that as a form of public transport Minitram could attract car users (although in a central-area system of the kind prepared for Sheffield these opportunities would seem to be limited) and stop roads encroaching even further on scarce space in cities. Minitram would provide a service at lower cost than urban motorways, with greater equity and less demolition, but an initial problem is the high capital cost of investing in guideways and automatic systems and this has been sufficient cause for the project to be abandoned for the present. Unless a public demonstration is set up, however, it will remain difficult to establish the potential of systems such as Minitram or Cabtrack. Some of these objections may be overcome in an interesting scheme based on an extended version of the Minitram being undertaken by the Tyneside Passenger Transport Executive (1973). Known locally as the Tyneside Metro, the plan involves a 34-mile, 43-station rapid-transit system which will provide a service from 06.00 hours to midnight between South Shields, Jarrow, Gateshead,

Tynemouth, Whitley Bay and, at the centre of the network, Newcastle. The trams will run at 2–7-minute intervals using 90-feet-long electric 'supertrams' with a 200-person capacity, and which can be linked together into groups of two or three during the peak hours. The Metro will provide the framework for an integrated bus and rail network which, it is hoped, will be good enough to lure commuters away from their cars and so help to ease traffic congestion, especially in the centre of Newcastle, allow pedestrianization of more shopping streets, and reduce the need for more urban road building. It is expected that the Metro will carry some 30 million passengers each year compared with 6 million a year on the existing rail network. This will be achieved at a lower cost than the existing operation. Only eight miles of the route involve construction of new track, some of it underground, especially in Newcastle and Gateshead, while the remainder will consist of upgraded British Rail suburban track.

Other guideway systems

There are also a number of other intermediate systems which use guideways such as those utilized by Cabtrack (these are not personal rapid-transit systems, but have a lower capacity than standard railway system). An example is Ford ACT, which uses rubber-tyred cars on a fixed guideway that has automatic control on line switching and the speed of the 24-person-capacity cars. Speeds of up to 30 miles per hour are possible with as little as two-second headways between vehicles. A second example is the Hawker Siddeley Minitram, which is a medium-capacity vehicle on a fixed guideway. It was developed in connection with the City of Westminster Study in 1970 and is somewhat similar to the Ford ACT and Cabtrack, despite differences in the methods of controlling the operation of the systems. Airtrans is a system being developed at Dallas/Fort Worth Airport, using rubber-tyred cars running on a guideway which will form a 13-mile network operated by 68 cars linking 28 stations, with each car having a capacity of 40 passengers. Many of these experimental systems are tested in environments with a large demand for relatively short-distance movement, such as airports or university campuses. They all require a fixed track, either elevated, in a tunnel or at ground level, which must be of substantial dimensions to carry the relatively large vehicle operated. It is therefore easy to incorporate them in new developments such as New Towns, airports or new university areas, but in existing central areas there are likely to be considerable difficulties of finding space and of reconciling the visual presence of the support structures with the city scene. In addition, while a fixed-route service can be provided at the times required by passengers, the degree of personal choice of route must be limited in systems of intermediate capacity, which depend on a number of people wishing to make the same journey.

Moving pavements

The demand for many short-distance movements in central areas need not only be met by new intermediate-scale rapid-transit networks but may be adequately provided for by passenger-conveyor or moving-pavement systems (Truelove, 1971; Richards, 1969). Many city-centre trips are best made on foot and this mode could be even more attractive if speeds could be increased. There are several examples of moving pavements, one of the best-known being the one at Bank underground station in London, and there are also a number of units at London Heathrow Airport. Proponents of large-scale provision of moving pavements suggest that their main advantage is that they benefit all travellers in the city centre, rather than a select group as is commonly the case if a new urban motorway is constructed. Because many intra-CBD trips are for firms' business purposes, a 50-per-cent reduction from 10 to 5 minutes for example, will generate substantial benefits for commercial and related activities. There will also be benefits accruing to shoppers and others who will not have to expend as much energy in getting from one destination to the next. Moving pavements eliminate waiting times completely, which gives them a considerable advantage over public vehicular systems, but they are only really feasible where high-density, short-distance movements of people predominate. This means that they must be used in busy shopping and commercial areas or they should focus on heavily used commuter rail termini such as Liverpool Street or Victoria in London. Perhaps the most likely application is a role as feeders to other transport facilities. Operating speeds vary: the Dunlop 'Starglide' moves at no more than two miles per hour, so that passengers can easily step on or off and frequent access points can be provided, and this counterbalances the disadvantage of low speed. The Dunlop 'Speedaway' system is more advanced and can operate at 10 miles per hour, but access is restricted to areas where acceleration and deceleration belts can be used to reduce the moving speed of the pavement (see *Traffic Engineering and Control*, vol. 13, 1971, pp. 558–9). It can operate over distances of 0.25–2.00 miles, it is silent and can run either above ground or within a glazed tube over or along main roads, or below ground. The capabilities of such a system have been examined for the central area of Liverpool where it would meet two requirements: to act as a distribution network for a park-and-ride system based on two 5,000-space car parks on the edge of the inner motorway; to serve as a movement system along a pedestrain precinct across the central area. Other studies have been for the La Défense scheme in Paris and for a link between London Bridge and Liverpool Street stations in central London, where there is a dense corridor of pedestrian movement between these two stations and three underground stations.

Innovations in public transport technology, some of which have been

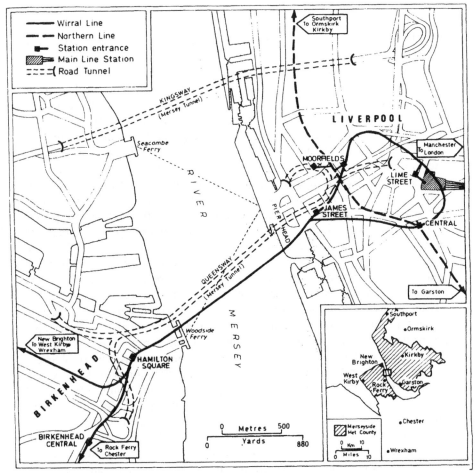

Figure 10.6 *The loop-and-link underground system on Merseyside.*

outlined in this section, only represent a medium- to long-term solution for specific kinds of movement problems in cities. During the interim, individual metropolitan areas must seek solutions for their existing public transport problems. A recent example is the opening on Merseyside of a loop-and-link underground railway system which joins together a number of rail termini which surround Liverpool city centre (figure 10.6). It represents a logical development of the pre-existing rail network and provides a rapid-transit service around the city centre which was formerly undertaken by bus services. It also improves interchange facilities for trips between various areas within the Merseyside network and outside it from other parts of the country.

While the logic of the new network is clear enough, it may well prove to be a good example of the difficulties of providing a new transport system on the basis of apparently reasonable passenger forecasts. When originally proposed back in the late 1950s, the very steep decline in city-centre employment upon which the new system would depend for a substantial part of its payload, was not fully anticipated. The Merseyside Transportation Study published in 1969 also confirmed the advantages of the loop and link, even though the commuters upon which the system would depend were already beginning to abandon central Liverpool. Since 1961 employment in the central area has fallen by some 50 per cent to approximately 90,000, of which about half is accounted for by office workers. Therefore, a system which is designed to convey upwards of 50,000 passengers each way during the peak hour under the Mersey alone, at present carries just 9,000 an hour and, even assuming that the current economic ills of central Liverpool are reversed, this is not expected to increase to more than 17,000 during the peak hour in 1980. The completed scheme has cost approximately £45m and, although 75 per cent of this is underwritten by the Department of the Environment, the remainder has to be found by the Merseyside Passenger Transport Executive via operating profits and a precept from the rates over nine years. Given the reduction in city-centre employment and the possibility of a further reduction in passenger levels, it seems likely that the system will operate at a deficiency which will have to be met entirely from the rates. The Passenger Transport Executive is attempting to counter the problem by reorganizing the bus/rail network feeding into the loop/link so as to minimize competition between the two modes and to ensure maximum ridership on the rail system. Further increases in the tolls for using the two Mersey road tunnels may also direct more commuters to the rail route, but the Passenger Transport Executive has so far failed to close the ferries across the Mersey, which are used by some 7,000 commuters each day and cost about 50 per cent less than the rail fare.

It therefore remains to be seen whether events have overtaken this ambitious plan for a city centre which has seen employment decline for more than a decade. Its supporters hope that these trends will now be reversed because the 'tube' will make it much easier for shoppers, tourists, businessmen and regular commuters to reach and use the city centre. No doubt, similar queries and uncertainties face the operation of the new Tyneside Metro which is the only other major new public transport scheme outside London. The Picc-Vic line in Manchester, which in some respects would have been the equivalent to the Liverpool scheme, has been abandoned on the grounds of costs, which in 1977 were estimated to exceed £100m. Such enforced abandonment may be more realistic than it seems, in view of the changing status of city centres in most large metropolitan areas in Britain.

Developments in private transport

Many of the rapid-transit systems outlined briefly here are more likely to become operational in the year 2000 than in 1980, provided that some of the ever-present environmental and land-use conflicts which they all seem to generate can be reconciled with the values of late-twentieth-century society. There is also an assumption here that future movements in cities will continue to be dominated by central areas and that they will be more public-transport-orientated than at present. This is certainly the view of a growing body of city authorities and transport planners, but this does not mean that attempts should not be made to improve the efficiency with which the conventional car uses space on urban highways and parking areas. Some of the possibilities have been considered by the Ministry of Transport, which showed that there is little to be gained from reducing the size of vehicles unless this is accompanied by changes in traffic conditions, particularly the segregation of smaller cars from larger vehicles. Average-sized cars, such as the Ford Escort or Austin Allegro, could move about a quarter as many people again if they were given traffic routes for their exclusive use. In the same road space, vehicles such as the Morris Mini and the Fiat 400 could increase capacity a little over 1.5 times. The 'optimum' vehicle could increase the capacity to just over twice the present level, but only by 10–15 per cent if operated in mixed traffic, which is the typical situation at present.

Various types of four-wheeled car suitable for use in cities and intended to derive the maximum benefit from minimum size were outlined in *Cars for Cities* (Ministry of Transport, 1967c). These were: a single-seater car suitable for commuter journeys or short, intra-central area trips (Citycar 1); a two-seater with staggered seating (Citycar 2); a two-seater with side-by-side seating (Citycar 2s); and a four-seater car (Citycar 4). Small city cars permit more vehicles to be parked in the same parking spaces and 'optimum' Citycar could more than double the capacity of parking spaces, e.g. five Citycar 2s could be parked in the space at present taken up by one conventional car. One of the main difficulties is similar to that confronted by new rapid-transit systems, and this is the need to provide new structures mainly above ground level, and the traffic segregation needed to make small cars a more effective way of accommodating trip demand on a free-for-all, personal basis. It may be possible to segregate lanes on existing urban roads, but this would present difficulties at junctions, as well as a reduction in the space available for larger vehicles.

The use of small cars in cities could extend beyond their use as privately owned vehicles and therefore allow more effective control on congestion or length and frequency of journeys. They could be provided as self-drive vehicles by local authorities or private companies, with the cars made available free of charge within controlled areas, to be driven anywhere within them but left in clearly defined parking zones from where they could be picked up by the next user. Alternative systems include payment for use of a car via a slot machine on the car; purchase of keys which, when used in a car, provide coded information relating to identity of the key-holder, length of trip undertaken, etc. – this can be used to calculate accounts forwarded each month to the owner of the vehicle (Richards, 1969, pp. 73–5). An example of the latter is a privately financed scheme using Witkars, which is currently operating in Amsterdam using five 'stations' located in the highly congested central area inside the Singelgracht (Bendixson and Richards, 1976). It is hoped to expand the network of stations to fifteen sites, but the actual impact of the scheme on traffic coming into central Amsterdam seems likely to be negligible unless it is supported by restraint on the use of conventional vehicles. Again, these systems would operate best on segregated roads, but their main disadvantages are the high maintenance costs arising from the widely different standards of driving by a large number of drivers. This cooperative approach to vehicle use has been more than matched by the growth of car-hire companies offering self-drive cars, vans and lorries. Intermittent car hire probably matches the demand patterns of individuals rather better than cooperative ownership.

Such difficulties could be overcome, however, by using battery-driven vehicles. There is considerable interest in developing this method of propulsion, given that cars which use petrol as their main source of energy are remarkably inefficient and oil supplies are not infinite (Independent Commission on Transport, 1974, pp. 49–55, 292–8). A typical car only uses 18 per cent of the energy in its fuel to overcome the main resistive forces, which are rolling resistance (6 per cent), aerodynamic drag (4 per cent) and braking (8 per cent) (K. Owen in *The Times*, 13 June 1975; see also Waters and Briggs, 1975). The remaining 82 per cent goes in losses caused by inefficiencies in the machine: 20 per per cent out of the exhaust, 12 per cent in radiation from the engine, 8 per cent in driving auxiliaries, 2 per cent in the transmission, and 40 per cent into the coolant. Some of these inefficiencies can be reduced by better engine design, more efficient cooling systems, more widespread use of radial-ply tyres, lighter cars or improved driving techniques. It may be better, however, to turn to other sources of propulsion, and the electric car, apart from saving oil, is almost silent, free of pollution and has much lower fuel costs than a petrol-engined vehicle. A recently introduced electric car is the Enfield 8000, which is undergoing detailed evaluation as the first British electric car in regular production. It is eight inches shorter than the Morris Mini, and with a turning circle of 25 feet it almost falls into the Citycar category already described and is seen as an answer to traffic congestion. Battery technology is still not far enough advanced, however, to produce compact batteries which also give long range and high speeds. But with a range of 55 miles and a top speed of 40 miles per hour provided by eight lead-acid batteries which can be charged overnight from a 13-amp socket, the Enfield could be well-suited to urban commuter travel. Silver batteries, a spin-off from space research, are much more compact and efficient than lead-acid batteries but are prohibitively expensive at the present time. The limitations of battery-powered cars could perhaps be improved by providing small auxiliary petrol engines (perhaps 200–300 cc), which could give the minimum constant power needed and could be used to charge the batteries. This would increase the safety of the car, its range and top speed, while at the same time reducing the consumption of petroleum. A battery-powered bus is at present undergoing trials in Manchester, and satisfactory development of this method of propulsion promises considerable environmental gains as well as changes in energy consumption. Even battery-powered bicycles have been mooted, and a prototype giving 30 miles travel (or 3 hours) from a 2p electricity charge (1976 prices) has been developed (*The Times*, 22 April 1976).

It has even been suggested that a return to former transport systems may also be necessary in the interests of reducing the environmental impact of transport. This has been suggested by the Motor Industry Research Association with reference to trolley buses, which over the past 30 years have been abandoned by all towns and cities in Britain because of the costly, unsightly and inflexible overhead wiring which they require (M. Bailey in *The Times*, 5 April 1975). But trolley buses are a clean and quiet public transport mode, and modern technology allows them to operate with only 25 per cent of the route system wired. These stretches should be located where several routes overlap, and wherever possible they could be placed away from environmentally sensitive areas. The trolley buses could draw off enough power from the overhead wires to propel them over the rest of their route. Such trolley-bus networks would be feasible in large cities and medium-sized towns. There has also been increasing interest in using bicycles for short-distance urban travel in order to save energy, to reduce accidents and improve traffic management (Institute of Transportation, 1973, pp. 159–240; Mitchell, 1973). There are almost as many bicycles as cars in Britain, about 12 million, and 8 million are in regular use, many of them by a growing number of adults. A measure of resurgence of interest in this form of transport is the cycle-route experiment in Portsmouth. The object is to see whether cyclists can be persuaded to desert busy and dangerous main roads by providing them with an unimpeded route through parallel back streets. A further objective is to see whether the provision of safer and more comfortable conditions for cycling generates additional cycling trips. In order to do this, cycle lanes are provided which are ten feet wide and follow the centre of narrow residential streets. Cars are allowed into these streets, but for access only. Initial surveys show a marked increase in the use of bicycles on the designated routes. The London Borough of Lewisham has also approved an experimental 4.5-mile cycle route which runs north–south from Lewisham through Catford to Beckenham Place Park, parallel to the A21.

Finally, in this brief review of some new transport technology, the capacity of urban highway systems could be extended even further by introducing automatically guided vehicle systems which can be transferred from automatic to personal operation. An example is Starrcar, an American minicar system, which can be driven along streets in the conventional way, or if the driver wishes to have a more direct and faster route, he can drive up a ramp to an elevated track where his vehicle joins a train of vehicles travelling at 60 miles per hour (Richards, 1969). The car is automatically controlled on the elevated track and ejected at the selected exit, where it can again transfer to normal operation. Each track can carry the equivalent of 15 lanes of motorway or 30,000 persons per hour. An alternative might be a three-lane motorway with one lane reserved for normal operation, and one switching lane on to which cars or buses move in preparation for moving on to the third automatically controlled lane of traffic. Careful and regular maintenance of vehicles will be an essential prerequisite of Starrcar or its equivalents and this may make it more appropriate for public rather than private transport operation.

Only a few of the alternative systems offered by new transport technology have been described here, but they provide some idea of the opportunities available for introducing new methods of movement in cities. It remains to be seen whether any of these prototypes/ideas will ever actually become operational in any significant way in our urban areas where the costs of reshaping the urban fabric may well outweigh the benefits subsequently endowed by the new transport system. It may be more reasonable, on both social and economic grounds, to concentrate on the existing transport infrastructure of British cities and to try and devise ways to ensure that it is made to work as efficiently as possible and in the interests of all users of the system.

Social priorities

The last point in the preceding section raises an issue which has gained credence in the transport literature during the last few years. As Hillman, one of the main protagonists of the social approach to transport planning, has pointed out, 'transport affects society in more ways than is implied by considerations of efficiency, economics and environment' (Hillman, 1975, p. 16; see also Hillman *et al.*, 1973; Hillman, 1974, pp. 162–7). These, of course, are precisely the themes which have dominated much of the discussion in this and earlier chapters of this book. It has come to be recognized that for too long the motor vehicle has been given priority irrespective of the high social costs which it generates. Hence in its recent outline of the objectives of transport policy the Department of the Environment (1967a, vol. 1, p. 12) has put forward the view, not previously articulated, that 'people have a right to expect a reasonable degree of mobility; and the social objective of transport policy is to ensure that this is available to all and not just to those who own cars'. It is important to distinguish here between the social *costs* of transport such as noise, dirt or air pollution, which are well-documented (and already discussed in this chapter), and the social *consequences* of transport. One of the most obvious examples of the latter is the out-migration of population since the war from inner-city to suburban residential areas, which has not been matched by similar movements of employment. Hence more people wish to travel longer distances to places of employment, and this has increased the demand for transport services adequate to meet 'peak hour' travel. Fare levels and subsidies have been raised to the point where they cover at least a part of the cost of providing these transport services (including roads), which are subsequently under-utilized in off-peak periods. Changes in fare levels are usually applied 'across the board' and therefore affect inner-city residents, who live much nearer to their employment and make comparatively short journeys. Their relative contribution to the costs of maintaining urban transport services is much smaller than that of longer-distance commuters, but they must still meet, usually from inferior resources, fare increases of a similar order.

This applies to both road and public transport travel but especially the latter, which has been prone to the effects of the spread of car ownership. Some of the consequences of this for public transport have already been outlined, but it must be borne in mind that in 1990 at least one-third of all households in Britain are still unlikely to own a car and about 50 per cent of the population will not have a driving licence. With the financial bases of public transport already undercut by the growth of car ownership, the potential mobility of the minority will be deprived further to even lower levels than they have been accustomed to, if emphasis continues to be given to planning for the car. In addition, if the car-owning majority continue to dominate our thinking about the spatial pattern of land use in cities, the non-car-owning households are likely to suffer to an increasing degree. Out-of-town shopping centres, larger schools or suburban industrial estates and office parks imply that greater, not less, mobility is required by all; indeed the advantages of being mobile have probably increased. The basic problem is that 'mobility becomes even more necessary: but command over it for the minority grows less' (DoE, 1976a, vol. 1, p. 12).

In a sense, 'mobility deprivation' in cities is a component of the social deprivation which is equated with particular socio-economic groups located in distinctive areas of our cities. It would be myopic to pretend, however, that most of the transport-related problems affect the residents of the inner city where social deprivation is concentrated. Many pensioners, for example, live in suburban areas, and all who do not own a car (the majority) must walk to obtain public transport services. The lower-density public transport network in most suburbs means that for many this may be a real hardship, while for those who find it difficult to walk for medical or other reasons, it becomes impossible to get to, let alone use, the bus or rail transport upon which they are almost totally dependent. Few can afford to call a taxi as and when required. Another city-wide social consequence of increasing car ownership and travel is the inability of children to move about on foot or bicycle with the freedom or without the fear of road accidents which reduces dependence on their parents for any sort of travel, whether to school or to see friends. Hillman *et al.* suggest (1976, pp. 17–21) that this is part of a more general loss of personal freedom directly arising from past attitudes to transport provision and planning in cities.

Identification of the problem is a little easier than prescribing solutions. Subsidies for public transport have frequently been mooted as a means of ensuring greater social equity from transport policies. In order to meet social priorities it is inadequate to give indiscriminate subsidies, however, because these tend to benefit the better-off, who travel most, rather than the less well-off minority (DoE, 1976a, vol. 2, pp. 29–52). Subsidies to suburban rail services, for example, benefit the middle-class medium- and long-distance commuters far more than the lower-class short-distance commuters, because they keep fares at artificially low levels and thus allow individual households to spend more on the consumption of, for example, land or housing. This problem is less acute in the case of bus services, because they are used by a wider cross-section of urban households, but it is still necessary to ensure that essential services are retained by providing selective help to bus services in particular areas. Individual local authorities have considerable discretion in this context with reference to local transport plans. Vehicle restraint, road pricing, taxation, as well as subsidies, may also effect changes with social implications, but it is as well to remember that 'no pricing system in transport is likely to be fully equitable as between all users. The best we can aim at is rough justice between main groups, tempered both by gradual implementation where individuals are most vulnerable and a systematically greater priority for mitigating the hardship likely to be caused to those least able to help themselves' (DoE, 1976a, vol. 1, p. 30).

It might also be helpful if transport policy were devised in a 'socially aware way' (Hillman *et al.*, 1976, p. 23), whereby priority is given to journeys on foot, to journeys by bicycle, to public transport for essential journeys over longer distances and, last rather than first on the list, to travel by car, which will always be the first choice mode for most journeys undertaken by those who have one. This is a rather radical reordering of contemporary priorities in urban transport and will require enormous changes in attitude, as well as in the infrastructure of cities, if it is to be achieved. Cities and their transport systems comprise a very large complex of fixed investment and it will not be possible to modify them very quickly. Social priorities for transport planning are undoubtedly important, but much will have to be achieved through modifying the existing transport infrastructure and its use rather than restructuring cities in the way implied by advocates of social transport planning.

Intra-urban travel and the future city

Much of the preceding discussion assumes that there will continue to be a growing demand for personal and goods movement within British cities and that this will be met by investment in new private and public transport facilities. Increased travel demands will arise from the growth of car ownership; improvements in the distribution of real and disposable income amongst the population; and changes in the amount of leisure time as average hours worked per week decrease or the number of days worked, already reduced to four per week in the United States, is reduced. It is also assumed that urban land-use patterns will remain much as they are now, with trip-generation characteristics, upon which plans and strategies for transport planning depend, very similar to those of today. Maintenance of the *status quo* could be unwise, however, because it could lead to a misallocation of investment and resources at a time when the optimism about social and economic growth characteristic of the 1960s has been replaced by the realization during the 1970s that this cannot take place indefinitely. Public investment in new road facilities was once easy to justify on socio-economic grounds, but following the fuel crisis in 1973 and the growing role of the conservation movements this acceptance of contemporary trends has been severely eroded. But it has not changed substantially the conventional wisdom about the relationships between urban form and travel. Any reappraisal of the future of transport in our cities ought not to exclude an allowance for the possible impact of travel substitution on future urban form and travel behaviour.

Movement substitution in cities involves the consumption or provision of a good, service, information or any other form of contact without the participants moving from one point in the city to another. Telecommunications are the key to this kind of travel substitution; their impact could invalidate the forecasts upon which present programmes for urban development and transportation are based. During the last decade advances in telecommunications have been rapid; data networks, high-speed facsimile devices, Confra-Vision, the video telephone, national and even international private telephone circuits, increased household and commercial penetration of the telephone, telex, remote-controlled typewriters and related devices, all create new opportunities for reducing the volume and pattern of urban travel. Harkness has commented that if 'a reduction in the ease of communicating electronically is not balanced by a corresponding facilitation of travel (or some other compensating measure) surely the percentage of contacts made electronically will increase' (Harkness, 1973, p. 57). From one point of view the emergence of telecommunications during the last fifteen years could be interpreted as an extension of the long-standing tendency for improvements in communication to lead to a more dispersed city form. This has already been traced in Chapter 1 where it was shown that each advance in transport technology has loosened the hold of transport as a constraint on the location of urban activities; other things such as cheap land, the ease of providing parking facilities, the attractions of low-density suburban housing or the desire to reside in areas concomitant with occupation status, have assumed greater importance. All these things have not reduced the necessity for travel in cities, but this is precisely what telecommunications technology offers. Some movement substitution has, of course, been taking place independent of new telecommunications: the use of freezers to store large quantities of discounted food products or fresh garden produce has reduced the number of household trips required to purchase perishable foods; visiting tailors, cosmetics representatives, insurance salesmen or door-to-door grocery vans, all contribute in small part to reduce the aggregate level of urban travel.

Such adjustments become insignificant when compared with the contribution which the telephone has already made. It is impossible to estimate how many trips the telephone has already saved by allowing friends and relatives to communicate without travelling or by allowing businessmen and others to undertake low-order discussion, to exchange information, or to arrange meetings. Perhaps 60–70 per cent of business meetings could be effectively substituted by telecommunications. Television has brought the theatre and sport into the home and has helped to reduced dependence upon living in cities where there is easy access to these activities. The use of the telephone to order groceries from supermarkets, which are subsequently delivered by goods vehicles, already takes place in the United States and could take place in British cities. Because comparison shopping is important for durable goods, there is probably a limit to the extent to which shopping trips can be substituted: it would require complex organization to permit an individual to sit at home in front of his television selecting a good such as a suit or a hi-fi set from those held in the stock lists of all relevant retailers in his city. Shopping trips also perform an important social role for many urban residents who would otherwise be at home for much of the time. Similarly, it is doubtful whether telecommunications will be an adequate substitute for social and recreational trips, in which the destinations and perceived benefits vary widely from one urban dweller to another. This leaves the journey to work as the most likely candidate for trip substitution and, given its regularity and significant contribution to the daily volume of travel in cities, the impact of telecommunications could be substantial.

Urban activities engaged in the production and distribution of tangible goods and services will still need to assemble a labour force to undertake the work. The only scope for a reduction in journey to work is through increased mechanization of production which will cut down the size of the labour force. It seems unlikely, for example, that individual employees will be able to control their part of the production process from a console in their own homes. On the other hand, computers are now widely used to monitor and forecast the inventories of stock of industrial and wholesale distributors, so that greater centralization of activities at a smaller number of locations has therefore been possible. This has been paralleled by the use of computers to plan efficient distribution networks and routes for freight vehicles. Suburban journeys to work are at least 50 per cent shorter in distance and time than trips to the CBD; a further reduction could occur if the trend towards concentration of industrial development at limited numbers of suburban locations were replaced by a more dispersed pattern of location.

Activities engaged in the exchange of intangibles such as information, ideas and transactional work are the most likely to contribute to substantial movement reduction. Face-to-face meetings are the basis for much of the contact amongst the largely office-based activities comprising this group. Because of this, office activities are still characterized, despite decentralization in recent decades, by concentration in central business districts where the time taken to get to meetings or to exchange important and confidential documents can be kept to a minimum. The decision-takers usually involved in transactional activities require a large number of support staff to collect, store, retrieve and process the information and other documentation upon which executives depend for efficient negotiation or the provision of a service to a client. Many of the support tasks performed in office premises, as well as meetings, are of a kind which can be undertaken without the need for personal face-to-face contact or proximity of premises. The Viewphone, for example, permits two individuals to have direct audiovisual contact; Confra-Vision allows groups of participants in, say, London and Glasgow, to hold a meeting in which they can see and hear each other without leaving their respective cities; documents can be transferred over long distances using facsimile devices; computers can be used to coordinate work fed in from remote terminals; typewriters can be operated on a remote basis, perhaps from an office-worker's home, with amendments and corrections conveyed by line printers or by dictation on the telephone.

The implications of this should be obvious: if employees or participants in an important meeting do not need to leave their respective offices or homes, the aggregate demand for travel will be reduced. With individuals able to work from home or at small neighbourhood workplace clusters linked to other parts of their organization elsewhere in the city by telecommunications, it is very likely that concentration of quaternary employment in the central business district will no longer be necessary. This would mean that investment in expensive public transport, particularly rail facilities, which service CBD workplaces, is unnecessary. Existing network capacities will remain adequate for sustaining those activities in the CBD for which travel substitution is not an attractive alternative. Some office-based activities, such as international banking or commodity trading, may resist any opportunities to reduce direct face-to-face meetings, because spatial separation of participants could be considered to undermine the confidence and mutual understanding necessary for reliable high-level decision-making. Equally, it has been suggested that retailing does not seem widely suited to the use of telecommunications and, although decentralization is certainly taking place in British cities, large department or chain stores and specialist retailers may well decide to continue operations in highly accessible central locations.

Massive investment in CBD-oriented rail or road facilities can then be diverted to providing efficient suburban and extra-suburban transport to handle new patterns of employment location. It should not be assumed that this means investment in high-standard motorways and similar urban roads; assuming, for example, that many workers for psychological and social reasons will not want to be encapsulated within their own homes, it will be possible to encourage patterns of workplace location which generate sufficient demand for rapid-transit or rail facilities to be provided. Employers may also prefer to keep as many workers on their own centralized premises as they require in the interests of effective management and control of output. It may even be possible to define route capacities and volumes servicing suburban employment concentrations very precisely by specifying that office-workers should always travel to their nearest neighbourhood centre (Harkness, 1973). Telecommunications would make this feasible by providing immediate audiovisual contact, information and document exchange with other self-contained centres within the city; so discouraging travel between them. Cities can therefore be envisaged in which the demand for journey-to-work travel can be manipulated using travel substitution to levels compatible with the needs, priorities and resources of individual cases; ranging from total to limited telecommunications substitution.

Telecommunications would have the added advantage of minimizing the environmental impacts of travel in cities. Most of the powerlines can be buried underground at reasonable cost when compared with most new rapid or personal transit of the kind described earlier in this chapter, all of which creates considerable environmental damage. Assuming that levels of travel would decline, air pollution and related environmental effects of transport would also be curtailed. Electrical energy used to power telecommunication systems could also be supplied by coal or nuclear-based generators, thus saving scarce fossil fuels such as oil and gas, especially oil, for non-transport use.

Perhaps shortages of energy supply for private transport will force society to look more seriously at travel substitution in cities than it has done to date. The technology already exists and, although it is unrealistic to expect the new patterns of behaviour and travel consequent upon intensive use of telecommunications to occur overnight, it must surely be wise to plan existing investment in transport infrastructure in a way which does not lead to redundant routes and stock in 20 or 30 years' time because of changes subsequently brought about by travel substitution. Contemporary or short-term transport investment should be made with a clear recognition that the future city could be less centralized than today and with a much more limited concern with providing transport facilities adequate for the purpose of handling peak-hour travel. The potential savings in transportation costs created by telecommunications substitution are enormous, perhaps several hundred million pounds for a city like London. Naturally, these savings will need to be counterbalanced by the cost of providing comprehensive telecommunications networks for inter and intra-firm contacts. More information is still needed about the size of these costs, as well as the social consequences of reduced travel in cities and diminished opportunities to develop a clear identity with a place of work or association with colleagues. All this assumes that decentralization of cities will continue, but land shortages in Britain may help reverse this trend. There is already a revival of interest in more effective regeneration of the inner cities, but even if this should happen it may not invalidate the case for reassessing assumptions about the transport movement and congestion likely to be created. Telecommunications may again offer an opportunity to reduce travel demands, even though participants will never be very far apart. Every city resident has the right to be able to move freely; the demand for movement is insatiable but attempts to meet it have, so far, created more imbalance and inequality than is now considered acceptable. A combination of careful land-use planning, efficient transport investment and telecommunications provision would move us nearer to a city form in which space, time and cost offered least resistance to the travel needs of the majority of the population, with minimum environmental damage and resource depletion.

11
The geographical perspective on urban movement

The form and content of this geographical interpretation of movement in cities has grown out of two major fields of writing and research: the literature of academic and practical transport planning, and that concerned with social and economic aspects of urban geography. We have adopted an integrative strategy towards these fields that has enhanced our understanding of the structure of travel within towns, its geographical implications, and the planning problems which it presents. Whatever the success of the book in communicating our spatial perspective on urban movement, there is no doubt that it is selective and only an introduction to the insights that each field can give to the other.

Being principally a geographical work, the book has repeatedly focused on the spatial patterns of movement and of transport services, and upon the distribution of land uses, urban activities and populations with which they are so strongly related. The objectives of the book have included not only an attempt to provide an improved spatial description of movement within cities, but also the pursuit of a deeper understanding of the factors generating such patterns. In other words, as well as answering the question, 'What is there there?', the more challenging supplementary, 'Why is it there?', has also been tackled. Furthermore, it is hoped that the book also reflects the tendency during the 1970s for human geographers to be concerned with 'What should be there?', partly by including critical evaluations of public policy-making and execution at national and local levels. Although an increasing number of geographers are seeking to demonstrate their subject's usefulness beyond its established role in a liberal education, urban transport and mobility questions have received only modest attention to date (but see Wheeler, 1973, especially pp. 163–80; it should be pointed out that the concern to recommend improved spatial distributions is not entirely novel – the late Dudley Stamp's work exemplified an earlier approach in, e.g., *Applied Geography*, 1960). Nonetheless an attempt has been made to illustrate the kinds of questions that geographers are prone to raise about spatial distributions, and the kinds of applied mobility research that they can readily and usefully undertake. Our initial belief that geographical research can play a constructive part in improving personal mobility and the operation of transport systems has been sustained and broadened throughout the writing of this book.

In this, the final chapter, some of the more important disciplinary and applied issues which we have encountered are reiterated and stressed. To be more specific about the ways in which geographers could more usefully contribute to applied transport studies, it is helpful to discuss separately three scales of activity. The first is at the aggregate level of all towns (the national scale) at which geographers can be more active in identifying failings or misdirections in public policy. Recent years have shown that the amount and structure of public expenditure devoted to urban transport is not necessarily related to a comprehensive assessment of its requirements and priority. Instead it appears to depend upon the health of the national economy, upon macro-economic policies such as those of price controls on local transport services, upon the policies which are adopted to accommodate or placate vested transport interests, and upon the sometimes simplistic reactions to perceived failings in the policies of the recent past. Changes in direction at the national level are preeminently matters of political decision, but in the transport field they often become apparent only by the changed emphasis common to a host of minor government directives and recommendations. Since there is no single national transport policy (DoE, 1976a), it is possible for all these minor adjustments to contradict each other.

Alterations are continually being made to the mandatory and discretionary requirements placed on local authorities by central government, and there are also many other changes, such as in the grants available to transport operators or highway authorities, which have important long-term transport policy implications as well as the fiscal and broader political objectives for which they are normally conceived. It is therefore particularly important for informed geographers, along with others, to monitor and evaluate, and where it is believed right, to challenge and criticize without delay the directions implied by a sequence of government directives and circulars. No exceptional role for geographers in this monitoring is claimed; transport issues are notable for the very wide range of vested interests and consumer groups that they affect (extending even to the Church of England; see Independent Commission on Transport, 1974. In 1975 the DoE listed 134 organizations concerned with research, political pressure or professional interests in the transport field; see Norman, 1977, p. 7). In comparison to many of these groups, however, geographers and other academic students of travel should have a more detached view of the functions, efficiency and equity of the urban facilities for travel and mobility. For a field in which particularist interests are so strong and so practised in the advocacy of their wishes, the value of commentators who stress the need for balanced, fair and long-term policies is clear.

There are already several examples of policy revisions that have resulted (at least in part) from the critical comments of transport students rather than the lobbying of vested interests. For over a decade they have argued that the transport professions and transport policies have been concerned with only selected types and purposes of movement, and that strongly biased investment programmes have not only encouraged but in some cases made more necessary longer, mechanically-assisted trips. Both the high energy consumption and the spatial and personal inequalities in access to more highly capitalized forms of travel have been submitted as reasons for querying these policies and trends. It has been argued that, according to local circumstances and the specific case, the provision of transport services and roads will sometimes repay improvement, but in other cases requires curtailment. The effect of land-use changes and of environmental changes on the total demand for 'obligatory' (or inelastic forms of) travel has not been sufficiently realized, and there has been little recognition that where circumstances warrant, it will be advantageous to plan through land-use and activity distributions for a lower volume of travel and mobility in present and future cities. Policy has perceptibly, if not very tangibly, moved in the direction urged by such criticism, as the explicit references to different mobility groups and to energy conservation in the latest White Paper on *Transport Policy* (Department of Transport, 1977) shows. But the assumption that increases in travel can be equated with increases in mobility benefits and are therefore desirable is still widespread among commentators and policy-makers (for a vigorous debate about the correct interpretation of increasing travel, see Foster, 1977, pp. 21–46).

One other contribution which geographers can make at the national scale has been extensively discussed in this book and is worth reiterating. This derives from their increasing competence in numerical analysis as well as their spatial interests, and relates to the design and practice of the transport study and planning process. Geographers have not only participated in the formulation of this process, and incidentally been party to the establishment of some of its more questionable assumptions and procedures, but they have also been among those offering the most trenchant and telling criticisms. The techniques used to derive estimates of future car-ownership levels in Britain as the basis for calculating future travel demand have received some detailed scrutiny which casts serious doubt on their validity (Adams, 1974). A great deal of inertia in the way in which transport studies are organized, implemented and presented has also been detected. The cost of transportation studies is not easy to ascertain but one estimate puts the total cost at between £20m and £25m, of which the various London Transportation Studies have already absorbed £1m (Atkins, 1977). In all, it is estimated that some £50m (at 1977 prices) has been expended on urban transport research since the emergence of sophisticated transport studies in the early 1960s (Bayliss, 1977). While it is a small sum relative to the overall expenditure on transport, Britain will spend about £15,000m (17 per cent of GDP) on all aspects of transportation in 1977, much of this within urban areas; it is difficult to gauge what benefits have accrued from these studies. Some would argue that by encouraging expensive capital projects, particularly on urban motorways, the authority attached to the recommendations of expensive research has been an undesirable and harmful influence. Clearly there is no scope for complacency since, according to some, 'it is difficult to avoid the conclusion that over the past ten or twenty years, apart from increased access to cars, the transport circumstances in urban areas have more often got worse than better' (Bayliss, 1977, p. 5). Such a view is perhaps unduly pessimistic and it may be truer to say that reducing differentials (during the post-war period) in income, education, job opportunities and working conditions have not been matched by similar reductions in differential access and engagement in travel. In an increasingly egalitarian society, the surviving differentials become more obvious and unacceptable.

Transportation studies generate rather unrealistically grand ambitious solutions for coping with anticipated travel demands during a target year, which may be 15, 20 or even 30 years beyond the base date. But there is growing scepticism about the reliability of this crystal-gazing; survey errors and assumptions about relationships between countless variables which may or may not be autocorrelated are incorporated into models which become paragons of perfection by the time final reports are produced (Atkins, 1977). Apart from the sheer complexity of the transportation planning process, which has tended to increase as we become more aware of the diversity of travel demand variables, it is now realized that perhaps the wrong questions are being asked. Issues such as equity of access to movement opportunities in cities, the relationship between transport and the environment, the right of individuals and groups to participate in decision-making about urban transport policies and facilities have all been alluded to at various places in this book, and it is these considerations, rather than efficient planning of urban movements, which now call the tune in transport planning. 'Traditional' transport planning procedures are really unable to accommodate these nuances, which are so often difficult to quantify while, in addition, they deal very superficially with behavioural variables. The introspection of the procedures by which transport plans and proposals are examined in public, such as motorway inquiries, has also been highlighted recently (Adams, 1977). To a degree it may also be true that the anticipated increases in travel which transport planning seeks to accommodate are avoidable if the facilities they recommend are not provided. There does seem to be a need for further studies of people's travel behaviour according to their location within cities and in response to changes in their personal circumstances (such as in home or workplace location or increases in income), or as a result of changes in transport services. This is yet another subject within urban transport studies that geographers can further by the development of methods and techniques and the exploitation of data.

Changing emphases in transportation planning are reflected by the decline in the number of transportation studies commissioned during the last five years. Some 99 towns and cities had been studied by 1975, and 52 final reports had been completed (Adams, 1977). The peak year of publication was 1967, when ten appeared, since when the number has fluctuated around five per annum. Only two reports appeared in 1975, in which year most continuing studies had reached the analysis stage, and only four were still at the survey stage. It seems likely therefore that the annual output of reports will continue to be low, relative to the peak in 1967. As these area-wide studies and their recommendations have declined in importance, urban transport decision-making has turned more towards small-scale and shorter-term proposals, to form an incremental rather than an analytical approach to transport planning. City centres have attracted most attention, with recent action being characterized by the creation of pedestrian precincts, public transport interchanges, discriminatory parking schemes, careful management of traffic flow, segregated bus lanes, and the adoption of environmental management schemes. A related trend is concerned with more efficient use of existing capital investment rather than the development of new facilities which may involve major structural changes in urban areas. This reappraisal has also been encouraged by a more pressing recognition of the need to plan the consumption of energy as finite oil resources become more expensive (Maltby, 1977; Willey, 1975).

The second scale of applied studies is at the level of the individual town or urban region. Urban geographers have been tracing the changing distributions of urban activities and populations within towns for a long time, but undoubtedly much more could be done to specify the travel and mobility implications of the changing composition and distributions of population, employment and activities within urban areas. Studies of the transport implications of decentralizing offices have shown what is possible, but they need to be supplemented by a greater variety and number of enquiries. The transport effects throughout the city, of widespread local changes, such as the decline in the number of shops or the amalgamation of doctors' practices, should also be studied more extensively. It would often be interesting and valuable to monitor the spatial effects of new urban transport measures such as the pedestrianization of principal shopping streets, the implementation of new parking controls, or the integration of the services provided by different modes of public transport.

Until the present, little use has been made of geographical techniques and expertise within the transport planning process, as have for example techniques from applied economics such as cost-benefit analysis. The steady development of network analytic techniques does promise, however, to be of great value to the redesign and development of public transport routes and of goods-distribution services and itineraries within the city. A recent review of the field indicates that the geographers, mathematicians and others who are developing these techniques have so far been most attracted to inter-regional and inter-urban transport problems, but their extension to urban transport problems involves no radical conceptual or methodological problems; it is the interest and in some cases the data which are lacking (Leinbach, 1976).

The third scale of applied research that we recommend is the local study of movement. Under this heading would come the mobility impacts of detailed changes in urban design and of traffic-management measures, as well as their influence upon individuals' travel decisions, including routing. Many topics which deserve fuller investigation at this scale fall into what has been called in a recent review, 'a social geography of travel' (Muller, 1976). As has been shown by its fuller development to date in the United States than in Britain, this focuses on the wider functions and purposes of personal movement, and the pronounced socio-spatial variations in opportunities for travel. It has been suggested that 'differences among social, economic and cultural groups can be heightened by transport networks ... because of extreme differences in mobility and in part because of symbolic barriers which transport facilities can represent' (Wachs, 1977, p. 107). The mobility problems of the disabled, of the elderly and the young, and of ethnic groups have already received some attention (C. S. Davies and Albaum, 1972; Perle, 1968; Falcocchio and Cantilli, 1974; Norman, 1977), but there is scope for more. A strengthened social geography of travel would also contribute to urban social and economic geography by developing a more thorough knowledge of the spatial implications of movement behaviour. The study of movement at the micro-scale of the housing estate, the neighbourhood, or the city block has been palpably neglected by most geographers. This may be explained by the lack of attention by local and central government to mobility at this scale. Individually, it must be agreed that each small-scale local issue has little impact, but the aggregate of these issues throughout the country affects majorities of the population. It is lamentable that geographers, with their attuned spatial understanding and their interest in spatial patterns wherever they occur, have not been more energetic in bringing safety, accessibility and local-mobility issues to more widespread attention. Improvements in the allocation of resources for various urban services and/or greater efficiency in travel could well result from studies of access to, for example, chemists' shops, personal medical services, or infant schools (another outline of these issues has recently appeared: Ambrose, 1977). Should not these studies be at least as common as studies of leisure travel or of pedestrian behaviour within shopping centres? It is our view that the travel implications of the reorganization of local educational, retail, personal and institutional changes which has been one of the hallmarks of the 1970s, is a badly neglected aspect of contemporary urban studies. Those with a penchant for or competence in behavioural studies can find a great deal of useful research in this area, although our own opinion is that the development and testing of elaborate or speculative conceptualizations of decision-making is not necessarily the form that this research should take, and is more likely to give applied returns in the long than the short term. In the first instance, at least, there is great scope for research which is conceptually straightforward.

At all three scales of study that have been outlined, disciplinary as well as applied goals will be approached. We have argued through extensive illustrations that intra-urban movements, far from being chaotic in pattern and so diverse as to be incomprehensible except in terms of each individual's (or individual firm's) requirements and location, are in fact responsive in an ordered way to the geographical environment. It has been found possible to generalize about the intra-urban distribution of movements, by considering theoretically the general distributions of the origins and destinations of different types of journeys, by examining diverse empirical evidence, and by modelling the movements which occur. Moreover, we believe that it is realistic to interpret the mosaic of the movement pattern in terms of a small number of organizing principles, among which spatial factors feature strongly. The friction of distance, or distance decay, is clearly a pervasive influence on people's travel, and other spatial principles such as the use of the nearest available activity-opportunity, or the diversion of movements into concentrated flows on high-speed and high-capacity links, are also omnipresent. All these spatial features, as well as the structuring of the movement pattern through time and according to purpose, provide innumerable insights into the spatial processes and the spatial organization of urban areas. These subjects are therefore not only fitted for geographical investigation, but will stimulate and help develop geographical knowledge and understanding. We have found urban movement to be a rich and productive field of study for the urban geographer, and conclude simply by advocating its continued development.

Bibliography

ABERCROMBIE, P. (1945) *Greater London Plan*. London, HMSO.
ABU-LUGHOD, J. L. and FOLEY, M. M. (1960) Consumer differences. In Foote (1960).
ADAMS, J. S. (1974) Saturation planning. *Town and Country Planning*, 42, 550–4.
ADAMS, J. S. (1977) The breakdown of transport policy. *New Society*, 40, 548–51.
AGE CONCERN (1978) *Profiles of the Elderly: 6, Their Mobility and Use of Transport*. Mitcham, Age Concern England.
ALDRIDGE, H. R. (1915) *The Case for Town Planning*. London, National Housing and Town Planning Council.
AMBROSE, P. J. (1968) An analysis of intra-urban shopping patterns. *Town Planning Review*, 38, 327–34.
AMBROSE, P. J. (1977) Access and spatial inequality. In Open University Fundamentals of Human Geography Course Team (1977) *Values, Relevance and Policy*. Milton Keynes, Open University Press, 93–123.
AMOS, G. (1975) The probelms of the elderly in inner cities. In Brand and Cox, (1975), 50–4.
AN FORAS FORBARTHA (1972) *Dublin Transportation Study*. Dublin, AFB.
AN FORAS FORBARTHA (1973) *Dublin Transportation Study: Technical Reports*. Dublin, AFB.
ANTONIOU, J. (1968) Pedestrians in the city. *Official Architecture and Planning*, 31, 1035–9.
ANTONIOU, J. (1969) Traffic noise. *Official Architecture and Planning*, 32, 162.
APPLETON, J. H. (1962) *The Geography of Communications in Great Britain*. University of Hull Press.
ASHWORTH, R. (1971) Delays of pedestrians crossing a road. *Traffic Engineering and Control*, 13, 114–15.
ATKINS, S. T. (1977) Transportation planning: is there a road ahead? *Traffic Engineering and Control*, 18, 58–62.
BAERWALD, J. E. (1973) Traffic safety: problems and solutions. In Rose (1973).

BAGWELL, P. S. (1974) *The Transport Revolution from 1770*. London, Batsford.

BARKER, T. C. and ROBBINS, M. (1963) *A History of London Transport, Vol. 1: The Nineteenth Century*. London, Allen & Unwin.

BARKER, T. C. and ROBBINS, M. (1974a) *A History of London Transport, Vol. II: The Twentieth Century to 1970*. London, Allen & Unwin.

BARKER, T. C. and ROBBINS, M. (1974b) *An Economic History of Transport in Britain*. 3rd edn. London, Hutchinson.

BARRELL, D. W. F. and HILLS, P. J. (1972) The application of cost benefit analysis to transport investment projects in Britain. *Transportation*, 1, 29–54.

BATTY, J. M., HALL, P. and STARKIE, D. N. M. (1974) The impact of fares-free public transport upon urban land use and activity patterns. In *Symposium on Public Transport Fare Structure*. Crowthorne, Transport and Road Research Laboratory, Supplementary Report 37UC.

BAUM, H. J. (1973) Free public transport. *Journal of Transport Economics and Policy*, 7, 3–19.

BAYLISS, D. (1975) TPPs and structure plans. *The Planner*, 61, 334–5.

BAYLISS, D. (1977) Urban transport research priorities. *Transportation*, 6, 4–17.

BEAVON, K. S. O. (1972) The intra-urban continuum of shopping centres in Cape Town. *South African Geographical Journal*, 52, 58–71.

BEAVON, K. S. O. (1974) Generalising the intra-urban model based on Lösch. *South African Geographical Journal*, 56, 137–54.

BEAVON, K. S. O. (1976) The Lösch intra-urban model under conditions of changing cost functions. *South African Geographical Journal*, 58, 36–9.

BEAVON, K. S. O. (1977) *Central Place Theory: A Reinterpretation*. London, Longman.

BECHT, E. J. (1970) *A Geography of Transportation and Business Logistics*. Dubuque, Iowa, W. C. Brown.

BECKMAN, M., MCGUIRE, C. B. and WINSTEN, C. B. (1956) *Studies in the Economics of Transportation*. Yale University Press.

BEESLEY, M. E. (1965) The value of time spent in travelling: some new evidence. *Economica*, 32, 175–85.

BEESLEY, M. E. (1973) *Urban Transport: Studies in Economic Policy*. London, Butterworth.

BEESLEY, M. E. and DALVI, M. Q. (1973) Journey to work and cost benefit analysis. In Wolfe, J. N. (ed.) *Cost Benefit and Cost Effectiveness*. London, Allen & Unwin.

BEESLEY, M. E. and DALVI, M. Q. (1974) Spatial equilibrium and the journey to work. *Journal of Transport Economics and Policy*, 8, 1–26.

BEESLEY, M. E. and FOSTER, C. D. (1965) The Victoria Line: social benefit and finances. *Journal of the Royal Statistical Society*, Series A, 128, 67–88.

BEESLEY, M. E. and KAIN, J. F. (1964) Urban form, car ownership and public policy: An appraisal of *Traffic in Towns*. *Urban Studies*, 1, 174–203.

BELL, W. (1958) Social life, life styles and suburban residence. In Dobriner, W. M. (ed.) *The Suburban Community*, New York, Putnam's, 228–47.

BENDIXSON, T. (1970) Results of Reading's bus lane experiment. *Traffic Engineering and Control*, 12, 188–9.

BENDIXSON, T. and RICHARDS, M. G. (1976) Witkar: Amsterdam's self-drive hire city car. *Transportation*, 5, 63–72.

BENEPE, B. (1965). The pedestrian in the city. *Traffic Quarterly*, 19, 28–42.

BERRY, B. J. L. (1959) Ribbon developments in the urban business pattern. *Annals of the Association of American Geographers*, 49, 145–55.

BERRY, B. J. L. (1963) *Commercial Structure and Commercial Blight*. University of Chicago Department of Geography, Research Paper no. 85.

BERRY, B. J. L. (1967) *Geography of Market Centers and Retail Distribution*. Englewood Cliffs, NJ, Prentice Hall.

BERRY, B. J. L. and KASARDA, J. D. (1977) *Contemporary Urban Ecology*. New York, Macmillan.

BETT, W. H. and GILLHAM, J. C. (1957) *Great British Tramway Networks*. London, Light Railway Transport League.

BIRD, J. H. (1963) *The Major Seaports of the United Kingdom*. London, Hutchinson.

BIRD, J. H. (1968) *Seaport Gateways of Australia*. Oxford University Press.

BIRD, J. H. (1971) *Seaports and Seaport Terminals*. London, Hutchinson.

BISHOP, D. (1975) User response to a foot street. *Town Planning Review*, 46, 31–46.

BLAZE, J. R. (1972) Restructuring freight transportation in Chicago. *Transportation Engineering Journal of the American Society of Civil Engineers*, 98, 577–84.

BLAZE, J. R. and RAASCH, N. (1970) *Planning for Freight Facilities*. Chicago Area Transportation Study.

BOAL, F. W. and JOHNSON, D. B. (1965) The functions of retail and service establishments on commercial ribbons. *Canadian Geographer*, 9, 154–69.

BOAL, F. W. (1970) Technology and urban form. In Putnam, R. G., Taylor, F. H. and Kettle, P. T. (eds) (1970) *A Geography of Urban Places: Selected Readings*. London, Methuen.

BOAL, F. W. (1976) Ethnic residential segregation. In Herbert and Johnston (1976), 41–79.

BOR, W. and ROBERTS, J. (1972) Urban motorway impact. *Town Planning Review*, 43, 292–321.

BORDASS, W. (1975) Minitram in Sheffield: a consultant's view. *Traffic Engineering and Control*, 16, 172–4.

BOTTON, C. G. and GROOME, D. J. (1969) Road traffic noise – its nuisance value. *Applied Acoustics*, 21, 279–96.

LE BOULANGER, H. (1971) Research into urban travellers' behaviour. *Transportation Research*, 5, 113–25.
BOUTON, M. J. (1973) Parking policy as an instrument of parking restraint. *Institution of Highway Engineers' Journal*, 20, 7–12.
BOWLBY, S. (1977) Why people move. In Open University Fundamentals of Human Geography Course Team (1977) *Spatial Analysis: II Movement Patterns*. Milton Keynes, Open University Press, 131–81.
BRAND, J. and COX, M. (eds) (1975) *The Urban Crisis: Social Problems and Planning*. London, Royal Town Planning Institute.
BRENNAN, T. (1948) *Midland City*. London, Dobson.
BRIERLEY, J. (1962) *Parking of Motor Vehicles*. London, Applied Science Publishers.
BRIGGS, A. (1968) *Victorian Cities*. Harmondsworth, Penguin.
BRITISH MARKET RESEARCH BUREAU (1970) *Shopping in the Seventies*. London, IPC Women's Magazines.
BRITISH ROAD FEDERATION (1973) *Basic Road Statistics 1972*. London, BRF.
BROWNING, C. E. (1964) Selected aspects of land use and distance from the city centre: the case of Chicago. *Southeastern Geographer*, 4, 29–40.
BRUCE, A. (1974) Why we shop where we do. *Built Environment*, 3, 280–4.
BRUTON, M. J. (1970) *Introduction to Transportation Planning*. London, Hutchinson.
BRUTON, M. (1975) Land-use and transport planning. *The Planner*, 61, 331–3.
BRYAN, M. (1973) Noise laws don't protect the sensitive. *New Scientist*, 27 September.
BUCHANAN, C. D. (1958) *Mixed Blessing: The Motor Car in Britain*. London, Leonard Hill.
BUCHANAN, C. and PARTNERS (1970) *Canterbury Traffic Study*. London, Buchanan and Partners.
BUCHANAN, C. M., COOMBE, R. D. and HEWING, R. B. (1974) Bus priority in inner London: 1. Comparison of alternative bus priority strategies. *Traffic Engineering and Control*, 15, 480–6.
BUCKLES, P. A. (1973) Journey times by bus and car in central London in 1972. *Traffic Engineering and Control*, 14, 337–9.
BULLOCK, N. et al. (1974) Time budgets and models of urban activity patterns. *Social Trends*, 5, 45–63.
BUREAU OF PUBLIC ROADS (1954) *Manual of Procedures for Home Interview Traffic Study*. Chicago, Public Administration Service.
BURNS, W. (1967) *Traffic and Transportation in Newcastle-upon-Tyne*. Newcastle-upon-Tyne Corporation.
BUSSIERE, R. (1970) *The Spatial Distribution of Urban Populations*. Paris, Centre de Recherche d'Urbanisme.

BUTTON, K. J. (1973) Motor car ownership in the West Riding of Yorkshire: some findings. *Traffic Engineering and Control*, 15, 76–8.

BUXTON, M. J. and RHYS, D. G. (1972) The demand for car ownership: a note. *Scottish Journal of Political Economy*, 19, 175–81.

CABINET OFFICE *The Re-organization of Central Government*. London, HMSO (Cmnd 4506).

CARLSTEIN, T., PARKES, D. and THRIFT, N. (eds) (1978) *Timing Space and Spacing Time*, London, Arnold.

CARP, F. M. (1971) Walking as a means of transportation for retired people. *Gerontologist*, 11, 104–11.

CARP, F. M. (1972) Transportation and retirement. *Transportation Engineering Journal of the American Society of Civil Engineers*, 98, part TE4, 787–98.

CARRUTHERS, W. L. (1962) Service centres in Greater London. *Town Planning Review*, 33, 5–31.

CARTER, H. (ed.) (1969) *Techniques in Urban Geography*. Aberystwyth, Institute of British Geographers Urban Study Group, mimeo.

CASETTI, E. (1967) Urban population density patterns: an alternative explanation. *Canadian Geographer*, 11, 96–100.

CATANESE, A. J. (1970) Commuting behaviour patterns of families. *Traffic Quarterly*, 24, 439–58.

CATANESE, A. J. (1972) *New Perspectives in Urban Transportation Research*. Lexington, Mass., Heath.

CENTRAL STATISTICAL OFFICE, *Annual Abstract of Statistics*. London, HMSO, annual.

CENTRAL STATISTICAL OFFICE, *Highway Statistics*. London, HMSO, annual.

CENTRAL STATISTICAL OFFICE, *Social Trends*. London, HMSO, annual from 1970.

CHAPIN, F. S. and LOGAN, T. H. (1969) Patterns of time and space use. In PERLOFF (1969), 305–32.

CHAPIN, F. S. and STEWART, P. H. (1959) Population densities around the clock. In Mayer, H. M. and Kohn, C. F. (eds) (1959) *Readings in Urban Geography*. University of Chicago Press, 180–2.

CHAPPELL, C. W. and SMITH, M. T. (1971) Review of urban goods movement studies. In Highway Research Road (1971), 163–81.

CHINITZ, B. (1960) *Freight and the Metropolis*. Cambridge, Mass., Harvard University Press.

CHISHOLM, M. D. I. (1970) Forecasting the generation of freight traffic in Great Britain. In Chisholm, *et al.* (1970), 431–42.

CHISHOLM, M. D. I. (1971) Freight transport costs, industrial location and regional development. In Chisholm and Manners (1971), 213–44.

CHISHOLM, M. D. I. and MANNERS, G. (eds) (1971) *Spatial Policy Problems of the British Economy*. Cambridge University Press.

CHISHOLM, M. D. I. and O'SULLIVAN, P. (1973) *Freight Flows and Spatial Aspects of the British Economy.* Cambridge University Press.
CHISHOLM, M. D. I. and RODGERS, H. B. (eds) (1973) *Studies in Human Geography.* London, Heinemann.
CHISHOLM, M. D. I., FREY, A. and HAGGETT, P. (eds) (1970) *Regional Forecasting.* London, Butterworth.
CHOUDUHRY, A. R. (1971) Park-and-ride as a modal choice for the journey to work. *Traffic Engineering and Control,* 13, 252–5.
CHRISTALLER, W. (1966) *Central Places in Southern Germany.* Englewood Cliffs, NJ, Prentice Hall.
CHRISTIE, A. W., PRUDHOE, J. and CUNDILL, M. A. (1973a) *Urban Freight Distribution: a Study of Operations in High Street, Putney.* Crowthorne, Transport and Road Research Laboratory, Report LR556.
CHRISTIE, A. W., BARTLETT, R. S., CUNDILL, M. A. and PRUDHOE, J. (1973b) *Urban Freight Distribution: Studies of Operations in Shopping Streets at Newbury and Cambereley.* Crowthorne, Transport and Road Research Laboratory, Report LR603.
CHRISTIE, A. W., CUNDILL, M. A., EDMONSON, D. R. and MCCARTHY, S. P. (1977) The Swindon freight study: 2, Assessment of goods vehicle controls. *Traffic Engineering and Control,* 18, 252–6.
CHU, C. (1971) A review of the development and theoretical concepts of traffic assignment and their practical applications to an urban road network. *Traffic Engineering and Control,* 13, 136–41.
CITY OF COVENTRY (1972) *The Coventry Transportation Study.* Coventry, Transportation Study Group.
CITY OF LEICESTER (1964) *Leicester Traffic Plan.* Leicester, City Planning Office.
CITY OF NORWICH and NORFOLK COUNTY COUNCIL (1969) *Norwich Area Transportation Survey: Interim Report.* Norwich, City of Norwich and Norfolk County Council.
CLARK, C. (1951) Urban population densities. *Journal of the Royal Statistical Society,* Series A, 114, 490–6.
CLARK, C. (1957–8) Transport: maker and breaker of cities. *Town Planning Review,* 28, 237–50.
CLARK, C. (1977) *Population Growth and Land Use,* 2nd edn. London, Macmillan.
CLARK, D. (1976) Telecommunications and office location: some implications for regional development policies. Paper presented to the IBG Urban Study Group Conference, Keele.
COATES, B. E., JOHNSTON, R. J. and KNOX, P. L. (1976) *Geography and Inequality.* Oxford University Press.
COBURN, T. M., BEESLEY, M. E. and REYNOLDS, D. J. (1960) *The London–Birmingham Motorway.* Crowthorne, Road Research Laboratory, Technical Paper 46.

COHEN, J. E., DEARNELEY, J. and HANLEY, C. E. M. (1955) The risk taken in crossing a road. *Operational Research Quarterly*, 6, 120–8.

COHEN, S. B. and LEWIS, G. K. (1967) Form and function in the geography of retailing. *Economic Geography*, 43, 1–42.

COLLINS, M. F. and PHAROAH, T. M. (1974) *Transport Organization in a Great City: The Case of London*. London, Allen & Unwin.

COMMISSION ON THE THIRD LONDON AIRPORT (1970) *Papers and Proceedings, Stage III: Research and Investigation*. London, HMSO.

COOPER, M. (1975) *Motorways and Transport Policy in Newcastle-upon-Tyne*. Newcastle, SOLEM Committee.

COPLEY, G. and MAHER, M. J. (1973) *Pedestrian Movements: A Review*. University of Leeds Institute of Transport Studies.

COSTINETT, P. J. (1973) Modal split: theory and practice. *Proceedings of the Fourth Symposium on Public Transport*. University of Newcastle-upon-Tyne.

COSTINETT, P. J., GREGG, D. J. and MCCULLUM, D. G. (1971) North Tyne Loop Study: the alternative systems investigated. *Traffic Engineering and Control*, 13, 98–101.

COTTEN, J. E., DEARNELEY, J. and HANLEY, C. E. M. (1955) The risk taken in crossing a road. *Operational Research Quarterly*, 6, 120 8.

CREIGHTON, R. L. (1970) *Urban Transportation Planning*. Chicago, University of Illinois Press.

CREYKE, T. M. R. (1971) Car parking demand in a metropolitan town centre. *Traffic Engineering and Control*, 13, 301–4.

CULLINGWORTH, J. B. (1959–60) Overspill in South East Lancashire: the Salford-Worsley Scheme. *Town Planning Review*, 30, 189–206.

CULLINGWORTH, J. B. (1960) The social implications of overspill: the Worsley Social Survey. *Sociological Review*, 8, 77–95.

CULLINGWORTH, J. B. (1972) *Town and Country Planning in Britain*, 4th edn. London, Allen & Unwin.

CULLINGWORTH, J. B. and ORR, S. C. (eds) (1969) *Regional and Urban Studies*. London, Allen & Unwin.

CURRAN, F. B. and STEGMAIER, J. J. (eds) (1958) Travel patterns in 50 cities. *Highway Research Board Bulletin*, 203, 99–130.

DAHYA, B. (1974) The nature of Pakistani ethnicity in British cities. In Cohen, A. (ed.) (1974) *Urban Ethnicity*. London, Tavistock, 77–118.

DANIELS, P. W. (1973) Some changes in the journey to work of decentralised office workers. *Town Planning Review*, 44, 167–88.

DANIELS, P. W. (1974) New offices in the suburbs. In Johnson, J. H. (ed.) (1974), *Suburban Growth: Geographical Processes at the Edge of the Western City*. London, Wiley, 177–200.

DANIELS, P. W. (1975a) Strategic office centres in London. *Town and Country Planning*, 43, 209–14.

DANIELS, P. W. (1975b) *Office location*. London, Bell.

DANIELS, P. W. (1977) Office location in the British conurbations: trends and strategies. *Urban Studies*, 14, 261–74.

DANIELS, P. W. (1978) Office dispersal and the journey to work in Greater London: a follow-up study, in DANIELS, P. W. (ed.) (1978) *Spatial Patterns of Office Growth and Location*. London, Wiley.

DAOR, E. and HATHAWAY, P. J. (1973) *Influence of Bus and Rail Accessibility on Car Ownership*. London, GLC Research Memorandum 390.

DAVIES, C. S. and ALBAUM, M. (1972) Mobility problems of the poor in Indianapolis. *Antipode Monographs in Social Geography*, 1, 67–87.

DAVIES, R. L. (1968) Effects of consumer income differences on the business provisions of small shopping centres. *Urban Studies*, 5, 144–64.

DAVIES, R. L. (1969) Effects of consumer income differences on shopping movement behaviour. *Tijdschrift voor Economische en Sociale Geografie*, 60, 111–21.

DAVIES, R. L. (1971) The urban retailing system of Coventry. University of Newcastle-upon-Tyne Department of Geography, Seminar Paper 15.

DAVIES, R. L. (1972) Structure models of retail distribution: analogies with settlement and urban land-use theories. *Transactions, Institute of British Geographers*, 57, 59–82.

DAVIES, R. L. (1973a) *Patterns and Profiles of Urban Consumer Behaviour*. University of Newcastle-upon-Tyne Department of Geography, Research Series No. 10.

DAVIES, R. L. (1973b) The location of service activities. In Chisholm and Rodgers, 125–71.

DAVIES, R. L. (1974) Nucleated and ribbon components of the urban retail system in Britain. *Town Planning Review*, 45, 91–111.

DAVIES, R. L. (1976a) *Marketing Geography*. Corbridge, Retailing and Planning Associates.

DAVIES, R. L. (1976b) A framework for commercial planning policies. *Town Planning Review*, 47, 42–58.

DAVIES, W. K. D. and MUSSON, T. C. (1978) Spatial patterns of commuting in South Wales, 1951–71: a factor analysis definition. *Regional Studies*, 12, 353–66.

DAWS, L. F. and BRUCE, A. J. (1971) *Shopping in Watford*. Garston, Building Research Station.

DAWS, L. F. and MCCULLOCH, M. (1974) *Shopping Activity Patterns: a Travel Diary Study of Watford*. Garston, Building Research Station.

DAWSON, R. F. F. (1963) Survey of commercial traffic in London. *Traffic Engineering and Control*, 5, 246–50.

DAWSON, R. F. F. (1971) *The Current Cost of Road Accidents in Great Britain*. Crowthorne, Road Research Laboratory, Report LR396.

DAY, J. R. (1973) *The Story of the London Bus: London and its Buses from Horse Bus to the Present Day*. London Transport.

DAY, R. A. (1973) Consumer shopping behaviour in a planned urban

environment. *Tijdschrift voor Economische en Sociale Geografie*, 64, 77–85.
DEMETSKY, M. J. and HOEL, L. A. (1972) Modal demand: a user perception model. *Transportation Research*, 6, 293–308.
DENDY MARSHALL, C. F. (1963) *History of the Southern Railway*, two vols. London, Ian Allan.
DEPARTMENT OF EDUCATION AND SCIENCE (1973) *School Transport*. London, HMSO.
DEPARTMENT OF EMPLOYMENT (1975) *Family Expenditure Survey, 1974*. London, HMSO.
DEPARTMENT OF THE ENVIRONMENT (1972) *New Roads in Towns*. London, HMSO.
DEPARTMENT OF THE ENVIRONMENT (1973a) *Passenger Transport in Great Britain, 1971*. London, HMSO.
DEPARTMENT OF THE ENVIRONMENT (1973b) Local transport grants, Circular 104/73.
DEPARTMENT OF THE ENVIRONMENT (1974) *Road Accidents in Great Britain 1974*. London, HMSO.
DEPARTMENT OF THE ENVIRONMENT (1975) *National Travel Survey 1972/73: Cross-sectional Analysis of Passenger Travel in Great Britain*. London, HMSO.
DEPARTMENT OF THE ENVIRONMENT (1976a) *Transport Policy: A Consultation Document*, two volumes. London, HMSO.
DEPARTMENT OF THE ENVIRONMENT (1976b) *British Cities: Urban Population and Employment Trends, 1951–71*. London, DoE.
DEPARTMENT OF THE ENVIRONMENT (1976c) *National Travel Survey 1972/73: a Comparison of the 1965 and 1972/73 Surveys*. London, HMSO.
DEPARTMENT OF THE ENVIRONMENT (1976d) *National Travel Survey 1972/73: Number of Journeys per Week by Different Types of Households, Individuals and Vehicles*. London, HMSO.
DEPARTMENT OF THE ENVIRONMENT and STEVENAGE DEVELOPMENT CORPORATION (1974) *Stevenage Superbus Experiment: Summary Report*. Stevenage Development Corporation.
DEPARTMENT OF HOUSING AND LOCAL GOVERNMENT, PLANNING ADVISORY GROUP (1965) *The Future of Development Plans*. London, HMSO.
DEPARTMENT OF SCIENTIFIC AND INDUSTRIAL RESEARCH, ROAD RESEARCH LABORATORY (1965) *Research on Road Traffic*. London, HMSO.
DEPARTMENT OF TRANSPORT (1977) *Transport Policy*. London, HMSO (Cmnd. 6836).
DESKINS, D. R. (1972) Race, residence and workplace in Detroit, 1880–1965. *Economic Geography*, 48, 79–94.
DEWDNEY, J. C. (1960) The journey to work in County Durham. *Town Planning Review*, 31, 107–24.

DICKEN, P. and LLOYD, P. E. (1978) Inner metropolitan industrial change, enterprise structures and policy issues: case studies of Manchester and Merseyside. *Regional Studies*, 12, 181–98.

DICKINSON, G. C. (1959) Stage coach services in the West Riding of Yorkshire between 1830 and 1840. *Journal of Transport History*, 4, 1–11.

DICKINSON, G. C. (1960) The development of suburban road passenger transport in Leeds, 1840–95. *Journal of Transport History*, 4, 214–23.

DICKINSON, G. C. and LANGLEY, C. J. (1973) The coming of cheap transport: a study of tramway fares. *Transport History*, 6, 107–27.

DICKINSON, R. E. (1964) *City and Region*. London, Routledge.

DIXON, O. M. (1972) Models of nodal regions. *Institute of British Geographers, Occasional Papers*, 1, 33–48.

DOMANSKI, R. (1967) Remarks on simultaneous and anisotropic models of transportation networks. *Papers and Proceedings of the Regional Science Association*, 19, 223–8.

DONALDSON, B. (1973) An empirical investigation into the concept of sectoral bias in the mental maps, search spaces and migration patterns of intra-urban migrants. *Geografiska Annaler*, 55B, 13–33.

DONNISON, D. (1975) The age of innocence is past: some ideas about urban research and planning. *Urban Studies*, 12, 263–72.

DOUGLAS, A. A. and LEWIS, R. J. (1970–1) Trip generation techniques. *Traffic Engineering and Control*, 12, 362–65, 428–31, 477–79, 532–55.

DOUGLAS, A. A. (1973) Home-based trip end models: a comparison between category analysis and regression analysis procedures. *Transportation*, 2, 53–70.

DOWNES, J. D. and WROOT, R. (1974) *1971 Repeat Survey of Travel in the Reading Area*. Crowthorne, Transport and Road Research Laboratory, Supplementary Report 43UC.

DOWNS, R. M. (1970) The cognitive structure of an urban shopping centre. *Environment and Behaviour*, 2, 13–39.

DREWETT, R., GODDARD, J. B. and SPENCE, N. (1975) What's happening to British cities? *Town and Country Planning*, 43, 523–30.

DUPUIT, J. (1844) De la mesure de l'utilité des travaux publics. Reprinted in Munby, D. (ed.), *Transport*. London, Penguin.

DYOS, H. J. (1955–6) Railways and housing in Victorian London. *Journal of Transport History*, 2, 11–21, 90–100.

DYOS, H. J. (1961) *Victorian Suburb: a Study of the Growth of Camberwell*. Leicester University Press.

DYOS, H. J. and ALDCROFT, D. H. (1971) *British Transport*, 2nd edn. Leicester University Press.

EDWARDS, S. L. (1969) Transport costs in the wholesale trades. *Journal of Transport Economics and Policy*, 3, 272–8.

ELLIS, H. (1959) *British Railway History*. London, Allen & Unwin.

ELLMAN, P. (1968) Commuting. In Royal Commission on Local

Government in England, *Research Studies 1 : Local Government in South East England.* London, HMSO.

EVANS, A. W. (1972) On the theory of valuation and allocation of time. *Scottish Journal of Political Economy.* 19, 1–17.

FAIRHURST, M. H. (1974) *The Influence of Public Transport on Car Ownership in London.* London Transport Operational Research Report No. R203.

FALCOCCHIO, J. C. and CANTILLI, E. J. (1974) *Transportation and the Disadvantage.* Lexington, Mass., Heath.

FARRINGTON, J. H. (1973) *Morphological Studies of English Canals.* University of Hull Occasional Papers in Geography No. 20.

FARRINGTON, J. H. (1973) *Morphological Stuies of English Canals.* University of Hull Occasional Papers in Geography No. 20.

FISHWICK, F. (1972) The influence of economic factors on car ownership in Britain. *Proceedings, Urban Traffic Research Seminar.* London, Planning and Transport Research and Computation Co. Ltd.

FOOTE, N. N. *et al.* (eds) (1960) *Housing Choices and Constraints.* New York, McGraw-Hill.

FOSTER, C. D. (1963) *The Transport Problem.* London, Blackie.

FOSTER, C. D. (1964) Can we afford Buchanan? *The Statist,* 26 February.

FOSTER, C. D. (1974) Transport and the urban environment. In Rothenberg and Heggie (eds) (1974), 166–87.

FOSTER, C. D. (ed.) (1977) *A Policy for Transport.* London, Nuffield Foundation.

FOSTER, C. D. and BEESLEY, M. E. (1963) Estimating the social benefits of constructing an underground railway in London. *Journal of the Royal Statistical Society,* Series A, Part I, 126, 46–92.

FRATAR, T. J. (1954) Forecasting the distribution of inter-zonal vehicular trips by successive approximations. *Highway Research Board, Proceedings,* 33, 376–84.

FREEMAN FOX, WILBUR SMITH and ASSOCIATES (1968a) *West Midlands Transport Study.* Birmingham, Freeman Fox *et al.*

FREEMAN FOX, WILBUR SMITH and ASSOCIATES (1968b) *London Transportation Study, Phase III.* London, GLC.

FREEMAN, T. W., RODGERS, H. B. and KINVIG, R. H. (1966) *Lancashire, Cheshire and the Isle of Man.* London, Nelson.

FRESKO, D., SHUNK, G. and SPIELBERG, F. (1972) Analysis of need for goods movement forecasts. *Journal of the Urban Planning and Development Division of the American Society of Civil Engineers,* 98, 1–16.

FRUIN, J. J. (1971) *Pedestrian Planning and Design.* New York, Metropolitan Association and Environmental Planner Inc.

FRUIN, J. J. (1972) Goods movement on urban transit systems. *Transportation Engineering Journal of the American Society of Civil Engineers,* 98, 617–32.

FULLERTON, B. (1975) *The Development of British Transport Networks.* Oxford University Press.
FULLERTON, B. and BULLOCK, K. M. (1968) *Accessibility to Employment in the Northern Region.* University of Newcastle-upon-Tyne Department of Geography.
GARNER, B. J. (1966) *The Internal Structure of Retail Nucleations*, Evanston, Ill., Northwestern University Studies in Geography, No. 12.
GETIS, A. (1963) The determination of the location of retail activities with the use of a map transformation. *Economic Geography*, 39, 14–22.
GILLESPIE, A. (1977) Journey to work trends within British labour markets, 1961–71. Paper presented to the Institute of British Geographers Urban Study Group Conference, King's College, London.
GLYN-JONES, A. (1975) *Growing Older in a South Devon Town.* University of Exeter.
GOELLER, B. F. (1971) Freight transport in urban areas: issues for research and action. In Highway Research Board (1971), 149–62.
GOELLER, B. F. (1972) Bibliography on urban commodity transportation. *Council of Planning Librarians Exchange Bibliography*, No. 276.
GOLLEDGE, R. V., RIVIZZIGNO, V. L. and SPECTOR, A. (1976) Learning about a city: analysis by multidimensional scaling. In Golledge, R. V. and Rushton, G. (eds) (1976) *Spatial Choice and Spatial Behaviour.* Columbus, Ohio State University Press, 95–116.
GOODMAN, W. I. (1972) Staggered work hours. *Traffic Engineering and Control*, 14, 283–4.
GODWIN, P. B. (1973) Some data on the effects of free public transport. *Transportation Planning and Technology*, 1, 159–74.
GOODWIN, P. B. (1975) Variations in travel between individuals living in areas of different population density. Planning and Transportation Research Corporation, Summer Meeting, University of Warwick.
GOVERNMENT STATISTICAL SERVICE (1977) *Transport Statistics, Great Britain 1965–75.* London, HMSO.
GRANT, B. E. and RUSSELL, W. J. (1973) *Opportunities in Automated Urban Transport.* Crowthorne, Transport and Road Research Laboratory.
GRAYSON, G. B. (1975) Child pedestrians. *New Behaviour*, 2, 180–1.
GREATER LONDON COUNCIL (1966) *London Travel Survey: Part II.* London, GLC.
GREATER LONDON COUNCIL (1968) *London Transportation Study.* London, GLC.
GREATER LONDON COUNCIL (1969) *Movement in London.* London, GLC.
GREATER LONDON COUNCIL (1970) *London Road Plans 1900–70.* GLC Intelligence Unit Research Report II.
GREATER LONDON COUNCIL (1973) *Pedestrianized Streets.* London, GLC.
GREATER LONDON COUNCIL (1975) *Free Fares.* GLC Intelligence Unit Research Bibliography No. 33.

GREYTAK, D. (1974) Central city access and the journey to work. *Socio-Economic Planning Sciences*, 8, 57–9.
GWILLIAM, K. M. (1964) *Transport and Public Policy*. London, Allen & Unwin.
GWILLIAM, K. M. (1976) Appraising local transport policy. *Town Planning Review*, 47, 26–42.
GWILLIAMS, K. M. and MACKIE, P. J. (1975) *Economics and Transport Policy*. London, Allen & Unwin.
HAGGETT, P., CLIFF, A. D. and FREY, A. (1977) *Locational Models*. London, Macmillan.
HAGGETT, P. and CHORLEY, R. J. (1969) *Network Analysis in Geography*. London, Edward Arnold.
HALL, P. G. (1962) *The Industries of London Since 1861*. London, Hutchinson.
HALL, P. G. (1964) The development of communications. In Coppock, J. T. and Prince, H. C. (eds) (1964), *Greater London*. London, Faber.
HALL, P. G. (1971) The spatial structure of metropolitan England and Wales. In Chisholm and Manners (1971), 96–125.
HALL, P. G. et al. (1973) *The Containment of Urban England*, vol. 1, *Urban and Metropolitan Growth Processes*. London, PEP.
HALL, P. G. (1974) The re-birth of the bus. *New Society*, 29, 798.
HAMMOND, A. (1973) The evaluation of amenity. Paper presented at PTRC Conference, Summer 1972. London, Planning and Transport Research and Computation Co. Ltd.
HANSEN, F. (1972) *Consumer Choice Behaviour: A Cognitive Theory*. London, Collier-Macmillan.
HARKNESS, R. C. (1973) Telecommunications substitutes in travel: a preliminary assessment of their potential for reducing urban transportation costs by altering office location pattern. Unpublished Ph.D. Thesis, University of Washington.
HASSALL, B. B. (1974) Pedestrian traffic generated by retail stores in central London. *Traffic Engineering and Control*, 15, 566–70.
HAY, A. (1973) *Transport for the Space Economy*. London, Macmillan.
HAY, A. and SMITH, R. H. (1970) *Inter-regional Trade and Money Flows in Nigeria, 1964*. Oxford University Press.
HELLEWELL, D. S. and LEDSON, K. M. (1975) To make a T.P.P: problems of interpretation. Regional Studies Association, North West Branch Conference, Manchester, April.
HELVIG, M. (1964) *Chicago's External Truck Movements*. University of Chicago Department of Geography, Research Paper 90.
HEMMENS, G. C. (1970) Analysis and simulation of urban activity patterns. *Socio-Economic Planning Sciences*, 4, 53–66.
HENSHER, D. A. (1975) Perception and commuter mode choice – an hypothesis. *Urban Studies*, 12, 101–4.

HERBERT, D. T. (1972) *Urban Geography: a Social Perspective.* Newton Abbot, David & Charles.

HERBERT, D. T. and JOHNSTON, R. J. (eds) (1976) *Social Areas in Cities: Spatial Processes and Form.* Chichester, Wiley.

HERMANN, P. E. (1968) *Forecasts of Vehicle Ownership in Counties and County Boroughs in Great Britain.* Crowthorne, Transport and Road Research Laboratory, Report LR200.

HEWITT, W. (1928) *Workplaces and the Movement of Workers in the Merseyside Area.* London, Hodder.

HIGHWAY RESEARCH BOARD (U.S.A.) (1971) *Urban Commodity Flow.* Washington, DC, Highway Research Board, Special Report 120.

HILLE, S. J. (1971) Urban goods movement research: a proposed approach. *Traffic Quarterly*, 26, 25–38.

HILLMAN, M. (1970) Mobility in new towns. Unpublished Ph.D. thesis, University of Edinburgh.

HILLMAN, M. (1974) Travel needs of individuals. In *Symposium on Public Transport Fare Structure.* Crowthorne, Transport and Road Research Laboratory, Supplementary Report 37UC.

HILLMAN, M. (1975) Social implications of transport planning. In Lester and Miller (1975), 16–25.

HILLMAN, M., HENDERSON, I. and WHALLEY, A. (1973) *Personal Mobility and Transport Policy.* London, PEP Broadsheet No. 542.

HILLMAN, M., HENDERSON, I. and WHALLEY, A. (1976) *Transport Realities and Planning Policy.* London, PEP Broadsheet No. 567.

HITCHCOCK, A. J. M., CHRISTIE, A. W. and CUNDILL, M. A. (1974) *Urban Freight: Preliminary Results from the Swindon Freight Survey.* Crowthorne, Transport and Road Research Laboratory, Supplementary Report No. 126UC.

HOLLY, B. P. and WHEELER, J. O. (1972) Patterns of retail location and the shopping trips of low-income households. *Urban Studies*, 9, 215–20.

HOLMES, E. H. (1973) The state-of-the-art in urban transportation planning and how we got there. *Transportation*, 1, 379–410.

HOLMES, J. (1968) An analysis of patterns of journey to work in a part of the Yorkshire, Nottinghamshire and Derbyshire coalfield. Unpublished M.A. thesis, University of Sheffield.

HORTON, F. E. and REYNOLDS, D. R. (1971) Effects of urban spatial structure on individual behaviour. *Economic Geography*, 47, 36–48.

HORWOOD, E. M. (1958) Center city goods movement: an aspect of congestion. *Highway Research Board Bulletin*, 203, 40–55.

HOUSE OF COMMONS, SELECT COMMITTEE ON EXPENDITURE (1972–3 session) *Minutes of Evidence on Urban Transport Planning.* Days 1–6.

HOWSON, H. F. (1967) *London's Underground.* London Transport.

HUDDERT, K. W. and ALLEN, B. L. (1972) Bus priority in Greater London: the general picture. *Traffic Engineering and Control*, 14, 324–26, 328.

HUDSON, R. (1975) Patterns of spatial search. *Transactions Institute of British Geographers*, 65, 141–50.

HUMPHRYS, G. (1962) The importance of commuters to their areas of residence. *Journal of the Town Planning Institute*, 48, 73–6.

HUMPHRYS, G. (1965) The journey to work in industrial South Wales. *Transactions Institute of British Geographers*, No. 36, 85–96.

HURST, M. E. E. (1969) The structure of movement and household travel behaviour. *Urban Studies*, 6, 70–82.

HURST, M. E. E. (1973) Transportation and the societal framework. *Economic Geography*, 49, 163–80.

HUTCHINSON, B. G. (1974) *Principles of Urban Transport Systems Planning*. New York, McGraw-Hill.

INDEPENDENT COMMISSION ON TRANSPORT (1974) *Changing Directions*. London, Coronet.

INSTITUTE OF TRANSPORTATION AND TRAFFIC ENGINEERING (1973) *Proceedings of the Pedestrian/Bicycle Planning and Design Seminar, 1972*. San Francisco, University of California.

IRWIN, N. A., DODD, N. and VON CUBE, H. G. (1961) Capacity restraint in assignment programs. *Highway Research Board Bulletin*, 297, Washington, D.C.

ISARD, W. (1956) *Location and Space Economy*. New York, Wiley.

JOHNSTON, R. J. (1966) The distribution of an intra-metropolitan central place hierarchy. *Australian Geographical Studies*, 4, 19–33.

JOHNSTON, R. J. and RIMMER, P. J. (1967) Some recent changes in Melbourne's commercial landscape. *Erdkunde*, 21, 64–7.

JONES, E. (1960) *A Social Geography of Belfast*. Oxford University Press.

JONES, P. M. (1978) Urban transport and land use planning: a unified approach. Paper presented at the Institute of Bitish Geographers Annual Conference, University of Hull.

JONES, R. (1967) Central place theory and the hierarchy and location of shopping centres in a city: Edinburgh. In Institute of British Geographers Study Group in Urban Geography, *Aspects of Central Place Theory and the City in Developing Countries*. Durham Conference Proceedings, unpaginated.

JOYCE, F. E. and WILLIAMS, H. E. (1972) On assessing the environmental impact of urban road traffic. *International Journal of Environmental Studies*, 3, 201–7.

KAHAN, S., BERKOWITZ, C. M. and KRAFT, W. H. (1976) Apparel-goods movement process. *Transportation Engineering Journal of the American Society of Civil Engineers*, 102, 507–24.

KAIN, J. F. (1967) Urban travel behaviour. In Schnore and Fagin (1967), 161–92.
KANSKY, K. J. (1967) Travel patterns of urban residents. *Transportation Science*, 1, 261–85.
KEEBLE, D. E. (1978) Industrial decline in the inner city and conurbation. *Transactions Institute of British Geographers*, NS3, 101–14.
KELLETT, J. R. (1969) *The Impact of Railways on Victorian Cities*. London, Routledge.
KIRWAN, R. M. (1969) Economics and methodology in urban transport planning. In Cullingworth and Orr (1969), 188–212.
KISSLING, C. C. (1973) Demand responsive buses: progress and prospects. *Traffic Engineering and Control*, 14, 132–4.
KNIGHT, T. E. (174) An approach to the evaluation of changes in travel unreliability: a safety margin hypothesis. *Transportation*, 3, 393–408.
KOFOED, J. (1970) Person movement research: a discussion of concepts. *Regional Science Association Papers*, 24, 142–55.
KUTTER, E. (1973) A model for individual travel behaviour. *Urban Studies*, 10, 235–58.
LAMB, G. M. (1970a) Introduction to transportation planning: 1. Context of transportation planning. *Traffic Engineering and Control*, 11, 422–5.
LAMB, G. M. (1970b) Introduction to transportation planning: 3. Trip generation. *Traffic Engineering and Control*, 11, 554–7.
LAMBE, T. A. (1972) Parking prices in the central business district. *Socio-Economic Planning Sciences*, 6, 133–44.
LANE, R., POWELL, T. J. and SMITH, P. P. (1971) *Analytical Transport Planning*. London, Duckworth.
LANGDON, M. E. (1971) The Cabtrack urban transport system. *Traffic Engineering and Control*, 12, 634–8.
LANSING, J. B. and HENDRICKS, J. (1967) How people perceive the cost of the journey to work. *Highway Research Record*, 197.
LATCHFORD, J. C. R. and DOBSON, C. G. (1964) *Road Traffic Generated by Households and Factories*. London, International Road Federation.
LATCHFORD, J. C. R. and WILLIAMS, T. E. H. (1965–6) Prediction of traffic in industrial areas. *Traffic Engineering and Control*, 7, 498–501, 566–8, 628–30, 679–81, 735–7.
LAW, C. M. and WARNES, A. M. (1976) The changing geography of the elderly. *Transactions Institute of British Geographers*, NS 1, 453–71.
LAWSON, J. M. (1967) An evaluation of two proposals for traffic restraint in central London. *Journal of the Royal Statistical Society*, 130, 327–77.
LAWTON, R. (1959) The daily journey to work in England and Wales. *Town Planning Review*, 29, 241–57.
LAWTON, R. (1963) The journey to work in England: forty years of change.

Tijdschrift voor Economische en Sociale Geografie, 54, 61–9.
LAWTON, R. (1968) The journey to work in Britain: some trends and prospects. *Regional Studies*, 2, 27–40.
LAWTON, R. (1977) People and work. In House, J. W. (ed.) (1977) *The U.K. Space*. London, Weidenfeld.
LEAKE, G. R. and GAN, R. C. H. (1973) Traffic generation characteristics of selected industrial groups. *Traffic Engineering and Control*, 15, 349–53.
LEE, C. E. (1972) *The Metropolitan Line*. London Transport.
LEE, C. (1973) *Models in Planning*. Oxford, Pergamon.
LEE, T. R. (1962) Brennan's law of shopping behaviour. *Psychology Reports*, 11, 662.
LEE, T. R. (1970) Perceived distance as a function of direction in the city. *Environment and Behaviour*, 2, 40–51.
LEEDS CITY COUNCIL (1969) *Planning for Transport: the Leeds Approach*. London, HMSO.
LEEMING, F. A. (1959) An experimental survey of retail facilities in part of north Leeds. *Transactions Institute of British Geographers*, 26, 133–52.
LEINBACH, T. R. (1976) Transportation geography I: networks and flows. *Progress in Geography*, 8, 177–207.
LESLEY, L. J. S. (1975) Bus priority and modal split. *Traffic Engineering and Control*, 16, 127–9.
LESTER, J. and MILLER, D. (eds) (1975) *The Urban Crisis: Transport Problems of Cities*. London, Royal Town Planning Institute.
LEVIN, P. H. and BRUCE, A. J. (1968) The location of primary schools – some planning implications. *Journal of the Royal Town Planning Institute*, 54, 56–66.
LEWIS, G. (1974) Pedestrian flows in the central area of Leicester: a study in spatial behaviour. *East Midland Geographer*, 6, 79–91.
LEWIS, R. (1977) Central place analysis, in Open University Fundamentals of Human Geography Course Team, *Spatial Analysis: Point Patterns*. Milton Keynes, Open University, 48–100.
LICHFIELD, N. and ASSOCIATES (1969) *Stevenage Public Transport Cost Benefit Analysis*. Stevenage Development Corporation.
LICHFIELD, N. and CHAPMAN, H. (1968) Road proposals for a shopping centre. *Journal of Transport Economics and Policy*, 2, 280–320.
LIEPMANN, K. K. (1944) *The Journey to Work*. London, Kegan Paul.
LISCO, T. E. (1968) The value of commuters' travel time: a study in urban transportation. *Highway Research Record*, 245.
LLEWELYN-DAVIES, WEEKS, FORESTER-WALKER, BOR, OVE ARUP AND PARTNERS (1971) *Motorways in the Urban Environment*. London, British Road Federation.
LOEWENSTEIN, L. K. (1963) The location of urban land uses. *Land Economics*, 39, 407–20.

LOGAN, M. I. (1968) Work–residence relationships in the city. *Australian Geographical Studies*, 6, 151–66.
LONDON AND CAMBRIDGE ECONOMIC SERVICE (1970) *The British Economy, Key Statistics: 1900–1970*. London, Times Newspapers.
LONDON BOROUGH OF GREENWICH (1972) *The Elderly in Greenwich: Report of a Survey*. LBG.
LONDON TRANSPORT EXECUTIVE (1956) *London Travel Survey 1954*. LTE.
LÖSCH, A. (1954) *The Economics of Location*. Yale University Press.
LOVEJOY, W. B. (1972) Urban commodity flow: suggested research projects. *Journal of the Urban Planning and Development Division of the American Society of Civil Engineers*, 98, 55–61.
LOWE, J. C. and MARYADAS, S. (1975) *The Geography of Movement*. Boston, Mass., Houghton Mifflin.
MACKAY, M. (1973) Traffic accidents: a modern epidemic. In Rose (1973).
MACKIE, P. J. and URQUHART, G. B. (1974) *Through and Access Commercial Vehicle Traffic in Towns*. Crowthorne, Transport and Road Research Laboratory, Supplementary Report 117UC.
MALTBY, D. (1970) Traffic at manufacturing plants. *Traffic Engineering and Control*, 12, 72–7.
MALTBY, D. (1973) Traffic models for goods, service and business movements to manufacturing establishments. *Transportation Planning and Technology*, 2, 21–39.
MALTBY, D. (1977) The implications of oil resource depletion for urban passenger transport investment. *Traffic Engineering and Control*, 18, 206–7.
MANSFIELD, N. W. (ed.) (1971) *Papers and Proceedings of a Conference on Research into the Value of Time*. London, Department of the Environment.
MARING, G. E. (1972) Pedestrian travel characteristics. *Highway Research Record*, 406, 14–20.
MARKIN, R. J. (1974) *Consumer Behaviour: a Cognitive Orientation*. London, Collier-Macmillan.
MARKOVITZ, J. K. (1971) Transportation needs of the elderly. *Traffic Quarterly*, 25, 237–53.
MARSHALL, J. U. (1969) *The Location of Service Towns*. Department of Geography, University of Toronto, Research Publication No. 3.
MARTENSSON, S. (1977) Childhood interaction and temporal organization. *Economic Geography*, 53, 99–125.
MARTIN, J. E. (1966) *Greater London: an Industrial Geography*. London, Bell.
MASSER, F. I. (1972) *Analytical Models for Urban and Regional Planning*. Newton Abbot, David & Charles.
MASSON, M. (1970) Study of goods movement in urban areas of Aix-en-Provence and Metz-Thionville. In OECD (1970), 23–40.
MATHEW, R., JOHNSON-MARSHALL AND PARTNERS (1974) *Minitram in Sheffield*. London, Robert Mathew *et al*.

MAXFIELD, D. W. (1972) Spatial planning of school districts. *Annals Association of American Geographers*, 62, 582–90.

MAY, A. D. (1975) Supplementary licensing: an evaluation. *Traffic Engineering and Control*, 16, 162–7.

MAY, T. (1972) Parking control for restraint in Greater London. *Greater London Council Quarterly Bulletin of the Intelligence Unit*, 19, 30–42.

MAYER, H. M. (1971) Changing urban structure and its implications for terminals and pickup and delivery problems in metropolitan areas. In Highway Research Board (1971), 110–20.

MCEVOY, D. (1968) Alternative methods of ranking shopping centres: a study from the Manchester conurbation. *Tijdschrift voor Economische en Sociale Geografie*, 59, 211–17.

MELLOR, J. and T. (1970) Manchester: surveys of a city's industry. *Journal of the Town Planning Institute*, 59, 384–8.

MENSINCK, T. G. M. (1973) Mode of travel to secondary schools. *Traffic Engineering and Control*, 15, 82–5.

MERCER, J. (1971) The Runcorn plan for rapid transit. Paper presented at Symposium for Public Transport Policy Makers, University of Newcastle-upon-Tyne, March 1971.

METRA CONSULTING GROUP LTD. (1970) *Study of Generation of Goods Vehicle Movements in Selected Town Centres* (a report to the GLC). London, Metra.

MEYBURG, A. H., DIEWALD, W. J. and SMITH, G. P. (1974) Urban goods movement planning methodology. *Transportation Engineering Journal of the American Society of Civil Engineers*, 100, 791–800.

MEYBURG, A. H. and STOPHER, P. R. (1976) Urban goods movement: a study design. *Transportation Engineering Journal of the American Society of Civil Engineers*, 102, 651–64.

MEYER, V. R., KAIN, J. F. and WOHL, M. (1965) *The Urban Transportation Problem*. Harvard University Press.

MILLWARD, C., COLEMAN, A. H. and DUNFORD, J. E. (1973) Passenger transport interchanges: theory and practice on Merseyside. *Traffic Engineering and Control*, 14, 578–80.

MILNE, R. (1974) Nottingham backs the bus. *Surveyor*, 143, 12–16.

MINISTRY OF TRANSPORT (1958) *Crush Hour Travel in Central London*. London, HMSO.

MINISTRY OF TRANSPORT (1963a) *Parking: The Next Stage*. London, HMSO.

MINISTRY OF TRANSPORT (1963b) *Traffic in Towns*. London, HMSO.

MINISTRY OF TRANSPORT (1963c) *The Transport Needs of Great Britain in the Next Twenty Years*. London, HMSO.

MINISTRY OF TRANSPORT (1964) *Road Pricing: The Economic and Technical Possibilities*. London, HMSO.

MINISTRY OF TRANSPORT (1965) *Advisory Memorandum on Urban Traffic Engineering Techniques*. London, HMSO.
MINISTRY OF TRANSPORT (1966) *Transport Policy*. London, HMSO (Cmnd. 3057).
MINISTRY OF TRANSPORT (1967a) *British Waterways*. London, HMSO (Cmnd. 3401).
MINISTRY OF TRANSPORT (1967b) *Restraint on Urban Traffic*. London, HMSO.
MINISTRY OF TRANSPORT (1967c) *Cars for Cities*. London, HMSO.
MINISTRY OF TRANSPORT (1967d) *Better Use of Town Roads*. London, HMSO.
MINISTRY OF TRANSPORT (1967e) *Public Transport and Traffic*. London, HMSO.
MINISTRY OF TRANSPORT (1967f) *Railway Policy*. London, HMSO (Cmnd. 3439).
MINISTRY OF TRANSPORT (1967g) *Transport of Freight*. London, HMSO (Cmnd. 3470).
MINISTRY OF TRANSPORT (1967h) *National Travel Survey 1964*, Part I, *Household Vehicle Ownership and Use*; Part II, *Personal Travel by Public and Private Transport*. London, Ministry of Transport Statistics Division.
MINISTRY OF TRANSPORT (1968a) *Transport in London*. London, HMSO (Cmnd. 3686).
MINISTRY OF TRANSPORT (1968b) *Traffic and Transport Plans*. Road Circular 1/68. London, HMSO.
MITCHELL, C. G. B. (1973) *Pedestrian and Cycle Journeys in English Urban Areas*. Crowthorne, Road Research Laboratory, Laboratory Report 497.
MITCHELL, C. G. B. (1977) *Some Social Aspects of Public Passenger Transport*. Crowthorne, Transport and Road Research Laboratory, Supplementary Report 278.
MITCHELL, C. G. B. and TOWN, S. W. (1977) *Accessibility of Various Social Groups to Different Activities*. Crowthorne, Transport and Road Research Laboratory, Supplementary Report 258.
MOGRIDGE, M. J. H. (1967) The prediction of car ownership. *Journal of Transport Economics and Policy*, 1, 52–74.
MOGRIDGE, M. J. H. (1973) *Car Ownership Forecasting in London*. London, GLC Research Memorandum 386, 4.
MORGAN, B. S. (1976) The bases of family status segregation. *Transactions Institute of British Geographers*, NS 1, 83–107.
MORGAN, J. N. (1967) A note on the time spent on the journey to work. *Demography*, 4, 360–2.
MORGAN, R., TRAVERS AND PARTNERS (1969) *Cambridge Transportation Study: an Interim Report on Existing Travel*. Cambridgeshire and Isle of Ely CC, Cambridge CB, and University of Cambridge.

MORTON, M. (1971) Public transport in North America. *Traffic Engineering and Control*, 13, 161.

MOSER, P. J. (1972) Aesthetic and ecological disharmonies of new highways. *Transportation*, 1, 55–67.

MOSES, L. N. and WILLIAMSON, H. E. (1963) Value of time, choice of mode, and subsidy issues in urban transportation. *Journal of Political Economy*, 71, 247–64.

MOULDER, J. S. (1971) Heavy vehicles in town centres. *Quarterly Bulletin of the Intelligence Unit of the GLC*, 15, 9–14.

MULLER, P. O. (1976) Transportation geography II: social transportation. *Progress in Geography*, 8, 208–31.

MUMFORD, L. (1964) *The Highway and the City*. London, Secker.

MUNT, P. W. (1970) Strategic cordons and screen line studies: general report for 1968–9. GLC Department of Planning and Transportation, Intelligence Unit, Research Memorandum, RM244.

NADER, G. A. (1969) Socio-economic status and consumer behaviour. *Urban Studies*, 6, 235–45.

NAKANISHI, C. (1973) Physical distribution in large cities. *The Wheel Extended*, 3, 12–20.

NASH, C. A. (1976) *Public versus Private Transport*. London, Macmillan.

NATIONAL ECONOMIC DEVELOPMENT OFFICE (1970) *Urban Models in Shopping Studies*. London, HMSO.

NATIONAL ECONOMIC DEVELOPMENT OFFICE (1971) *The Future Pattern of Shopping*. London, HMSO.

NEILSON, G. K. and FOWLER, W. K. (1972) Relation between transit ridership and walking distances in a low density Florida retirement area. *Highway Research Record*, 403, 26–34.

NEWLING, B. E. (1966) Urban growth and spatial structure: mathematical models and empirical evidence. *Geographical Review*, 56, 213–25.

NIKOLAIDIS, G. C. (1975) Quantification of the comfort variable. *Transportation Research*, 9, 55–66.

NOISE ADVISORY COUNCIL (1972) *Traffic Noise: the Vehicle Regulations and their Enforcement*. London, HMSO.

NOISE ADVISORY COUNCIL (1973) *A Guide to Noise Units*. London, HMSO.

NORMAN, A. (1977) *Transport and the Elderly*. London, National Corporation for the Care of Old People.

NORTH WEST SPORTS COUNCIL (1972) *Leisure in the North West*. Salford, NWSC.

NOTESS, C. B. (1973) Life-style factors behind modal choice. *Transportation Engineering Journal of the American Society of Civil Engineers*, 99, 513–20.

NOTESS, C. B. and PAASWELL, R. E. (1972) Demand activated transportation for the elderly. *Transportation Engineering Journal of the American Society of Civil Engineers*, 98, 807–22.

O'DELL, A. C. and RICHARDS, P. S. (1971) *Railways and Geography*. London, Hutchinson.
OFFICE OF POPULATION CENSUSES AND SURVEYS (1974) *Availability of Cars*. London, HMSO.
O'FLAHERTY, C. A. and PARKINSON, M. H. (1972) Movement on a city centre footway. *Traffic Engineering and Control*, 13, 434–8.
OGDEN, K. W. (1977) Modelling urban freight generation. *Traffic Engineering and Control*, 18, 106–9, 117.
OI, W. Y. and SHULDINER, P. W. (1962) *An Analysis of Urban Travel Demands*. Chicago, Northwestern University Press.
O'MALLEY, B. and SELINGER, C. S. (1972) Staggered work hours in Manhattan. *Traffic Engineering and Control*, 13, 418–23.
OPINION RESEARCH CENTRE (1971) *Report on London Commuters*. London, ORC.
ORGANIZATION FOR ECONOMIC COOPERATION AND DEVELOPMENT, CONSULTATIVE GROUP ON TRANSPORTATION RESEARCH (1970) *The Urban Movement of Goods*. Proceedings of the Third Technology Assessment Review. Geneva, OECD.
O'SULLIVAN, P. (1970) Forecasting interregional freight flows in Great Britain. In Chisholm *et al.* (1970), 443–50.
OWEN, W. (1956) *The Metropolitan Transportation Problem*. Washington, D. C., Brookings Institution.
OXLEY, P. R. (1970) Dial-a-ride demand activated public transport. *Traffic Engineering and Control*, 12, 146–8.
PAIN, G. M. (1967) *Planning and the Shopkeeper*. London, Barrie & Rockliff.
PAINE, F. T., NASH, A. N., HILLE, S. J. and BRUNNER, G. A. (1967) *Consumer Conceived Attributes of Transportation*. University of Maryland Dept. of Business Administration.
PARKER, H. R. (1962) Suburban shopping facilities in Liverpool. *Town Planning Review*, 33, 197–223.
PARKER, J. (1967) *Transport Interchanges*. London, Report on Winston Churchill Travelling Fellowship.
PARKER, J. and HOILE, C. (1975) Central London's pedestrian streets and ways. *Greater London Intelligence Quarterly*, 33, 16–26.
PARR, J. B. (1978) Models of the central place system: a more general approach. *Urban Studies*, 15, 33–50.
PARR, J. B., DENIKE, K. G. and MULLIGAN, G. (1975) City-size models and the economic base: a recent controversy. *Journal of Regional Science*, 15, 1–8.
PARRIS, H. (1965) *Government and the Railways*. London, Routledge.
PERKIN, H. (1970) *The Age of the Railway*. Newton Abbot, David & Charles.
PERLE, E. D. (1964) *The Demand for Transportation: Regional and Commodity Studies in The United States*. University of Chicago, Department of Geography, Research Paper 95.

PERLE, E. D. (1968) Urban mobility needs of the handicapped: an exploration. In Horton, F. E. (ed.) (1968) *Geographic Studies of Urban Transportation and Network Analysis*. Evanston, Ill., Northwestern University Studies in Geography No. 16, 20–41.

PERLOFF, H. S. (ed.) (1969) *Quality of the Urban Environment*. London, Resources for the Future and Johns Hopkins.

PICK, F. (1926–7) Growth and form in modern cities. *Journal of the Institute of Transport*, 8, 163–9.

PLOWDEN, W. (1971) *The Motor Car and Politics*. London, Bodley Head.

POCOCK, D. C. D. (1968) Shopping patterns in Dundee: some observations. *Scottish Geographical Magazine*, 84, 108–16.

POST OFFICE (1972) *Transport, the Urban Environment and Telecommunications*. London, Post Office.

POTTER, S. (1978) *Modal Conflict at the Mezzo Scale*. Milton Keynes, Open University New Towns Study Unit.

PRED, A. (ed.) (1977) Planning-related Swedish geographic research. *Economic Geography*, 53, 99–221.

PREST, A. R. and TURVEY, R. (1965) Cost-benefit analysis: a survey. *Economic Journal*, 75, 683–735.

PROUDFOOT, M. J. (1937a) City retail structure. *Economic Geography*, 13, 425–8.

PROUDFOOT, M. J. (1937b) The outlying business centres of Chicago. *Journal of Land and Public Utility Economics*, 13, 57–70.

PURCELL, R. H., FIRTH, J. N., CUNDILL, M. A. and CHRISTIE, A. W. (1977) The Swindon freight study: 1 Objectives, surveys and model building. *Traffic Engineering and Control*, 18, 162–6.

QUARMBY, D. (1967) Choice of travel mode for the journey to work. *Journal of Transport Economics and Policy*, 1, 273–314.

RAE, J. B. (1968) Transportation technology and the problem of the city. *Traffic Quarterly*, 22, 299–314.

REDDING, B. G. (1970) Traffic generated by hotels in central London. *Traffic Engineering and Control*, 12, 366–8.

REDDING, B. G. (1972) Traffic generated by industrial premises in London. *Traffic Engineering and Control*, 14, 170–3.

RICHARDS, B. (1969) *New Movement in Cities*. London, Studio Vista.

RICHARDS, B. (1976) *Moving in Cities*. London, Studio Vista.

RICHARDSON, H. W. (1973) Theory of the distribution of city sizes: review and prospects *Regional Studies*, 7, 239–51.

RICHBELL, L. E. and VAN AVERBEKE, B. A. (1972) Bus priorities at traffic control signals. *Traffic Engineering and Control*, 14, 70–5.

ROBERTSON, J. J. S. (1971) Urban traffic policy for goods vehicles: notes on a visit to the Continent. *Quarterly Bulletin of the Intelligence Unit of the Greater London Council*, 17, 27–36.

ROBSON, B. T. (1975) *Urban Social Areas*. Oxford University Press.
RODGERS, H. B. (1965) *Overspill in Winsford: a Social and Economic Survey of the Winsford Town Expansion Scheme*. Winsford Urban District Council.
ROLPH, I. K. (1929) *The Location and Structure of Retail Trade*. Washington, DC, Bureau of Foreign and Domestic Commerce, Domestic Commerce Series No. 8.
ROSE, D. M. (1963) *Commercial Traffic Generation in City Centres*. University of Birmingham Department of Transportation, Research Bulletin 5.
ROSE, J. (ed.) (1973) *Wheels of Progress: Motor Transport, Pollution and the Environment*. London, Gordon & Breach.
ROSS, H. R. (1971) Future developments in personal transit. In *Rapid Transit Vehicles for City Services*. London, Institute of Mechanical Engineers, 38-53.
ROTHENBURG, J. G. and HEGGIE, I. G. (eds) (1974) *Transport and the Urban Environment*. London, Macmillan.
ROWLEY, G. and TIPPLE, G. (1974) Coloured migrants within the city: an analysis of housing and travel preferences. *Urban Studies*, 11, 81-90.
ROYAL INSTITUTE OF PUBLIC ADMINISTRATION (1973) *Modal Choice and the Value of Time*. University of Reading Local Government Operational Research Unit, Report C143.
RUSHTON, G. (1972) Map transformations of point patterns: central place patterns in areas of variable population density. *Regional Science Association Papers*, 28, 140-56.
RUSSELL, W. (1972) Architectural and environmental studies of an automated public transport system in an urban context. In Anderson, J. E. *et al.* (eds) (1972), *Personal Rapid Transit*. Minneapolis, University of Minnesota.
SAUNDERS, L. (1972) *The Characteristics and Impact of Traffic Generated by Chelsea Football Stadium*. GLC Research Memorandum 344.
SAVAGE, C. I. (1974) *An Economic History of Transport*, 3rd edn. London, Hutchinson. (First published 1959; 2nd edn 1966.)
SCHAEFFER, K. H. and SCLAR, E. (1975) *Access for All: Transport and Urban Growth*. London, Penguin.
SCHNORE, L. F. and FAGIN, H. (eds) (1967) *Urban Research and Policy Planning*. Beverly Hills, Sage.
SCOTT, P. (1970) *Geography and Retailing*. London, Hutchinson.
SEALY, K. R. (1966) *Geography of Air Transport*. London, Hutchinson.
SEALY, K. R. (1976) *Airport Strategy and Planning*. Oxford University Press.
SEKON, G. A. (1938) *Locomotion in Victorian London*. Oxford University Press.
SELF, P. (1972) The Department's transport philosophy. *Town and Country Planning*, 40, 351-4.
SEXTON, B. H. (1969) Traffic noise. *Traffic Quarterly*, 23, 427-39.
SHARP, C. (1967) *Problems of Urban Passenger Transport*. Leicester University Press.

SHARP, C. (1973) *Living with the Lorry: a Study of Goods Vehicles in the Environment*. London, Road Haulage Association.

SHERWOOD, P.T. and BOWERS, P. H. (1970) *Air Pollution from Road Traffic: a Review of the Present Position*. Crowthorne, Transport and Road Research Laboratory.

SILK, J. (1971) Search behaviour: general characteristics and review of the literature in the social sciences. University of Reading Geographical Papers, No. 7.

SIMMONS, J. (1966) The pattern of the tube railways in London, *Journal of Transport History*, 7, 234–40.

SIMMONS J. W. (1964) *The Changing Pattern of Retail Location*. University of Chicago. Department of Geography, Research Paper No. 92.

SLEEMAN, J. F. (1961) The geographical distributiion of motor cars in Great Britain. *Scottish Journal of Political Economy*, 8, 71–81.

SLEEMAN, J. F. (1969) A new look at the distribution of private cars in Britain. *Scottish Journal of Political Economy*, 16, 306–18.

SLEVIN, R. and COOPER, A. E. (1973) Minibus and dial-a-ride: initial experience with the Abingdon experiment. *Traffic Engineering and Control*, 14, 586–9.

SLEVIN, R. and OCHAJNA, S. (1974) *Dial-a-Ride in Maidstone*. Cranfield.

SMAILES, A. E. and HARTLEY, G. (1961) Shopping centres in the Greater London area. *Transactions Institute of British Geographers*, 29, 201–13.

SMEED, R. J. (1961) The traffic problem in towns. *Manchester Statistical Society*. February.

SMEED, R. J. (1968) Traffic studies and urban congestion. *Journal of Transport Economics and Policy*, 2, 33–70.

SMITH, D. M. (1977) *Human Geography: a Welfare Approach*. London, Edward Arnold.

SMITH, W. AND ASSOCIATES (1961) *Future Highways and Urban Growth*. New Haven, Conn.

SMITH, W. AND ASSOCIATES and P. E. CONSULTING GROUP LTD (1977) *Hull Freight Study: Collection of Data and Construction of Computer Model*. Crowthorne, Transport and Road Research Laboratory, Supplementary Report 315.

SMITH, W. S. (1971) State of research and data on urban goods movements and some comments on the problem. In Highway Research Board (1971), 60–74.

SMYTH, R. (1974) G.L.C. land use trip generation studies. *Greater London Council Intelligence Unit Quartery Bulletin*, No. 26.

SOCIAL AND COMMUNITY PLANNING RESEARCH (1973) *Travel Unreliability Study: a Report on the Pilot Stage*. London, SCPR.

SOUTH EAST LANCASHIRE AND NORTH EAST CHESHIRE TRANSPORTATION STUDY (1971) *Planning Data: Existing and Forecast*. Manchester, SELNEC.

SPAIN, B. (1965) *Vector Analysis*. London, Van Nostrand.
STARKIE, D. N. M. (1967a) Modal split and the value of time: a note on 'idle' time. *Journal of Transport Economics and Policy*, 1, 216–19.
STARKIE, D. N. M. (1967b) *Traffic and Industry: A Study of Traffic Generation and Spatial Interaction*. London School of Economics, Geographical Paper No. 3.
STARKIE, D. N. M. (1970) Commercial vehicles in urban transportation. *Journal of the Institute of Highway Engineers*, September, 19–25.
STARKIE, D. N. M. (1973a) Transportation planning and public policy. *Progress in Planning*, 1, 313–89.
STARKIE, D. N. M. (1973b) *Transportation and Public Policy*. Oxford, Pergamon.
STARKIE, D. N. M. and JOHNSON, D. M. (1975) *The Economic Value of Peace and Quiet*. Farnborough, Saxon House.
STEEL, M. A. (1965) Capacity restraint: a new technique. *Traffic Engineering and Control*, 7, 381–4.
STOPHER, P. R. (1968) Predicting travel mode choice for the journey to work. *Traffic Engineering and Control*, 9, 436–9.
STOPHER, P. R. (1969) A probability model of travel mode choice for the journey to work. *Highway Research Record*, 57–65.
STOUFFER, S. A. (1948) Intervening opportunities: a theory relating mobility and distance. *American Sociological Review*, 5, 845–67.
STUTZ, F. P. (1973) Distance and network effects on urban social travel fields. *Economic Geography* 49, 134–44.
STUTZ, F. P. (1976) *Social Aspects of Interaction and Transportation*. Washington, DC, Association of American Geographers.
SZALAI, A. (ed.) *The Use of Time: Daily Activities of Urban and Suburban Populations in Twelve Countries*. The Hague, Mouton.
SZUMELUK, K. (1968) Central place theory: I, a review. Centre for Environmental Studies Working Paper, No. 2.
TAAFFE, E. J., GARNER, B. J. and YEATES, M. H. (1963) *The Peripheral Journey to Work*. Evanston, Ill., Northwestern University Press.
TAAFFE, E. J. and GAUTHIER, H. L. (1973) *Geography of Transportation*. Englewood Cliffs, NJ, Prentice Hall.
TANNER, J. C. (1962) Forecasts of the future number of vehicles in Great Britain. *Roads and Road Construction*, 40, 263–74.
TANNER, J. C. (1963) Car and motorcycle ownership in the counties of Great Britain in 1960. *Journal of the Royal Statistical Society*, series A, Part II, 276–84.
TANNER, J. C. (1965) Forecasts of vehicle ownership in Great Britain. *Roads and Road Construction*, 43, 341–7, 371–6.
TAYLOR, M. A. (1965) Travel surveys by home questionnaire. *Traffic Engineering and Control*, 7, 206–12, 218.
TAYLOR, M. A. (1968) *Studies of Travel in Gloucester, Northampton and Reading*.

Crowthorne, Road Research Laboratory, Report LR141.

TAYLOR, P. J., and PARKES, D. N, (1975) A Kantian view of the city: a factorial ecology experiment in space and time, *Environment and Planning A*, 7, 671–88.

THOMAS, C. J. (1974) The effects of social class and car ownership on intra-urban shopping behaviour in Greater Swansea. *Cambria*, 1, 98–126.

THOMAS, E. N., HORTON, F. E. and DICKEY, J. W. (1966) *Further Comments on the Analysis of Non-residential Trip Generation*. Evanston, Ill., Northwestern University Transportation Centre.

THOMAS, R. (1968) *Journeys to Work*. London, PEP Broadsheet No. 504.

THOMAS, R. W. (1977) An interpretation of the journey to work using entropy maximising methods. *Environment and Planning A*, 9, 817–34.

THOMSON, J. M. (1967) Traffic restraint in central London, *Journal of the Royal Statistical Society*, Series A, 130, 327–67.

THOMON, J. M. (1969) *Motorways in London*. London, Duckworth.

THOMSON, J. M. (1973) Halfway to a motorized society. In Cullingworth, J. B. (ed.) (1973) *Problems of an Urban Society: Vol. 3, Planning for Change*. London, Allen & Unwin.

THOMSON, J. M. (1974) *Modern Transport Economics*. London, Penguin.

THOMSON, J. M. (1977) *Great Cities and Their Traffic*. London, Gollancz.

THORPE, D. (1968) The main shopping centres of Great Britain in 1961: their locational and structural characteristics. *Urban Studies*, 5, 165–206.

THORPE, D. (ed.) (1974) *Research into Retailing and Distribution*. Farnborough, Saxon House.

THORPE, D. and NADER, G. A. (1967) Customer movement and shopping centre structure: a study of the central place system in northern Durham. *Regional Studies*, 1, 173–91.

TOWNROE, P. M. (1974) *Social and Political Consequences of the Motor Car*. Newton Abbot, David & Charles.

TOWNSLEY, C. H. (1974) *Traffic Generation of Central London Offices*. GLC Research Memorandum 399.

TOYNE, P. (1971) Customer trips to retail business in Exeter. In Gregory, K. J. and Ravenhill, W. L. D. (eds) (1971) *Exeter Essays in Geography*. University of Exeter Press, 237–56.

TRAFFIC RESEARCH CORPORATION LTD (1968) Goods movement forecasting procedures. *West Yorkshire Transport Study, Working Paper*, 4. Leeds, TRC.

TRAFFIC RESEARCH CORPORATION LTD (1969a) *Merseyside Area Land Use/Transportation Study*. Liverpool, TRC.

TRAFFIC RESEARCH CORPORATION LTD (1969b) The goods vehicle survey. *Merseyside Area Land Use/Transportation Study, Technical Report*, 3, Liverpool, TRC.

TRAFFIC RESEARCH CORPORATION LTD (1969c) The development of

goods vehicle forecasting procedure. *Merseyside Area Land Use/Transportation Study, Technical Report*, 7. Liverpool, TRC.

TRANSPORT AND ROAD RESEARCH LABORATORY (1970) *A Review of Road Traffic Noise*. Crowthorne, TRRL, Report LR352.

TRANSPORT AND ROAD RESEARCH LABORATORY (1975) *Pedestrians and Shopping Centre Layout: A Review of the Current Situation*. Crowthorne, TRRL, Report LR577.

TRANSPORT AND ROAD RESEARCH LABORATORY (1977) *The Management of Urban Freight Movements*. Crowthorne, TRRL, Supplementary Report 309.

TRAVERS MORGAN, R. and PARTNERS (1968) *Belfast Transportation Plan*. London, Travers Morgan.

TRENCH, S. and SLACK, J. H. (1973) Nottingham's new transport policy. *Traffic Engineering and Control*, 15, 200–4.

TRENCH, S. and SLACK, J. H. (1974) Restraining the automobile in Nottingham. *Ekistics*, 37, 392–7.

TRENCH, S. and SLACK, J. H. (1978) The Nottingham zone-and-collar experiment. *Traffic Engineering and Control*, 19, 240–1.

TRIPP, H. A. (1938) *Road Traffic and its Control*. London, Edward Arnold.

TRIPP, H. A. (1942) *Town Planning and Road Traffic*, London, Edward Arnold.

TRUELOVE, P. (1971) Moving pavements. *Official Architecture and Planning*, 34, 763–6.

TULPULE, A. H. (1969) *Forecasts of Vehicles and Traffic in Great Britain, 1969*. Crowthorne, Transport and Road Research Laboratory, Report LR288.

TULPULE, A. H. (1971) Trends in transport of goods in Britain. In *Proceedings of the PTRC Symposium on Freight Traffic Models, Amsterdam, May 1971*. London, Planning and Transport Research and Computation Co.

TULPULE, A. H. (1973) *Forecasts of Vehicles and Traffic in Great Britain, 1972 Revision*. Crowthorne, Transport and Road Research Laboratory, Report LR546.

TYNESIDE PASSENGER TRANSPORT EXECUTIVE (1973) *A Plan for The People*. Newcastle, TPTE.

ULLMAN, E. L. (1957) *American Commodity Flow*. Seattle, University of Washington Press.

URBAN MOTORWAYS COMMITTEE (1972) *New Roads in Towns*. London, HMSO.

VICKERMAN, R. (1972) Demand for non-work travel. *Journal of Transport Economics and Policy*, 6, 176–210.

VINING, D. R. and STRAUSS, A. (1977) A demonstration that the current deconcentration of the population in the United States is a clean break with the past. *Environment and Planning A*, 9, 751–8.

WABE, J. S. (1966) An econometric analysis of the factors affecting the journey to work in the London Metropolitan Region and their significance. Unpublished D.Phil. thesis, University of Oxford.

WABE, J. S. (1967) Dispersal of employment and the journey to work. *Journal of Transport Economics and Policy*, 1, 345–61.

WACHER, T. R. (1970) The effects of rapid transit on urban property development. *Chartered Surveyor*, 102, 420–8.

WACHS, M. (1977) Transportation policy in the eighties. *Transportation*, 6, 103–19.

WAGNER, F. P. (1973) The impact of the shorter working week on transportation. *Traffic Engineering*, 43, 12–13.

WALKER, G. (1947) *Road and Rail*, 2nd edn. London, Allen & Unwin.

WALTERS, A. A. (1961) The theory and measurement of private and social cost of highway congestion. *Econometrica*, 19, 676–99.

WALTERS, A. A. (1975) *Noise and Prices*. Oxford, Clarendon Press.

WARD, S. and ROBERTSON, T. S. (eds) (1973) *Consumer Behaviour: Theoretical Sources*. Englewood Cliffs, NJ, Prentice Hall.

WARNER, B. T. (1973) Transport planning for management in the 1970s: moving goods in urban areas. *Chartered Institute of Transport Journal*, 35, 231–6.

WARNER, S. L. (1962) *Stochastic Choice of Mode in Urban Travel: a Study in Binary Choice*. Chicago, Northwestern University Press.

WARNES, A. M. (1969) Changing journey to work patterns: an indicator of metropolitan change. In Carter (1969), unpaginated.

WARNES, A. M. (1970) Early separation of homes from workplaces and the urban structure of Chorley, 1780–1850. *Transactions of the Historic Society of Lancashire and Cheshire*, 122, 105–34.

WARNES, A M. (1972) Estimates of journey-to-work distances from census statistics. *Regional Studies*, 6, 315–26.

WARNES, A. M. (1975) Commuting towards city centres: a study of population and employment density gradients in Liverpool and Manchester. *Transactions Institute British Geographers*, 64, 77–96.

WARNES, A. M. (1977) The decentralization of employment from the larger English cities. London, Department of Geography, King's College, Occasional Paper No. 5.

WARNES, A. M. and DANIELS, P. W. (1978) Intra-urban shopping travel: a review of theory and evidence from British towns. Paper presented to the Institute of British Geographers' Annual Conference, University of Hull.

WARNES, A. M. and DANIELS, P. W. (1980) Spatial aspects of an intra-metropolitan central place hierarchy. *Progress in Human Geography*, 384–406.

WATERS, M. H. L. and BRIGGS, M. (1975) *A Review of Market Prospects for Battery Electric Road Vehicles: Parts I and II*. Crowthorne, Transport and Road Research Laboratory, Reports LR630 and LR631.

WATSON, P. L. (1972) *An Annotated Bibliography on Urban Goods Movement*.

Evanston, Ill., Northwestern University Transportation Center.
WATSON, P. L. (1974) *Behavioural Models of Modal Choice.* Lexington, Mass., Heath.
WEBB, B. and WEBB, S. (1920) *English Local Government: the Story of the King's Highway.* London, Longmans.
WEBBER, M. M. (1968/9) Planning in an environment of change. *Town Planning Review,* 39, 179–95, 277–95.
WEBSTER, F. V. (1972) *Priority to Buses as Part of Traffic Management.* Crowthorne, Transport and Road Research Laboratory, Report LR448.
WEBSTER, F. V. (1977) *Urban Passenger Transport: Some Trends and Problems.* Crowthorne, Transport and Road Research Laboratory, Report 771.
WEEKLY, I. G. (1956) Service centres in Nottingham, a concept in urban analysis. *East Midland Geographer,* 6, 41–6.
WESTERGAARD, J. (1957) Journeys to work in the London region. *Town Planning Review,* 28, 37–62.
WHEELER, J. O. (1968) Work-trip length and the ghetto. *Land Economics,* 44, 107–12.
WHEELER, J. O. (1972) Trip purposes and urban activity linkages. *Annals of the Association of American Geographers,* 62, 641–54.
WHEELER, J. O. (1970) The structure of metropolitan work trips. *Professional Geographer,* 23, 152–8.
WHEELER, J. O. (ed.) (1973) Transportation geography: Societal and policy perspectives. Special issue of *Economic Geography,* 49, pp. 95–184.
WHEELER, J. O. (1974) *The Urban Circulation Noose.* Belmont, Cal., Duxbury.
WHEELER, J. O. and STUTZ, F. P. (1971) Spatial dimensions of urban social travel. *Annals of the Association of American Geographers,* 61, 371–86.
WHITE, H. P. (1962) *A Regional History of the Railways of Great Britain, Vol. II, Southern England,* London, Phoenix.
WHITE, H. P. (1973) *Greater London (A Regional History of the Railways of Great Britain, Vol. III)* Newton Abbot, David & Charles.
WHITE, P. R. (1974) Use of public transport in towns and cities of Great Britain and Ireland. *Journal of Transport Economics and Policy,* 8, 26–39.
WILLEY, W. E. (1975) Transportation planning and the energy crisis. *Traffic Quarterly,* 29, 273–84.
WILLIAMS, P. M. (1969) Low fares and the urban transport problem. *Urban Studies,* 6, 83–92.
WILLIAMS, T. E. H. and LATCHFORD, J. C. R. (1966) *Prediction of Traffic in Industrial Areas.* London, Printerhall.
WILLMOTT, P. (1973) Car Ownership in the London Metropolitan Region *GLC Intelligence Unit Quarterly Bulletin,* No. 23
WILSON, A. G. (1974) *Urban and Regional Models in Geography and planning.* London, Wiley.
WILSON, F. R. (1967) *Journey to Work: Modal Split.* London, MacLaren.

WILSON, G. (1971) *London United Tramways: a History 1894–1937*. London, Allen & Unwin.

WINSON, C. B. (1967) Regression analysis versus category analysis. In *Proceedings of Seminar on Trip End Estimation*. London, Planning and Transport Research and Computation Co. Ltd.

WISEMAN, R. F. (1975) Location in the city as a factor in trip-making patterns. *Tijdschrift voor Economische en Sociale Geographie*, 66, 167–77.

WOLKOWITSCH, M. (1973) *Géographie des transports*. Paris, Armand Colin.

WOOD, R. T. (1967) Tri-State Transportation Commission's freight study programme. *Highway Research Record*, 165, 89–95.

WOOD, R. T. (1970) Measuring urban freight in the Tri-State Region. In OECD (1970), 61–82.

WOOD, R. T. (1971) Structure and economics of intra-urban goods movement. In Highway Research Board (1971), 22–33.

WOOTTON, H. J. and PICK, G. W. (1967) A model for trips generated by households. *Journal of Transport Economics and Policy*, 1, 137–53.

WROE, D. C. L. (1973) The elderly. *Social Trends*, 4, 23–34.

Index

Abingdon, 284
Accessibility, 75, 187, 189, 241, 327
  to employment, 117, 141, 209
  of facilities, 71
  importance, 36
  of opportunities, 129
  to shops, 33, 165
Accidents, 14, 50, 308, 309, 343
Adolescents, personal movement trends, 38, 52
Affluence, 4, 36–42, 157–71
Airtrans, 334
Aix-en-Provence, France, 87–8
Alweg, 331
Amsterdam, Netherlands, 339
Area licensing, 290
Atmospheric pollution (see Pollution, atmospheric)
Auto-taxis, 329

Barrows, 79
Battery-powered vehicles, 339
Behavioural models and studies, 204–13
Belfast, N. Ireland, 213
Bicycle, 12, 55, 340, 343 'see also Journey modes)
Birkenhead, 5, 7, 10, 83–4, 108 (see also Liverpool)
Birmingham, 4–6, 11, 72–4, 88, 103–4, 138, 140, 271
Black Country (see West Midlands)
Board of Trade, 232
Bournemouth, 13
Bradford, 6
Brighton, 13, 159, 166, 171
Bristol, 140, 294
British Electric Traction Company, 11
British Railways, 13, 179, 180, 231, 236, 258, 333

British Transport Commission, 235, 236
Bulk commodity flows (see Freight movements)
Buses, 10–14, 27–8, 71, 82, 177–81, 193, 209, 211, 234, 266, 271–6
  bus companies, 10–14, 249
  bus lanes, 276–9
  bus priority schemes, 203, 269, 273, 276–9
  bus-rail interchanges, 283, 334
  busways, 279–81
  express, 273–4
  guided, 334

Cabs (see Taxis)
Cabtrack, 332–4
Calibration, in models, 219, 224
Cambridge, 28, 32, 59
Canal network, 4–5
Captive transport users, 197, 223–69
Cardiff, Wales, 154
Cars, 13–15, 50, 52, 53, 73, 155, 176–9, 238, 307
  access to, 37, 52, 223, 268, 342
  car ownership, 36, 38–42, 60–1, 68–9, 144–5, 164, 235
  car parking (see Parking)
  car sharing or car pools, 290
  efficiency of, 339
  future of, 225–8, 338–41
Category analysis, 186, 189, 190
Cedar Rapids, Ohio, 119
Central business district (CBD), 15, 110, 119, 139, 141, 143, 147, 206, 266, 335, 346, 353
Charing Cross (see London, Charing Cross)
Cheap fares (see Fares)
Chicago, Illinois, 79, 110–11, 138, 145–6
  Chicago Area Transportation Study, 77, 78, 146

388 MOVEMENT IN CITIES

Children, 34, 36, 48–52, 132–3, 270, 343
  (*see also* Journey distances; Journey purposes; Journey rates)
Cities:
  commodity movements within (*see* Freight movements)
  development of, 1, 9, 11, 147, 175–81, 322
  foot, 2–3
  functions, 2, 4, 307–8
  future of, 338–41, 344–8
  inner, 33, 35, 73, 119, 122, 131, 144, 342–3, 348
  motorways, 307, 315–21
  movement in (*see* Journey . . .)
  organization or structure of, 2, 4, 9, 51
  and population, 18, 132, 143
  size of, 2–4, 8, 14–15, 123, 125, 171
  synchronization of activities, 55, 65–6, 298, 300
  transport policy for, 229–62
  (*see also* Central business district, CBD)
City of London, 5, 9, 10
  Bank, 5
  Barbican, 324–5
  (*see also* London)
Citycar, 338–9
Cleveland, Ohio, 81
Coaches, 12, 82, 177, 178, 179
Commerce, 15
  commercial ribbons, 14–15, 250
  location of (*see* Shopping centres; Shops)
Commodity movements (*see* Freight movements)
Commuters, 10, 13, 125, 136–56
Commuting (*see* Journey distances; Journey frequencies; Journey purposes)
Concessionary fares (*see* Fares)
Congestion, 2, 15, 32, 117, 203, 205, 241, 242, 263–8
  in central London, 9–10, 230, 299
  congestion pricing, 286–301, 315
  of housing, 13
Contra-flow lanes, 276
Conurbations, 59, 71, 84, 87, 89, 97, 103, 104, 124, 145, 148–9, 171, 208, 235, 236
Cost benefit analysis, 66, 80, 184, 246, 274, 275, 291, 302–5, 311, 321
Costs, 86, 207–12, 221, 287
  energy, 176, 351
  operating, 53, 205, 265

  of parking, 199, 208, 221, 271
  of public transport, 37, 201, 243, 266, 284
  of transport studies, 351–2
  of urban movement, 122, 299, 342
Coventry, 166, 167, 169, 171, 180, 297, 298, 323
Crawley, 171
Cumbria, 306
Cut-and-cover construction, 9–10

Daily movement patterns (*see* Journey timing; Time budgets)
Dalls, Texas, 334
Data (*see* Movement data)
Decatur, Illinois, 282
Decentralization, 14, 42, 149, 348
  of employment, 140, 147
  of industry and commerce, 15, 148, 329
  of offices, 155–6, 353
  of population, 39, 40, 147
  of retailing, 175, 347
Density:
  of employment, 138, 142–3, 145
  of population, 128–30, 136, 138, 143, 181, 307
  of urban development, 2, 152
Department of the Environment, 61, 150, 176, 179, 203, 236, 250, 253, 254, 257
Department of Transport, 261
Department stores, 89, 100, 194
Detroit, Michigan, 118
Dial-a-bus, 282
Dial-a-ride, 281–4
Distance minimization, 118–20 (*see also* Journey distance)
Distribution:
  of employment, 136–41, 154
  of population, 132
Dublin, Ireland, 276
Duisburg, Germany, 270
Dundee, 167
Dunlop 'Speedway', 'Starglide', 335
Duorail, 331–2

Ealing, 9
Edinburgh, Scotland, 279
Education (*see* Journey distances; Journey frequencies, Journey purposes)
Eighteenth century transport, and the growth of cities, 1

INDEX

Elderly people 46–9, 51–3, 59, 60, 122, 223, 245, 343
Electric car, 339–40
Electric trams, 7, 11
Electrification, 9–13, 230, 234
Employment, 136–56
   distribution, 136–41, 154
   part-time, 121, 161
   relationship with freight-trip generation, 191–6
   trends, 147–56
   (see also Journey distances; Journey purposes; Journey rates)
Energy:
   costs, 176
   resources, 339, 348, 351
Environmental issues, 225, 229, 243, 306, 308–27, 338, 348
   environmental areas, 327–9
   environmental evaluation, 321
Ethnic segregation (see Segregation, ethnic)
Evaluation procedures, 222–5, 321
Evolution, of transport, 1–11, 147–8, 175, 229, 230, 232, 237, 307, 338–41
Exmouth, Devon, 51
Express bus services (see Buses, express)

Familism, 34, 35
Family (see Life cycle)
Family status segregation (see Segregation, family status)
Fares, 6–8, 181, 269–72, 293
   concessionary, 52
Feeder services, to railways, 5
Ferry services, 5, 218
Flexitime, 55, 65–6, 298, 300, 326
Foot cities, 2–3
Foot traffic (see Journey modes; Pedestrians; Walking)
Ford A.C.T., 334
Fratar technique, 210, 214, 215
Freight movements, 4, 12, 20, 77, 103–8, 192
   classification of, 84–95
   cost benefit comparisons with passenger travel, 80
   extra-urban factors, 96
   intra-urban, 17, 78
   questions concerning, 110–13
   timing, 55, 60, 65, 99–102
   trip modelling, 109–12

Future, of the city, 338–41, 344–8

Germany, West, 42–7, 178, 270, 332
Gillingham, 13
Glasgow, Scotland, 5, 7, 10, 13, 72, 181, 235, 242
Gloucester, 28
Goods movement (see Freight movements)
Goods vehicles, 176–8 (see also Lorries; Vans)
Government:
   intervention, 229, 242, 257–60
   local, 13, 236, 241, 251, 255–7
   regulation, 229–62
   responsibilities, 252–4
   transport policy, 223, 229–62
Gravity model, 213, 216–18, 271
Great Eastern Railway, 13
Great Northern Railway, 9
Great Western Railway, 9
Greater London Council (GLC) (see London, Greater London Council)
Growth factor model, 213–15
Guided buses, 334

Hampstead Garden Suburb, 282
Hanover, Germany, 271
Harlow, 213, 282
Harrow, 9, 155
Highway city, 3
Holidays, 63 (see also Journey purposes)
Hornsey, 9
Horse-drawn vehicles, 4, 12
   omnibus, 5–6, 9, 232
   trams, 6–8
Household, 26, 34–7, 119, 189–90
   car ownership, 36, 223
   declining size of, 40
Housewives, 44–7, 52, 59, 159, 161–2, 164, 245 (see also Activity patterns)
Housing, 13, 33–5
Housing and Transport Ministries, 28
Hull, 7, 242

Industrial revolution, 1
Industrial trip generation (see Freight movement)
Inner city (see Cities, inner)
Intervening opportunities model, 213, 218–20
Italy, 178, 276, 293
Ithaca, New York, 204

Journey destinations (*see* Journey purposes or destinations)
Journey distances, 2, 5, 12, 13, 42, 61, 66, 69, 70, 74, 81, 175
  commercial, 97–9
  educational, 126
  longer distance journeys, 116, 121, 125, 132–5, 154
  shopping, 157, 162, 165–8, 171, 345
  short distance journeys, 59, 90, 97, 109, 116, 120, 126–7, 130, 145
  work, 35, 123, 126, 137, 149, 152
  (*see also* Time)
Journey frequencies (*see* Journey (or trip) rates)
Journey locations, 123–6, 149–53, 164
  central area, 130
  inner city, 35, 129–31
  neighborhood, 125–32, 157
  suburban, periphery, 14–15, 125, 133–4
Journey modes, 37, 42, 54, 69, 71–6, 82, 117, 122, 175–81, 195–213
  bus, 61, 71, 74, 82, 124, 133, 155, 234, 266, 271–6
  car, 13–15, 60, 69–72, 74, 124, 133, 144, 164, 235, 307, 338
  cycle, 12, 55, 61, 72, 75, 76, 82, 132, 176, 340, 343
  taxi, 81, 82, 123, 179, 182
  train, 8–13, 61, 71, 75, 124, 133, 155, 230, 236
  vans and lorries, 14, 80–2, 177
  walking, 2, 5, 55, 59, 61, 69, 74–6, 124–8, 223–4, 322, 343
Journey origins, 30–31, 131, 187
  home-based, 32, 47, 134
  non-home based, 30, 51, 57, 58
Journey purposes or destinations, 1, 5, 14, 17–33, 37–41, 53, 57, 75, 124, 172–3
  change mode, 30, 31
  daily variation, 62
  education, schools, 20, 28, 29, 31, 32, 46–8, 56, 132–3
  entertainment, leisure, 20, 28, 29, 46–8
  freight movements, 103–8, 111, 195
  higher education, 45
  home, return to, 27–9, 31, 48, 56
  medical facilities, 20, 51
  multi-purpose, 22, 94
  offices, 155–6
  parks, 48–50, 310
  personal business, 28–30, 47, 51, 57
  shops, retail land uses, 20, 28, 29, 31, 47, 48, 51, 56, 157–74
  social, 26, 31–2, 126
  work, 2, 8, 15, 20, 28–31, 35, 51, 56, 73, 136–56, 212
  worship, religion, 21
  (*see also* Freight movements)
Journey (or trip) rates, 14, 33, 42, 53, 54, 116, 127, 129
  and car ownership, 66, 68
  commercial, 87–91, 93, 97
  educational, 69
  and household size, 66
  and journey distance, 69
  shopping, 68, 69, 157, 161–3, 166–9
  social, 26, 30–2
  vans and lorries, 81
  work, 26–7, 33
Journey timing or temporal distribution, 47, 54–66, 115–17, 124, 162
  diurnal patterns, 32, 46, 55–9, 100–1, 158
  of freight movements, 55, 60, 99–102
  seasonal patterns, 32, 63–4, 99, 116
  weekly patterns, 60–3, 133–4, 164, 167–8
  (*see also* Life cycle)

Lake District, 65
Land use distributions and patterns, 112, 139
  change in, 301–2
Land use-transportation studies and plans, 28, 110, 176, 223, 243, 251, 257, 261
Least effort principle, in routeing, 118–21, 129
Leeds, 7, 159, 276, 298
Leicester, 15, 213
Licensing, of vehicles, 68–9, 223, 342
  supplementary, 224, 289–92
Life cycle, 42–7, 116, 122–3, 128–9, 164
  characteristics and trends of stages in, 38–41
  defined, 36–7
  effects on transport, 34, 36, 64–6
  school children and the elderly, 34, 36, 46–53, 59, 60, 122
Life styles, 35
Linear regression models, 85, 87, 109, 111, 186, 188–9
Liverpool, 5, 10, 13, 72, 83, 84, 102, 106–8, 148, 152, 325
  Edge Hill Station, 9
  Loop and Link, 336–7

Speke, 140, 278–9
(*see also* Merseyside)
Local government (*see* Government, local)
Local Government Acts, 229, 238, 249, 255
London, 5, 7–8, 83, 84, 96, 100, 126, 131, 142–3, 148, 230–6
  Blackheath, 9
  Charing Cross, 102, 138
  Circle Line, 9, 10
  The City, 5, 9, 10, 324–5
  City and South London Railway, 10
  Cockfosters, 13
  Edgware, 13
  Euston Station, 13
  Golders Green, 12
  Greater London Council (GLC), 224, 236, 237
  Greenwich, 9, 51
  Hammersmith, 82
  Holborn, 194
  Hounslow, 13, 138
  King's Cross Station, 13
  Lewisham, 9
  Liverpool Street Station, 13, 335
  London and Birmingham Railway, 9
  London and Greenwich Railway, 9
  London Bridge Station, 9, 335
  London County Council (LCC), 231–4
  London General Omnibus Company, 6
  London Passenger Transport Act, 1933, 234
  London Passenger Transport Board, 234
  London Traffic Act, 1924, 233
  London Transport, 179, 180, 233, 236
  London Transport Board, 236
  London Transport Executive (LTE), 26, 235, 236
  Northern Line, 13
  Oxford Street, 194, 277
  Paddington, 5
  Piccadilly Line, 13
  Putney, 101
  Royal Commission on London Traffic, 232
  Strand, 194
  Vauxhall Bridge Road, 276
  Victoria Line, 224, 299, 303, 304, 331
  Victoria Station, 194, 335
  Waterloo and City Line, 10
  Wembley, 82
  West End, 5, 9, 10
  Westminster, 138, 334
  Woolwich, 9

Lorries, 14, 80–1 (*see also* Freight movements)
Los Angeles, California, 138
Lyons, France, 332

Manchester, 72, 142–6, 148, 152, 153, 194, 251, 276
  Liverpool Street Station, 9
  Market Place, 8
  Market Street, 8
  Oldham Street Station, 9
  Piccadilly/Victoria Line, 299, 237
  Princess Road, 278
Marseilles, France, 276
Mass transit, 244, 330–1
Medway Towns, 85, 86, 193
Melbourne, Australia, 109, 131
Mersey, 102, 108, 337
Merseyside, 83–4, 87, 89–93, 97, 99, 104, 108, 213
  Passenger Transport Executive, 284, 337
  Transportation Study, 337
  (*see also* Liverpool)
Metropolitan Railway, 9, 234
Migration, 1
Milan, Italy, 276
Milk deliveries, 89–90
Miners, journey to work, 154
Ministry of Housing and Local Government, 252
Ministry of Transport, 60, 179, 205, 223, 224, 233, 236, 239–41, 244, 255
Minitram, 332–4
Mobility, 51–3, 131, 223, 230, 254
  as a commodity, 307
  difficulties and handicaps, 342–3
  personal, 123–7, 342
  residential, 35
Modal choice, or split, 61, 71, 75, 76, 117, 182, 196–213, 268, 298
  analysis, 211–13, 268, 298
  binary choice, 198, 206–7, 211–13
  factors in, 199
  theory in, 196–204
  time factor in, 205–13, 217, 220
  (*see also* Journey modes)
Modelling (*see* Trip modelling)
Monorail, 331–2
Mopeds, 176
Motor bus; Motor car; Motor van (*see* Buses, Cars, Vans, respectively)
Motorway city, 15
Motorways, 15, 153, 236, 315–21 (*see also* Roads)

Movement data, 25–6, 112, 176, 189, 204
  (*see also* Surveys; Trip modelling)
Movement patterns (*see* Journey timing or temporal distribution; Synchronization, of urban activities)
Movement substitution, 344–5
Movements (*see* Journey . . .)
Moving pavements, 335–7
Municipalization, 7

Nashville, Tennessee, 81
National Bus Company, 249, 259
National Travel Survey, 60–1, 66–7, 69, 71, 74, 75
Neighborhood, journeys within (*see* Journey location)
Netherlands, 178, 339
New York, New York, 51, 77, 231
Newbury, 83, 101
Newcastle-upon-Tyne, 10, 13, 27–8, 276, 294, 334 (*see also* Tyneside)
Nineteenth century:
  and decentralization trends, 15, 147–8
  as golden era of transport, 1, 175, 237, 307
  urban transport during, 2–11, 229, 230, 232
Noise, 310–15
  control, 312, 314, 315
  measurement, 313
  origin, 312–13
Noise Advisory Council, 314
Non-home based trips (*see* Journey origins)
Non-residential trip generation, 191–6
North of England, 7
Northampton, 28, 29, 31, 32, 55, 56, 58, 59, 138–9
Norwich, 28, 29, 31, 32, 57–9, 89, 91, 102, 153
Nottingham, 281
  Victoria Centre, 279
  zone and collar scheme, 285–6

Offices:
  floorspace, 217, 290
  workers, 155–6, 289–91
  (*see also* Central business district)
Omnibus, 6, 232
Operating costs (*see* Costs, operating)
Opportunity costs, 205, 287, 307
Origin-destination surveys, 30–2, 136
Outer Metropolitan Area, 159, 171, 180

Paris, France, 138, 276
  La Defense, 335
Parking, 199, 208, 221, 271, 315
  controls on, 224, 290, 293–8
  integrated, 296–8
  meters, 294
  park-and-ride, 284–6
  spaces for, 15
  strategies for, 290
Parks, 48–50, 310
Part-time work, 121, 161
Passenger Transport Authorities and Executives, 246, 249, 269, 337
Passenger transport statistics, 175–81
Passengers, 13, 92, 330
  on buses, 181
  versus freight movements, 80
Peak-loads, 9, 32, 54–9, 65, 77, 81, 99–100, 123, 132, 203, 264, 271–3, 299
Pedestrians, 322–7
  access, 318, 320, 323
  accidents, 14, 50, 308, 309
  conflict with vehicles, 14, 308, 319, 324
  density, 132, 323, 326
  in shopping centres, 322
  (*see also* Foot cities; Journey modes; Walking)
Pensioners (*see* Elderly people)
Perception gap, 200
Personal mobility, 51–3, 123–7
  constraints on, 50, 307
Personal rapid transit, 331–2
Petrol delivery, 100
Philadelphia, Pennsylvania, 100, 118, 199, 201
Pipelines, 79–80
Planning, 181–6, 213, 223–5, 229, 235, 237, 349, 352
  blight, 14–15
  gain, 326
  and urban freight movements, 110–13
  (*see also* Cities, organization or structure of; Town and Country Planning)
Planning Advisory Group, 252
Policy makers, 223, 311, 351
Pollution, atmospheric, 309–11, 348
Population density, 128–130, 136, 138, 143, 181, 307
Population distribution, 18, 132
Postal services, 79
Prediction, of travel patterns, 181–2 (*see also* Cities, future of; Planning; Trip modelling)

INDEX

Pricing (*see* Roads, pricing)
Public transport, 4, 5, 11, 42, 60, 71, 155, 175, 192, 196
  decline of, 51, 52, 179–80, 223–4
  economics of, 37, 122, 199, 237, 269–72
  in inner-city areas, 33, 73
  and land-use distribution, 301–5
  peak hour capacity, 65, 265–8
  trends in, 38–41, 268–86
  (*see also* Journey modes)

Racial segregation (*see* Segregation, ethnic)
Railways, 12–13, 87, 133, 230, 236
  companies, 232
  construction, 13
  effect on city size, 5, 6, 147
  intra-urban, 8–11
  steam, 9, 10, 13
  underground, 9, 10, 13, 230, 231, 234
  underground freight, 79
  (*see also* Journey modes)
Rapid transit, 331, 332, 335, 338 (*see also* Underground railways)
Reading, 28, 47, 59, 76, 138–9, 276
Recreational journeys, 63 (*see also* Journey purposes)
Refuse collection, 90, 93, 95, 97
Research:
  on commuting patterns, 150
  on urban transport, 33–4, 109, 111–13
  (*see also* movement data; Surveys; Trip modelling)
Residences, 33–5
  choice, 117–18
  (*see also* Households; Segregation, residential)
Retail delivery (*see* Freight movements)
Retail functions (*see* Journey distances; Journey purposes; Shopping centres; Shops)
Retired population (*see* Elderly people)
Ribbon development, 14–15, 250
Road Research Laboratory (*see* Transport and Road Research Laboratory)
Roads, 236, 241
  construction of, 235, 237, 240, 241, 316
  pricing, 286–301, 315
  (*see also* Motorways)
Rome, Italy, 293
Royston, Hertfordshire, 13
Rubber city, 3
Rugby, 13
Runcorn, 280–1

Rush-hour, 54–9, 65, 77, 81, 99, 123, 132, 203, 231, 264, 271–3, 299

Safege, 331
Safety, 14, 308
  promotion of, 309
  (*see also* Accidents)
Salt Lake City, Utah, 81
San Diego, California, 81
School children (*see* Children)
School trips (*see* Journey purposes)
School transport, free, subsidized, 50–1
Scooters, 176
Scotland, 5, 7, 10, 13, 72, 181, 235, 253, 242, 279
Seasonal patterns, of movement (*see* Journey timing or temporal distribution)
Sectoral patterns, in cities, 34, 131
  of movement, 147
Segregation:
  ethnic, 35, 118
  family status, 34
  functional, 5, 123
  residential, 33, 34, 117–18
  vehicle/pedestrian, 309, 323–4
Sevenoaks, 13
Severance, of pedestrian movement, 311, 320
Sheffield, 6, 15, 192, 276, 333
Shirley Highway, Washington, D.C., 273–4
Shopping centres, 135, 159–60, 165, 170
Shops, distribution of, 2, 33–6, 158–67, 171
  (*see also* Journey purposes)
Singapore, 290
Smog, 310
Social geography, of travel, 354
Socio-economic status, 35, 75, 129, 131, 157–71, 199, 221, 230, 307, 342
  indices of, 188, 190
Solihull, 74, 104, 105, 140
South East Lancashire North East Cheshire conurbation (SELNEC), 217, 221, 249
  (*see also* Manchester)
Stage-coach services, 1, 4, 5
Staggered working hours (*see* Flexitime)
Statistics (*see* Research; Surveys)
Steam carriages, 6
Steam trains, 9, 10, 13
Steam trams, 7
Stetchford, 105
Stevenage, 48, 126
  superbus experiment, 274–6
Streets (*see* Roads)

Structure, of cities (see Cities, organization or structure of)
Students, 43–7, 59 (see also Children; Journey purposes, education)
Subsidies, 50–1, 255, 290
Suburbs, 14, 41, 49, 123, 134, 147–8
  employment densities, 138, 140, 142–4, 149
  (see also Journey location)
Supplementary licensing, 224, 289–92
Surrey, 306
Surveys, 19, 24–9, 48, 51, 58, 60, 84, 90, 96, 97, 100, 108, 111–13, 126, 158–9, 169, 271
Sutton Coldfield, 74, 104, 105, 140
Swansea, 154, 159, 167, 169–73
Synchronization, of urban activities, 55, 65–6, 300, 326

Taxis, 81, 82, 123, 179, 182
Technological change, 1, 4, 15, 147–8, 175, 229–30, 232, 237, 307, 329–41
Teenagers (see Adolescents)
Telecommunications, 79, 344, 345, 346, 348
Temporal co-ordination, of activities (see Journey timing; Synchronization, of urban activities; Time)
Time, 12, 15, 24–5, 155–6, 217, 221–2
  and modal choice, 207, 287
  value of, 205, 208–12
  (see also Journey distances; Journey timing; Modal choice, time factor in)
Time budgets, 23–4, 47 (see also Activity patterns)
Tokyo, Japan, 93, 103, 310, 332
Town and Country Planning Acts:
  1947, 250
  1968, 311
Town centres (see Central business district)
Tracked city, 3
Traffic, 178, 179, 242, 246, 264, 293, 308, 327–9
  control density, 178
  distributional models, 328
  restraint, 222, 246, 269, 275, 286–8
  speeds, 222
  surveys, 223–5
  zones, 222
Trains (see Journey modes; Railways)
Trams and tramways, 6–8, 13, 176, 177, 230, 232, 234, 276, 334

Transport(ation):
  innovations, 268–86
  investments, 236, 237, 244–6, 353
  legislation, 229–62
  management network, 306–55
  planning, 229, 236, 250–2, 314, 341, 349, 352
  policy (see Government, transport policy)
  private, 12–15, 175, 176, 181, 196–9, 223, 338–41
  social priorities for, 341–4
  statistics (see Surveys; Research)
  studies, 182–6, 222–5, 352, 353
  technology (see Technological change)
  and urban growth, 1–11
  users, 175–81, 197, 223, 269
Transport Acts:
  1947, 235, 241, 242
  1953, 235, 241
  1962, 236
  1960, 247–9
Transport and Road Research Laboratory, 63, 96, 113, 126, 243, 291
Transport Policy, 1976 Discussion Document, 257–62
Transport Policy and Programme (TPP), 255–7
Transport Supplementary Grants, 255
Travel behaviour (see Activity patterns; Journey . . .)
Tri-State Transportation Commission (USA), 78
Trip modelling, 85, 181–6
  ability to reproduce past trends, 224
  assignment, 220–2
  definition, 182–4
  distributional, 213–20
  of freight movements, 109–12
  generation at department stores, 194
  generation at manufacturing plants, 192
  generation at supermarkets, 196
  (see also Research; Surveys)
Trips (see Journey . . .)
Trucks (see Freight movements; Lorries; Vans)
Tube railways (see Railways, underground)
Tunnels, 10, 102, 218
Turnpike trusts, 237
Turnpikes, 237–8
Tyneside, Tyne-Wear, 181
  Metro, 333, 337

INDEX

Underground railways (*see* Railways, underground)
Unreliability, of services, 200–3
Urban . . . (*see* Cities)
Urbanism, 34

Value, of time, 205, 208–12 (*see also* Journey distances; Journey timing; Modal split, time factors in; Time)
Vans, 14, 80–1
Vehicle/pedestrian segregation, 309
Vehicles (*see* Buses; Cars; Lorries; Taxis; Vans)

Wales, 86, 153, 154, 253, 255
Walking, 27, 42, 55, 61, 75, 76, 123, 126–8, 223–4, 322, 343 (*see also* Foot cities; Journey modes; Pedestrians)
Walsall, 74, 103–5, 140
Warehouse, 14, 90, 290
Washington, D.C., 273–4
Watford, 158–67, 171

Weather conditions, 60, 63, 65 (*see also* Journey timing, seasonal patterns)
West Germany, 42–7, 178, 270, 332
West Midlands, 7, 59, 73, 74, 81, 89–93, 97, 99, 103–5, 138, 213, 271 (*see also* Birmingham)
Westinghouse, 331
Witkar, 339
Wolverhampton, 74, 103–5, 140
Woodside, 7
Workplaces:
　distribution of, 136
　separation from residences, 2, 5, 7, 15, 156–7
　(*see also* Journey distances; Journey purposes; Journey (or trip) rates)
Wuppertal, West Germany, 332

Zone and collar scheme, 285–6
Zones, 59, 106, 108, 150, 189, 190, 209, 210, 213–15, 220, 222
　concentric arrangement, 184–6